MAN IN BIOSPHERE
A CASE STUDY OF SIMILIPAL BIOSPHERE RESERVE

MAN IN BIOSPHERE

A Case Study of Similipal Biosphere Reserve

Editor

Samit Ghosal

Anthropological Survey of India
Kolkata - 700016

Published by

GYAN

Gyan Publishing House
5, Ansari Road
Daryaganj, New Delhi-110002
Phone: 011-47034999, 9811692060
E-mail: books@gyanbooks.com

Distribution Network
gyanbooks.com
India, USA, Canada, UK, Australia, France

ISBN: 978-93-6280-314-6 (PB)

First Published, 2024

Printed at: Gyan Press, Delhi.

Man in Biosphere: A Case Study of Similipal Biosphere Reserve
Editor: Samit Ghosal

CONTENTS

Foreword by **Prof. K.K. Misra** 7

Preface 9

List of Contributors 15

1. Introducing the Buffer Area Village 17
Surajit Kumar Pal, Amitabha Mitra, Rabi Ranjan Biswas, Arup Ray, Gautam Kumar Mallik and *Anupam Dutta*

2. Watersheds Management in Similipal 77
Surajit Kumar Pal

3. Religion and Worldview of the Buffer Zone Villages 125
Arup Ray

4. Impact of Development Plans and Programmes in the Buffer Zone Villages 161
Gautam Kumar Mallik

5. Ethno-Medical and Health Culture of Buffer Area Villages of Similipal Biosphere Reserve 209
Anupam Dutta

6. Folk Music and Dance: The Rhythm of a Cultural Milieu in Similipal 253
Amitabha Mitra

7. Tourism in Similipal: An Anthropological Observation 267
Rabiranjan Biswas

8. Maps of Buffer Villages 307
P.K. Guha

9. Introducing the Core Area Villages 315
Samit Ghosal, P.K. Guha and *Tamal Maity*

10. Indigenous Management of Land and Forest Resources in the Core Area Villages — 323
Samit Ghosal

11. Land Use Changes: A Study in the Core Villages — 347
P.K. Guha

12. Study on Material Culture in the Core Area: An Eco-Cultural Dimension — 429
Tamal Maity

13. Study on Society and Polity in the Core Area Villages of Similipal — 457
Tamal Maity

Photographs — 487
References — 491
Index — 497

प्रो. कमल कांत मिश्रा
निर्देशक
PROF. K. K. MISRA
Director
TEL : 91-33-2286 1761/33/1697
Fax : 91-33-2286 1799/1665
Email : director@ansi.gov.in / hohq@ansi.gov.in
ansihead@gmail.com
Website : www.ansi.gov.in

भारतीय मानवविज्ञान सर्वेक्षण
संस्कृति मंत्रालय
भारत सरकार
भारतीय संग्रहालय परिसर
27, जवाहरलाल नेहरू रोड कोलकाता - 700 016
ANTHROPOLOGICAL SURVEY OF INDIA
MINISTRY OF CULTURE
GOVERNMENT OF INDIA
INDIAN MUSEUM COMPLEX
27, JAWAHARLAL NEHRU ROAD, KOLKATA - 700 016

Foreword

The Anthropological Survey of India during the 11th Five Year Plan launched a research project '*Man and Environment : Biosphere Reserve Study*' with the basic objective of understanding the Man – Nature relationships, and pattern and extent of natural resource utilisation by the people living inside the Biosphere Reserves. As a sequel to this objective, the present volume *Man in Biosphere: A Case Study of Similipal Biosphere Reserve* contains research reports derived from a two months long intensive field research undertaken by the members of the Central Regional Centre, Nagpur and Head Office, Kolkata of the Anthropological Survey of India under the leadership of Dr Samit Ghosal, Anthropologist (Cultural), as the Core Member and Principal Investigator, who is also the editor of this volume.

The whole idea of demarcating biosphere reserves is to reconcile the conservation of biodiversity with its sustainable use. Importantly, biosphere reserves serve three complimentary and mutually reinforcing functions, *viz.* to conserve ecosystem, foster sustainable economic and social development, and fulfill logistical function of research and information sharing. The Similipal Biosphere Reserve is one such important biosphere reserves in India situated in the state of Odisha, covering a total area of 5324 sq. km. of which the core area of the reserve occupies 845 sq. km., while the buffer zone covers 1905 sq. km.

The Similipal Biosphere Reserve has a varied history of its exploitation, first by the former feudal state of Mayurbhanj in Odisha, and later by the local population due to their growing number in and around the reserve. As indicated earlier, biosphere reserves play a pivotal role in balancing critical environmental and human factors. If the Similipal Biosphere Reserve is not sufficiently protected, the whole of Odisha is likely to suffer from climatic disaster. Similipal guides the drainage system and perennial water supply of this part of Odisha to a great extent, because three major rivers, the Budhabalang, the Gangahar and the Khairi originate from Similipal and meet the Bay of Bengal in the east.

The reports presented in this volume provide a lucid and detailed account of the biodiversity and cultural diversity of the Similipal Biosphere Reserve. They also explore various facets of human interaction with nature, from forest economy to religion.

These reports would definitely help in understanding the Similipal Biosphere Reserve holistically, so that its resources, both natural and cultural, can be conserved for the posterity. In recent years, the study has become more relevant due to the relocation of a few studied villages from the core area of the Reserve, as the reports become an archive, a repository of agricultural and cultural biodiversity and traditional knowledge of the people who once lived there. I am sure this volume will be of great help to scholars, planners, administrators and environmentalists.

Prof. Kamal K. Misra
Director

Preface

The programme of Biosphere Reserve was initiated under the 'Man and Biosphere' (MAB) programme by UNESCO in 1971. Biosphere Reserves are areas of terrestrial and coastal ecosystem promoting the conservation of biodiversity with its sustainable use. Biospheres Reserves are designed locally recommended nationally and recognised internationally but remain under the sovereign jurisdiction of the states where they are located. The programme recognises the local communities as an integral part of ecosystem and promotes ecologically compatible socio-economic development of the local communities. The human dimensions of the biosphere reserves make them special from other protected areas, such as tiger reserves, national parks and sanctuaries.

Similipal was declared as a Biosphere Reserve on 22nd June 1994, it is located in the Mayurbhanj district of Odisha in the Mahanandian biogeographic region. The name 'Mayurbhanj' derived from Mayuras and Bhanjas, the two clans who ruled over the tract nearly for 1000 years till its merger in the Odisha in 1949. Similipal derived its name from the magnificent *Simul* (cotton) tree.

Similipal Reserve Forest was declared a tiger reserve in 1973, having a core area of 845.75 sq. km. and buffer area of 1904.25 sq. km. National park was created in 1980 with the core area of the tiger reserve as national park and the buffer area of the tiger reserve as the sanctuary. Protected area of 2750 sq. km. of erstwhile Similipal Reserve Forest is the conglomeration of project tiger, national park and sanctuary. Similipal was declared as the eighth biosphere reserve of India in 1994 with an area of 4,374 sq. km. Which was later extended to 5324 sq. km. with 1904 sq. km. buffer and 2575 sq. km. transition area. The Similipal Biosphere Reserve has three well-defined and well-demarcated zones. Core area – A legally protected area where bio-diversity is preserved. In case of Similipal it is the four core area villages. Buffer zone – An area where sustainable use of natural resources are permitted. In Similipal there are 61 village situated at the buffer area. Transition zone – Where all the stake holders work together to maintain the balance between conservation and economic development based on the cultural values of the local communities. Similipal has about 1200 villages in the transition zone of the biosphere reserve. Internationally designed Man and Biosphere (MAB) programme cannot be adopted in toto in Indian situation, as human intervention is more in India than any biosphere reserves located in other parts of the globe. An approach based on the design suitable to Indian condition has to be developed and implemented. Conservation Programmes in special protected areas like BRs cannot be achieved without the involvement of local people. Anthropologists have been advocating the importance of the participation of local people in the conservation and management of biosphere reserve. Centrality of the programme rests on the sustainable use of natural resources in the BRs. They show human concern and urges human involvement. The demarcation of the BRs in core, buffer and transition zones gives opportunities for human activities so that sustainability is maintained by balancing the interest of both man and nature.

Any development programme in the BRs should take local people into confidence and that should be compatible with the culture and value system of the local people. Expertise of anthropologists come handy in this sphere. Anthropological research documents local knowledge in conservation and propagate the concept of biodiversity held by the local people. In addition action oriented anthropologists act as advocates and allies of local people against exogenous factors.

The objective of the study was to generate primary database of all four-core area villages and nine buffer area villages selected on the basis of geographical location, remoteness and community representation using anthropological tools and techniques. Data have been collected through case studies, interviews, open-end questionnaires, schedules and personal observation. Thrust areas have been demography, population structure, marital status, family composition, occupation, land holding and so on and so forth. Four core villages are situated in the Gugudia gram panchayat. Nine buffer villages are distributed amongst the three gram panchayats, namely Gudgudia, Astakumar and Sarispal in the

Jasipur and Bangiriposi Block respectively. The nine buffer area villages studied are; Gudgudia, Kumari, Nenjaghosra, Chakundakocha, Garh-Similipal, Kuljhari, Saruda, Basilakacha, and Durdura (Birhor Settlement). The core villages covered are Kabatghai, Jamunagarh, Jenabil and Bakua. The following twelve communities are covered in the buffer area; Kol (Ho), Kharia, Bhatudi, Bhumij, Munda, Santal, Gaud, Kurmi-Mahato, Teli, Kaibarta, Sadgop and Birhor/Makdiya. In the core area communities covered are Ho (Kol), Santal and Bethudi.

Beside generating a holistic data base of selected buffer and core villages and an ethnos-cape for the entire buffer region, spot studies on some rehabilitated / re-settled village – a few issues have been identified for in depth studies.

The book consists of mainly two parts. In the first part case studies of the villages situated in the buffer area are given. In the second part the case studies of the four villages located in the core area are given. The *Preface* is a general outline of the Similipal Biosphere Reserve written by Dr. Samit Ghosal, the core member and principal investigator of the project, wherein Dr. Ghosal has presented the concept and design of the study.

In the first chapter Dr. Surajit Pal, Amitabh Mitra, R.B. Biswas, Arup Ray, Gautam Mallik and Anaupam Dutta have presented an *Introduction to the Buffer Area of the Biosphere Reserve*. Dr. Surajit Pal in the second report *Watershed Management in Similipal Biosphere Reserve* has given a detailed outline of the watershed system of Similipal Biosphere Reserve. He also pointed the need for its proper utilization so that misuse of water resources could be minimised. In the third chapter Arup Ray has dealt in detail on *Religion and World View of the Buffer Zone Villages of the Similipal Biosphere Reserve* where he has tried to explore and explain the intricate yet organic relationship between man, nature and spirit in understanding the world view of the different tribal groups of the Similipal Biosphere Reserve. He has also given a detailed account of festival and folk taxonomy and perception of tribal people about different natural phenomenon. Gautam Kumar Mallik in fourth report entitled *Impact of Development Plan and Programme : A Study in the Buffer Zone Villages of the Similipal Biosphere Reserve* has given a good account of the developmental programmes being undertaken among the different tribal groups of Similipal Biosphere

Reserve. Anupam Dutta in his report on *Ethno Medical and Health Culture of the Buffer Area of the Similipal Biosphere Reserve : A Case Study* has given a detailed account of the traditional medicine used by the population in this area. In the sixth chapter *Folk Music and Dance: Rhythm of Cultural Milieu in Similipal Blosphere Reserve* has been presented by Amitabha Mitra where he has outlined different forms of folk music and dances. R.R. Biswas has presented a report on *Tourism in Similipal Reserve : An Anthropological Observation* where he has discussed the development of tourism in Similipal Biosphere Reserve. Dr Samit Ghosal, Tamal Maity and Dr. P.K. Guha have given an *Introduction to the Core Area of the Biosphere Reserve,* Dr Samit Ghosal in his report on *Indigenous Management of Land and Forest Resources in the Core Villages of Similipal Biosphere Reserve* has given an account of the various methods used by the villagers in the core area for utilising their agricultural and forest resources for their survival. Dr P.K. Guha discussed the *Land* Use *Changes in the Core Villages of Similipal Biosphere Reserve* where he has discussed in detail the land use pattern on plot to plot basis of the four core villages namely, Bakua, Jenabil, Jamunagarh and Kobatghai. Tamal Maity in his report *on Study of Material Culture in the Core Area of Similipal Biosphere Reserve: An Eco-Cultural Dimension* has presented a detailed account of the material culture which is found in core villages. He has also written another paper on *Society and Polity of the Core Villages of Similipal Biosphere Reserve* where he has given a detailed account of the Nature–Man relationship and importance of polity and law in maintaining harmony among the different communities of the core villages of Similipal Biosphere Reserve.

The study on the project was conducted for a period of two months from November to December, 2006 by setting camp at Jashipur. This gives us the opportunity of understanding the people living outside the biosphere reserve as well as the people living inside the biosphere reserve. At the end of the study the draft report was presented at a seminar held in Bhubaneswar in April 2007 before finalising the report, which was attended, by the local scholars and forest officials.

In March, 2010 Jenabil village was relocated outside core area to a place called Ambadiha. The study was the last account of the village.

The authors are sincerely grateful to the Director, Anthropological Survey of India for permitting to undertake the present study in Similipal. They are also grateful to the officials particularly the Similipal Biosphere Reserve authorities for their cooperation in conducting fieldwork in the said location. Last but not least, thanks are due to the local people without whose cooperation this study could not have been possible.

Samit Ghosal

List of Contributors

1. **Dr Samit Ghosal**, Anthropologist (Cultural) , Anthropological Survey of India, Central Regional Centre, C.G.O. Complex, Block 'A', Seminary Hills, Nagpur - 440006.
2. **Dr Surajit Kumar Pal**, Former Assistant Anthropologist (Cultural), Nalini Appt.5, Flat-1, Shivam Kunj, Barde Nagar, Borgaon, Nagpur - 440013.
3. **Dr P.K. Guha**, Cartographer, Anthropological Survey of India, 27 Jawaharlal Nehru Road, Kolkata - 700016.
4. **Shri Amitabha Mitra**, Former Research Associate (Cultural), Duplex No. 18, Shivam Kunj, Barde Nagar, Borgaon, Nagpur - 440013.
5. **Shri Tamal Maity**, Research Associate (Cultural), Anthropological Survey of India, Central Regional Centre, C.G.O. Complex, Block 'A', Seminary Hills, Nagpur - 440006.
6. **Shri Rabiranjan Biswas**, Research Associate (Cultural), Anthropological Survey of India, 27 Jawaharlal Nehru Road, Kolkata - 700016.
7. **Shri Arup Ray**, Research Associate (Cultural), Anthropological Survey of India, 27 Jawaharlal Nehru Road, Kolkata - 700016.
8. **Shri Gautam Kumar Mallik**, Research Associate (Cultural), Anthropological Survey of India, 27 Jawaharlal Nehru Road, Kolkata - 700016.
9. **Shri Anupam Dutta**, Research Associate (Cultural), Anthropological Survey of India, Eastern Regional Centre, EN-79, Sector - V, Salt Lake , Kolkata - 700091.

1

Introducing the Buffer Area Village

Surajit Kumar Pal, Amitabha Mitra, Rabi Ranjan Biswas, Arup Ray, Gautam Kumar Mallik* and *Anupam Dutta

Discourses on the study of biosphere reserves in Indian context are relatively a recent phenomenon in social science disciplines especially in human science like anthropology and geography. It provides some basic insights as wells as broader understandings about man and environment and their relationships.

Biosphere bears several meaning. In conventional notion, one implies it only flora and fauna. But in anthropological sense, it does not refer to as flora and fauna only. Rather, it deals not only with the flora and fauna, human beings and other living and non-living organisms but also their cohesiveness that form a united whole. That means all are integral part of the system wherein man can never be isolated from its jurisdictions. Hence in biosphere study man gets special significance under whom future conservation and management of it largely depend.

In recent time, the geographical area of any biosphere reserve is commonly divided by the forest officials into three parts viz. core, buffer and transitional zones. This is done mainly from administrative and conservation point of view. But in anthropological perspective such zonations are redundant, rather untenable and misnomer. This type of classification does not carry any proper meaning of our understanding about biosphere in terms of relationship among all animate and inanimate including humans found within and outside its boundary. Anthropology, as a discipline is, in fact, based on holistic approach that bears the essence of interpreting the casual relations of all integral parts.

Thus, the present study on the Similipal Biosphere Reserve (SBR), where a sizeable human population belonging to different communities are found to live, has been made under anthropological perspective with a view to highlight several aspects primarily related to human beings whose livelihood is governed by forests as well. Since they are in the midst of forest ecology and environment, a brief outline about the SBR needs to be narrated to some extent.

Similipal Biosphere Reserve

The Similipal Biosphere Reserve is one of the prominent biosphere reserves in India, declared in 1994 by Govt. of India, located in the central part of erstwhile Mayurbhanj state, now a district of Orissa. Prior to declaration as a biosphere reserve some of its area became a Tiger Reserve (1973), a Sanctuary (1978) and also a National park (1980). These three are constituted to form this biosphere, now functioned as biosphere reserve as well as tiger reserve. The SBR, practically a hilly region, familiar as the Similipal Hill Range/Similipal Hills, lie between 20°-17′ and 22°-34′ north latitude and 85°-40′ and 87°-10′ east longitude covering an area of 5324 sq. kms. with innumerable crests, valleys, perennial streams, nalas, and rivers, surrounded by dense forests having rich biodiversity, wildlife and diverse ethnic groups dominated mainly by the adivasis (tribals). It influences the climatic conditions of adjoining eastern states namely West Bengal, Orissa, Bihar and Jharkhand. Its forests and wildlife are a rare asset to our nation, attract large number of nature lovers, tourists, environmentalists, scientists etc. of both foreign and Indian origin.

Like any other biosphere reserve the SBR has also three demarcated zonations - Core (845.75 sq. kms.), buffer (1904 sq. kms.) and transitional (2575 sq. kms). All the zones/areas are having human habitations. The core area has four villages, 61 villages in buffer area and 1200 villages are in the transitional area. The total human population in and around the SBR is about five lakhs and majority of them are directly and indirectly depended on it for their sustenance.

The SBR region witnesses a subtropical climate with three distinct seasons - summer, rainy and winter. The climate is characterized by mild summer of three months' duration from March to June, with temperature of less than 40° C; well distributed rainfall during south-west monsoon, conspicuously observed during June to September and

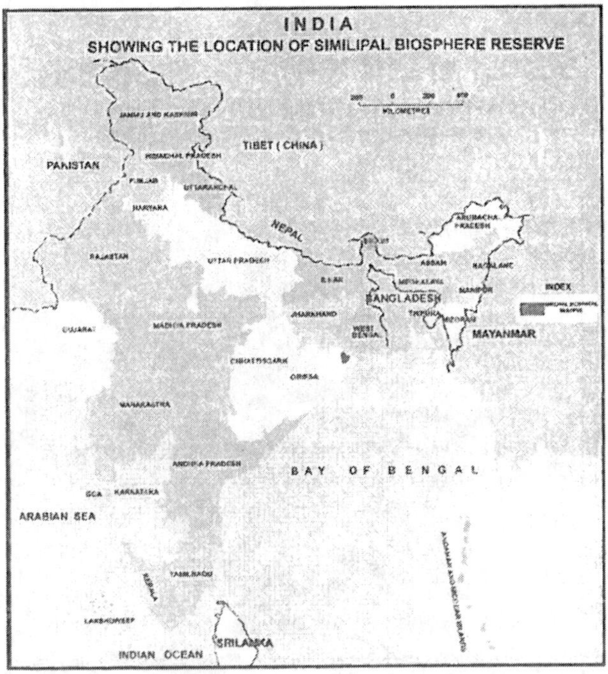

Map 1. Location of Similipal Biosphere Reserve

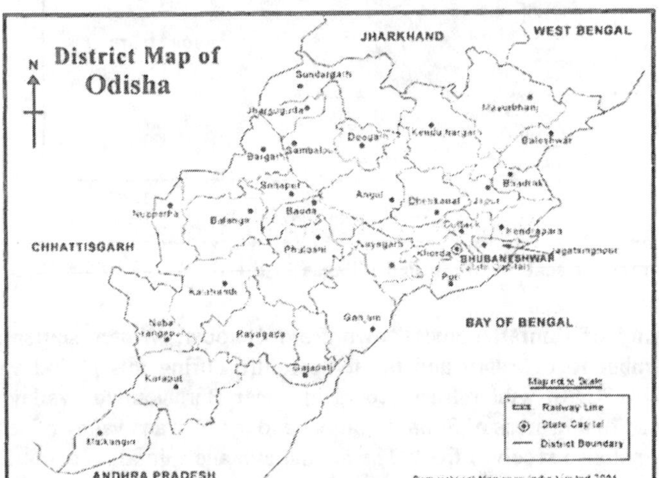

Map 2. Map of Mayurbhanj District

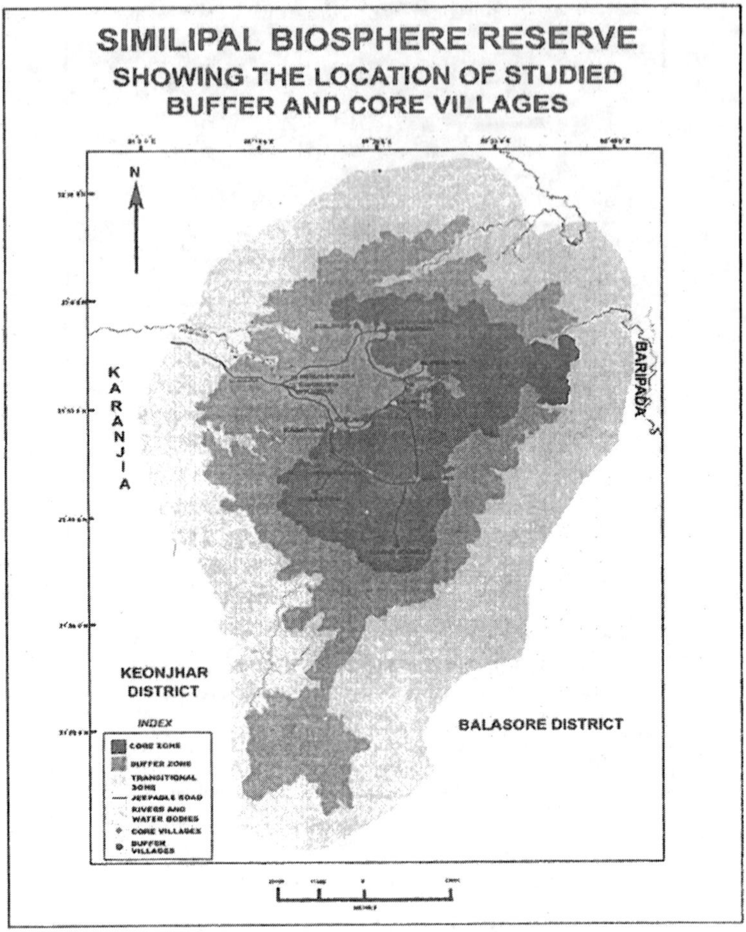

Map 3. Location of studied villages in the Buffer and Core Area

intensity of rainfall comes down from October. Winter starts from November to February and the temperature during this period varies 5°C to 8°C. In winter frost occurs in upper Barhakamuda valley and other adjoining areas of South Similipal and the Nawana valley of central Similipal also receives frost. The annual average rainfall is about 2200 mm. that influences the perennial character of watersheds available in the area.

Study villages and methodology

The villages were selected in such a way that could fulfill the objective of the study. The study stresses more importance in relation to ethnoscape, livelihood pattern vis-à-vis economic life of the people as these are vital in the arena of present biosphere study. Thus on this line of understanding each village bears certain features that reflect a kind of relations between the people and their immediate environment. That means village is regarded as a unit of study and the people are the centre of whole discourse in the present context.

Considering the vast geographical area of the SBR and from feasibility point of view, man power and for time constraints the present exercise is exclusively confined to eight buffer zone villages (Gudgudia, Lanjighosra, Kumari, Koljhari, Chakundakocha, Gad Similipal, Saruda, Basilakocha) and one transitional village (Durdura Makdiya or Mankhedia Tanda or segment) through which aforenoted aspects can reflect. The study was conducted during November and December 2006. The above mentioned villages were selected randomly from different directions namely east, west, north west, north east, but not from the core zone (having only four villages - Kabatghai, Jamunagarh, Jenabil and Bakua, located in the southern portion of the SBR primarily inhabited by the tribals) as this part was covered by other team of the SBR study, with the purpose of highlighting the ethnoscape, socioeconomic condition as well prevailed in the area.

In collections of field information within a stipulated period anthropological tools intermingled with survey method were adopted as this methodology was considered ideal for application in biosphere study. Because the objectivity of study requires such methodology through which a macroscopic outline about the people, their economy, culture and relationship with nature can be reflected and gets more attention for further interpretation and study particularly from conservation point of view.

Every selected village is having certain features like location, administrative boundary, ethnic composition, available civic amenities etc. Of all nine villages three (Gudgudia, Kumari and Gad Similipal) are multiethnic, fall under two Gram Panchayats namely Gudgudia and Astakunar (or Astakumar) where both tribal and caste population are seen to coexist and they maintain a harmonious relations among themselves as well as among other population groups living in their neighbourhood. The remaining six villages are of uniethnic type, distributed in Gudgudia (Lanjighosra), Barheipani (Koljhari and

Chakundakocha), Astakunar (Gad Similipal and Saruda), Paudagarh (Durdura) Gram Panchayats of Jashipur Block and Sarispal (Basilakocha) Gram Panchayat of Bangriposi Block of Mayurbhanj district of Orissa. While under biosphere administration all aforementioned villages barring Basilakocha (which falls in Baripada forest division) are situated under the jurisdiction of Karanjia forest division.

The study villages are, by and large, located in hilly and plain terrains, populated mainly by the tribals namely Kolha/Ho, Khadia, Bathudi, Mahali, Munda, Santal/Majhi, Makdiya/Birhor. However some caste groups such as Teli, Goud/Maghdha Goud, Lohar/Bindhani, Kudmi Mahato, etc. are also found to live with the tribals in some villages who are numerically less in number than the tribals. Among all these castes the Goud and the Kudmi Mahato are numerically more than other castes as seen in Gudgudia and Gad Similipal village. The features of each village, existing amenities and ethnic composition along with administrative boundary are listed in table 1, 2 and 3 for ready reference.

Table 1. Characteristic Feature of the Study Villages

Sl. No	*Village*	*Features*
1.	Gudgudia	Multiethnic, both tribes and castes coexist with arbitrary boundary demarcation of community hamlets, locally known as *sahi* or *sai* located in western direction of the SBR, not far away from core zone villages, a prominent centre for social, economic, education, political activities, having larger valley area meant for agriculture, quite proximity to the Khairi river vis-à-vis watershed
2.	Lanjig-hosra	Uniethnic, inhabited by the tribal group, habitation is in the hilly slope, moderate valley area meant for agriculture, also fall under the Khairi watershed, located in western side of the SBR
3.	Kumari	Multiethnic, inhabited by both tribal and caste population, quite proximity to the Khairi river, larger valley area (plain in nature) used for agriculture, habitation area demarcated by community hamlets, also located in western direction of the SBR
4.	Koljhari	Uniethnic, inaccessible and located in north-west direction, inhabited by a tribal group, surrounded by the hills and Barehipani Nala that create a natural barrier for people's movement, habitation is in the foot hills, isolated from other neighbouring villages due to lack of communication, limited agricultural area

(contd...)

Sl. No	*Village*	*Features*
5.	Chakun-dakocha	Uniethnic, also inaccessible and tiny type, surrounded by the hills and Barehipani Nala, located in north-west direction, very small habitation in the foot hills inhabited by the tribal people, devoid of agricultural field and no spatial communication,
6.	Gad Similipal	Multiethnic, located in north-eastern side, surrounded by forest, hills and nalas, habitation is in hillocks and foothills, inhabited by tribe and caste population with demarcation of community hamlets, larger valley exists for terrace cultivation and also some scope of future expansion of agriculture
7.	Saruda	Uniethnic, located in north-eastern direction, encircled by both hills and rivers (Budhabalanga and Saruda rivers), inhabited by the tribal people who are Christian by religion that leads them to get more exposure to the outer world and by virtue of Christianity their outlook is being changed, agriculture takes place in flat upland and Saruda basin
8.	Basilakocha	Uniethnic, a tiny and plain village located in eastern side of the SBR, very inaccessible and surrounded by dense forest and the Budhabalanga and Khadkai rivers, inhabited by the tribal people divided by both Hindu and Christian groups, limited agricultural field quite proximity to both rivers, no scope for further expansion of agriculture and habitation, despite its inaccessibility people are very conscious about their right on land and forest and they are also in the ambit of constant exposure to various market and urban centres like Bangriposhi, Bisoi, Baripada
9.	Durdura	Multiethnic, a plain village, located in north east direction of Jashipur, closed to N.H. 6, well communicated, inhabited by both tribe and caste population, having a separate segment for Makdiya (Birhor) settlement, vast agricultural field tilled mainly by the Santal and other communities

Table 2. Amenities Prevailed in Study Villages

Sl. No.	*Village*	*Amenities*	*Remarks*
1	Gudgudia	School - Middle* and Secondary, Post Office, Police Outpost, Anganwadi+, Allopathic Dispensary, Ayurvedic dispensary, Veterinary dispensary, Cooperative Centre, Youth Club @, Panchayat Office, Forest Office cum Eco-tourism Complex, Mobile Health Unit, Weekly Market, Grocery Shop, Grain Reservoir, Tubewells, Road linked with main arterial road leading to Jashipur and other core and buffer zone villages of the SBR	Hostel facility available for the girl students (VI and VII std), expenditures are borne by the Govt. + Meant for the children of villages of Gudgudia and Lanjighosra @ Now defunct but its house has been used as "Unnayana Mandir" set up by the Sambandh, a NGO working as development agency in tribal districts of Orissa
2	Lanjighosra	Tubewell, Wells; linked with Gudgudia – Jashipur Road and Lanhighosra – Barehipani Road	
3	Kumari	Middle School* (I – VII), Anganwadi, Tubewells, Grocery Shop. Linked with Gudgudia – Kondibil – Bakua Road	Located within the boundary of Saharpat village
4	Koljhari	Tubewell, Well, Anganwadi meant for both villages of Koljhari and Chakundakocha	

(contd...)

Sl. No.	*Village*	*Amenities*	*Remarks*
5	Chakun-dakocha	No basic amenities available	
6	Gad Similipal	Tubewell, Wells, Anganwadi, Linked with Nawana – Lulung Road	
7	Saruda	School*, Anganwadi, Church. Well	*EGS type
8	Basilakocha	Tubewell, linked with Bangriposi by forest road	
9	Duradura Makdiya Settlement	School*, Anganwadi, Farm tank, Tubewell, Wells, Linked with NH 6 – Chahala – Barehipani Road	*EGS type

The amenities listed in the above table are the outcome of some sort of development activities taken place over the years as part of people's well-being programmes implemented by the state. But in recent time implementation of various welfare schemes have been standstill because of declaration of biosphere reserve. Whatever the infrastructures created under various development programmes during the period of last few decades by the local Block Development offices for providing opportunities to the people to earn minimum livelihood needs in such an isolated location are being damaged or not functioning properly due to lack of proper maintenance, regular inspection of the development officials as well. Biosphere reserve rules and regulations have eventually enhanced people's agonies and woes. And there is no immediate relief from such situation that the people can lead a kind of modest living in years to come.

Table 3. Ethnic Composition in Study Villages

Block	*Gram panchayat*	*Village*	*Community*	
			Tribe	**Caste**
			Kolha/Ho	Goud
			Khadia	Lohar/ Bindhani
		Gudgudia	Mahali	Teli
			Bathudi	Kaibarta
	Gudgudia			Sadgop (Gobora caste)
Jashipur		Lanjighosra	Kolha/Ho	Nil
		Kumari	Bathudi	Goud
			Bhumij	
		Koljhari	Kolha/Ho	Nil
	Barheipani	Chakundakocha	Kolha/Ho	Nil
		Gad Similipal	Kolha/Ho	Kudmi Mahato
	Astakunar		Munda	Goud
			Santal/ Majhi	
		Saruda	Munda	Nil
	Paudagarh	Durdura	Makdiya/ Birhor	Lohar/ Bindhani
			Santal	
Bangriposhi	Sarispal	Basilakocha	Santal	Nil

Among all study villages Gudgudia stands as a prominent centre in whole of the SBR in terms of social, economic, healthcare, educational, cultural and political activities because of its location, exposure and availability of various infrastructures like schools (middle and secondary), forest beat office, eco-centre, dispensary, police out post, weekly market etc. Due to such facilities people of this village get more exposure to the outer world that help them to exert influence the people of other neighbouring villages in the above mentioned spheres.

It is mentioned that the SBR is having a kind of road network (gravel type) that connects majority of the buffer villages. The people can move from one village to other village, either by bicycle, two wheeler, jeep, truck or on foot despite undulating terrain and dense forest. Of these conveyances bicycle is regarded an affordable conveyance to the people. But majority of the families even do not have such type of transport and the people commonly move from one place to other place on foot. Anyway this opportunity provides some scope of interactions among various communities living in and around the biosphere.

Since after declaration of biosphere almost all development-cum-welfare activities have become standstill that creates problem not only to maintain existing amenities available in different villages but also to extend development activities to many other villages which are hitherto devoid of minimum civic facilities. The people have their grievances, but they are helpless to resolve their problem due to biosphere reserve. The dearth of basic amenities compels them to lead impoverished life.

Population in SBR

Existence of human population in any biosphere reserve or any wildlife sanctuary is a universal phenomenon and India is no exception to this fact. There are a large number of human populations in different zones of SBR living since centuries back. As the present study confined to the buffer area (where altogether 61 villages is located) it needs to highlight some salient features related to decadal population growth which is pertinent in the sphere of conserving biodiversity and maintaining ecological balance in the area. Whether the existing population pressure is threat to wildlife and ecological balance that may be judged by the ecological scientists.

If we look into the decadal population growth during the span of 50 years (1961 - 2001) it is found that no significant increase is visible (table 4 and chart 1). Such low growth rate may take place among the population due to outbreak of diseases (fatal malaria, tuberculosis, diarrhoea etc), child mortality, starvation death and limited survival needs (because of devices, used by the people are commonly eco-friendly and traditional in nature, for harnessing natural resources for livelihood that limit the consumption needs of the people).

Table 4. Village wise Population Growth and Decadal Variation

Sl. No.	*Village Name*	*Area (In Hectre)*	*1961*	*1971*	*1981*	*1991*	*2001*
1	Asabani	112.32	28	62	138	113	259
2	Bad-Kasira	97.77	105	135	150	178	230
3	Bad-Uski	135.3	111	114	170	213	241
4	Barigaon	136.01	95	130	154	129	182
5	Bharrachua	35.98	26	49	46	63	90
6	Bilapagha	84.71	147	177	180	216	267
7	Chandikhaman	58.26	78	78	82	115	158
8	Gudgudia	159.52	274	274	354	453	526
9	Jhatipani	N/A	75	75	75	75	75
10	Khariadunguri	142.05	65	70	83	79	110
11	Khejuria	171.43	193	269	245	296	418
12	Kohla	160.84	107	136	155	180	227
13	Kondabil	182.9	184	242	247	299	372
14	Kuanribil	265.66	162	242	304	297	346
15	Kumari	168.84	191	192	267	263	293
16	Kusumi	129.95	150	159	188	210	232
17	Nenjaghosra	112.55	61	61	61	88	85
18	Nuniaguda	111.32	59	83	105	141	182
19	Routola	154.49	113	149	194	226	259
20	Sankasira	102.11	101	139	142	169	184
21	San-Uski /Sanasoki	82.05	132	157	194	162	233
22	Saharpat	148.01	166	189	225	238	298
23	Astakunar	432.9	187	270	275	341	445

Sl. No.	*Village Name*	*Area (In Hectre)*	*1961*	*1971*	*1981*	*1991*	*2001*
24	Balarampur	357.75	157	208	255	298	324
25	Bandribasa	96.85	128	143	163	186	240
26	Barheipani	283.69	247	286	355	414	473
27	Barsia	360.52	174	237	240	329	403
28	Burhabalanga	315.53	86	150	175	199	249
29	Chahala	Un inhabited					
30	Chakundakocha	25.83	17	59	32	23	29
31	Daldali	Un inhabited					
32	Dhudruchampa	Un inhabited					
33	Garh-Similipal	405.14	121	255	297	345	417
34	Gopinathpur	476.6	47	74	123	159	217
35	Haladia	N/A	67	83	89	113	132
36	Jadadihi	46.73	65	61	78	101	126
37	Kiajhari	63.25	99	128	129	150	185
38	Kukurbhuka	874.3	109	172	173	233	300
39	Kuljhari/Koljhari	71.51	28	40	80	59	78
40	Lembhujharan	153.68	337	50	82	112	123
41	Makabari	267.03	77	150	146	176	241
42	Nawana/Naana	539.74	153	185	263	188	305
43	Nikhirda	304.46	23	23	35	34	60
44	Phulbari/Phulbadia	50.27	66	74	68	68	80
45	Rajupal/Rajapal	130.2	83	115	118	111	148
46	Saruda	177.31	61	94	92	136	139
47	Charbandha	118.17	133	283	246	308	449
48	Dantiakocha	43	198	61	60	71	89
49	Alapani	72.03	60	76	89	147	113
50	Amdapani	79.9	137	82	66	80	182
51	Barubeda/Baribeda	69.02	63	276	83	157	162

Sl. No.	*Village Name*	*Area (In Hectre)*	*1961*	*1971*	*1981*	*1991*	*2001*
52	Basilakacha	17.52	4	24	20	35	57
53	Bhoduakocha	56	12	14	17	17	23
54	Chakidi (Chakidihipanipal)	56.74	361	96	84	111	155
55	Jamtolia/Jamtalia	80.74	68	96	63	82	107
56	Jerkeni/Jarkani	60.71	43	50	36	62	88
57	Khadighati	29.95	18	33	43	62	78
58	Kukurbhuka	77.87	93	85	122	120	147
59	Kusumtota	86.42	40	96	42	41	65
60	Phuljhari/Fuljhara	75.86	27	51	80	98	138
61	Purunapani	273.32	43	183	171	173	242
	Total	9382.61	6255	7545	8249	9542	12076

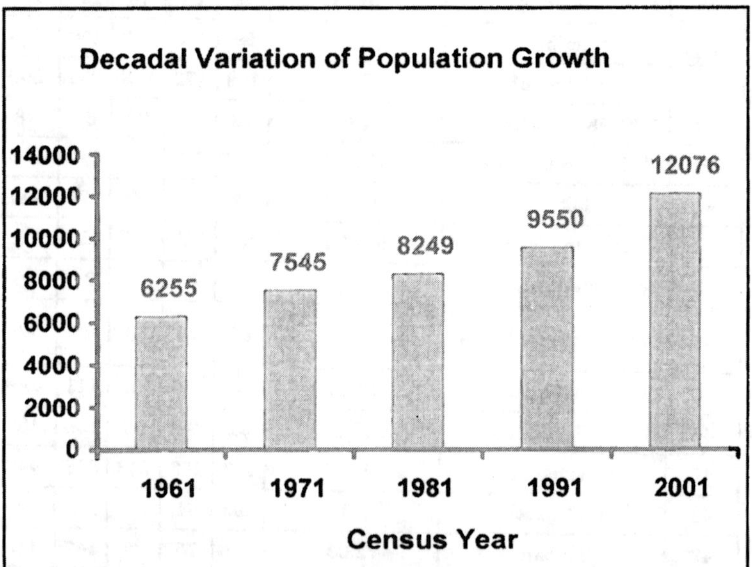

Chart 1

Population Structure in Study Villages

The population details have been made in study villages that may be useful for examining the future trend of growth rate and its impact on ecological balance of the area. It is observed that like the whole buffer area, the population variation during last 50 years (1961 - 2001) and even in 2006 enumeration conducted by An.S.I team in study villages is also low (Table 5, 6 and chart 2). There is no sharp gender difference noticed in respect of various age groups of population. Of the total population in nine villages the male and female percentage is 53.00 and 47.00 respectively. The chart shows that on and average only 200 individuals are seen to increase in every ten years of duration. Thus population increase in such location is not so alarming as found elsewhere that may bring any adverse situation in respect of maintaining ecological balance of the area in immediate future.

The study also indicates that the old age population is considerably very low in proportionate to other cluster of population (table 7 and chart 3). That means life expectancy beyond 60 years of age is limited. Because at this stage harnessing of natural resources and performing agricultural activities for maintaining livelihood is very strenuous job. The people since their childhood to old age are usually engaged in manual works for sheer existence in such location. Food they consume is inadequate to some extent in proportionate to physical work that creates problem to maintain good health till to old age. Moreover, the area is devoid of proper healthcare facilities. Hence majority of the people are unable to survive above 60 years of age.

The people have also been classified by religion. The study reveals that the area is predominantly a Hindu area. Of the total population around 78.00 per cent are the Hindu, followed by Christian (17.23 per cent) and Indigenous tribal Religion (about 5.00 per cent). Christianity and tribal religion prevail among the tribal population, but majority of them follow Hinduism. Even those tribal people claim as animist, may be treated as Hindu whose way of life is based on Hindu principle only. The caste groups are exclusively Hindu by religion (table 8 and chart 4). The religious structure of various ethnic groups, thus clearly indicates that people in SBR, whether tribes or castes, are basically Hindu and part of the larger Hindu society.

Ethnoscape

The term ethnoscape means that in a particular geographical space or in an ecological set up where different ethnic groups are distributed and how each group occupies their space for sharing resources for survival.

Table 5. Population in Study Villages (1961 - 2001)

Sl. No.	Village Name	Area (In Hectre)	1961		Total	1971		Total	1981		Total	1991		Total	2001		Total
Jashipur Block			*M*	*F*		*M*	*F*		*M*	*F*		*M*	*F*		*Male*	*Female*	
Gudgudia G.P.																	
1	Gudgudia	159.52	152	122	274	152	122	274	181	173	354	243	210	453	299	227	526
2	Nenjaghosra /Lanjighosra	112.55	28	33	61	28	33	61	26	35	61	38	50	88	45	40	85
3	Kumari	168.84	101	90	191	98	94	192	138	129	267	131	132	263	128	165	293
Barheipani G.P.																	
1	Kuljhari /Koljhari	71.51	14	14	28	27	13	40	37	43	80	31	28	59	40	38	78
2	Chakundakocha	25.83	7	10	17	26	33	59	13	19	32	12	11	23	12	17	29
Astakunar G.P.																	
1	Garh-Similipal /Gad Similipal	405.14	58	63	121	141	114	255	147	150	297	168	177	345	219	198	417
2	Saruda	177.31	35	26	61	48	46	94	46	46	92	65	71	136	70	69	139

(contd..)

Sl. No.	*Village Name*	*Area (In Hectre)*	*1961*		*Total*	*1971*		*Total*	*1981*		*Total*	*1991*		*Total*	*2001*		*Total*
Bangriposi Block																	
	Sarispal G.P.																
1	Basilakacha /Basilakocha	17.52	3	1	4	13	11	24	12	8	20	16	19	35	27	30	57
	Total	1138.22	398	359	757	533	466	999	600	603	1203	704	698	1402	840	784	1624

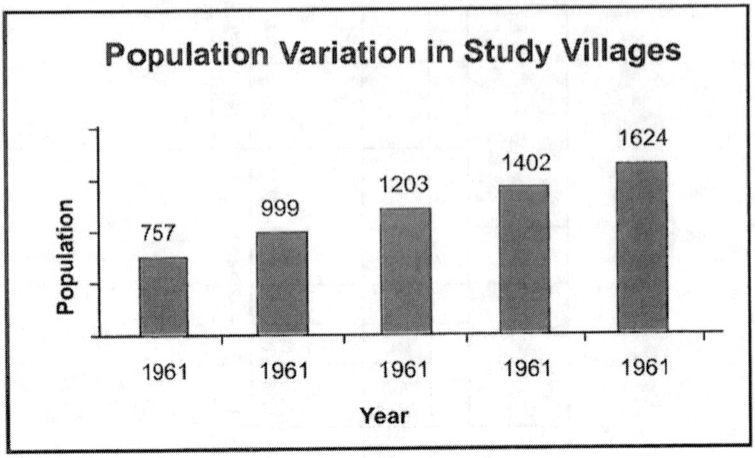

Chart 2

Table 6. Population distribution in study villages (2006)

Sl.No.	*Village*	*Male*	*Female*	*Total*
		225	209	434
1.	Nenjaghosra/ Lanjighosra	53	27	80
2.	Kumari	147	145	292
3.	Kuljhari/ Koljhari	43	44	87
4.	Chakundakocha	24	27	51
5.	Garh-Similipal/ Gad Similipal	107	81	188
6.	Saruda	75	60	135
7.	Basilakacha/ Basilakocha	31	33	64
8.	Durdura Tanda (Mankdiya tanda)	37	31	68
	Total	742	657	1399

Table 7. Age and Sex Wise Population in Study Villages

		Age Group																												
		0-5		*10-Jun*		*15-Nov*		*16-20*		*21-25*		*26-30*		*31-35*	*36-40*		*41-45*		*46-50*			*51-55*		*56-60*		*61+*		*Total*		
Village Name	*Community*	*M*	*F*	*M*	*F*	*M*	*F*	*M*	*F*	*M*	*F*	*M*	*F*	*M*	*F*	*M*	*F*	*M*	*F*	*M*	*F*	*M*	*F*	*M*	*F*	*M*	*F*	*M*	*F*	*Total*
	Kol/Ho	14	12	11	19	17	10	8	5	7	7	5	7	6	8	6	10	10	4	2	2	1	1	2	4	0	3	89	92	181
	Kharia	13	9	10	3	7	12	6	8	11	5	2	9	6	5	8	6	4	4	3	3	4	1	1	1	1	0	76	66	142
	Mahali	0	1	0	0	1	0	0	1	1	0	0	0	0	0	0	0	0	0	0	1	1	1	1	1	1	0	5	5	10
	Bhatudi	1	0	1	1	1	0	0	0	0	0	0	0	0	1	1	0	0	0	0	0	0	0	0	0	0	0	4	2	6
	Goud	6	3	2	2	2	2	3	2	1	1	2	4	3	3	2	0	1	5	2	1	2	0	0	2	2	2	28	27	55
	Lohar /Bindhani	3	1	4	2	0	2	0	0	0	1	1	1	0	2	1	0	1	0	0	0	0	0	0	1	1	0	11	10	21
	Teli	0	0	0	0	0	0	1	1	1	0	1	0	0	0	0	1	1	0	0	1	1	1	1	0	1	1	7	5	12
	Kaibarta	1	0	1	0	0	0	0	0	0	0	0	1	1	0	0	0	0	0	0	0	0	0	0	0	0	0	3	1	4
	Sodgop (Gobora Caste)	1	0	0	0	0	0	0	0	0	1	0	0	1	0	0	0	0	0	0	0	0	0	0	0	0	0	2	1	3
Gudgudia	Total	39	26	29	27	28	26	18	17	21	15	11	22	17	19	18	17	17	13	7	8	9	4	5	9	6	6	225	209	434
Nenjaghosra /Lanjighosra	Kol/Ho	15	4	8	1	8	1	1	6	5	2	1	5	4	3	4	3	2	2	3	0	1	0	0	0	1	0	53	27	80

(contd...)

Village Name	*Community*	*Age Group* 0-5		10-Jun		15-Nov		16-20		21-25		26-30		31-35	36-40			41-45		46-50		51-55		56-60		61+		Total		
		M	*F*	*M*	*F*	*M*	*F*	*M*	*F*	*M*	*F*	*M*	*F*	*M*	*F*	*M*	*F*	*M*	*F*	*M*	*F*	*M*	*F*	*M*	*F*	*M*	*F*	*M*	*F*	*Total*
Lanjighosra	Bathudi	11	10	10	14	8	17	12	17	11	6	8	10	8	9	6	9	7	5	6	2	1	2	3	2	4	1	95	104	199
	Bhumij	8	2	6	3	2	2	3	6	8	5	6	2	2	2	1	3	4	1	1	2	1	0	0	0	1	1	43	29	72
	Goud	1	2	0	3	0	1	2	2	3	0	1	1	1	0	0	0	0	0	0	2	0	0	0	1	1	0	9	12	21
Kumari	Total	20	14	16	20	10	20	17	25	22	11	15	13	11	11	7	12	11	6	7	6	2	2	3	3	6	2	147	145	292
Kuljhari/Koljhari	Kol/Ho	8	14	11	8	2	1	4	2	3	2	1	5	6	3	4	2	1	1	1	1	0	0	0	2	2	3	43	44	87
Chakundakocha	Kol/Ho	8	8	1	3	1	0	2	5	2	5	3	2	1	1	2	3	3	0	1	0	0	0	0	0	0	0	24	27	51
Garh-Similipal /Gad Similipal	Kudmi Mahato	1	3	4	3	2	1	7	1	3	1	1	1	2	3	3	4	3	3	4	2	2	2	1	1	3	2	36	27	63
	Kol/Ho	10	8	14	8	8	5	6	3	4	5	2	3	4	1	2	6	5	0	1	2	1	1	1	1	3	1	61	44	105
	Munda	0	2	0	1	1	0	0	0	0	0	0	1	0	0	1	0	0	0	0	0	0	0	0	0	0	0	2	4	6
	Santal /Majhi	1	0	0	1	1	0	0	0	0	0	0	1	1	0	0	0	0	0	0	0	0	0	0	0	0	0	3	2	5
	Goud	0	3	3	0	1	0	0	0	0	0	0	0	0	1	1	0	0	0	0	0	0	0	0	0	0	0	5	4	9
	Total	12	16	21	13	13	6	13	4	7	6	3	6	7	5	7	10	8	3	5	4	3	3	2	2	6	3	107	81	188
Saruda	Munda	17	8	9	10	7	4	8	12	9	10	7	2	3	1	1	6	5	2	5	2	0	1	2	0	2	2	75	60	135
Basilakacha /Basilakocha	Santal /Majhi	10	7	3	7	0	2	3	2	1	3	2	6	5	1	2	0	0	0	0	3	2	1	2	1	1	0	31	33	64
Durdura Tanda	Makdiya /Birhor	7	3	4	4	2	3	6	3	1	3	5	4	3	1	2	4	3	2	2	1	1	0	0	3	1	0	37	31	68
136 100	102 93 71 63	72 76	71 57	48	65	57		45	47	57		50	29	31		25	18	11		14		20		25	16	742			657	1399

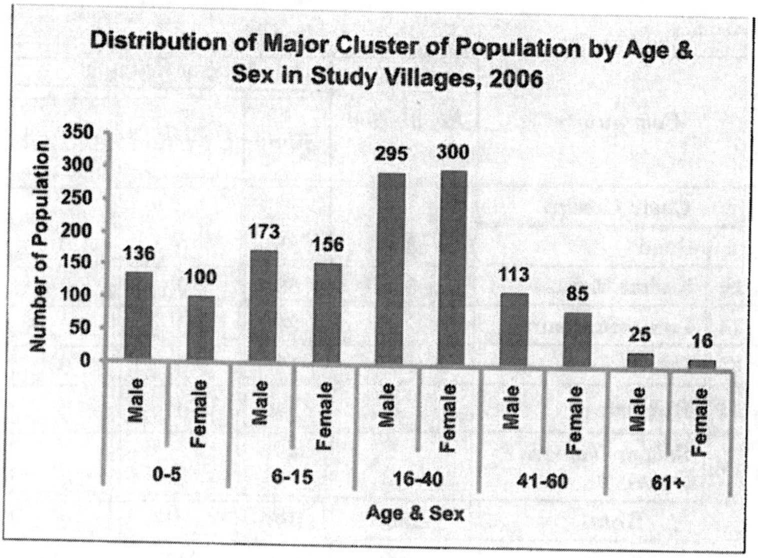

Chart 3

Table 8. Community Wise Population Distribution by Religion

	Community	*Population*	*Type of Religion*		
			Hindu	*Christian*	*Tribal*
	Tribal Groups				
1	Kol/Ho	504	429	75	0
2	Kharia	142	142	0	0
3	Bhatudi	205	205	0	0
4	Bhumij	72	72	0	0
5	Munda	141	0	141	0
6	Santal/Majhi	69	44	25	0
7	Mahali	10	10	0	0
8	Makdiya/Birhor	68	0	0	68
	Total	1211	902	241	68

(contd...)

Community		Population	Type of Religion		
			Hindu	Christian	Animist
Caste Groups					
9	Goud	85	85	0	0
10	Kudmi Mahato	63	63	0	0
11	Lohar/Bindhani	21	21	0	0
12	Teli	12	12	0	0
13	Kaibarta	4	4	0	0
14	Sodgop (Gobora Caste)	3	3	0	0
Total		188	188	0	0
Grand Total		1399	1090	241	68

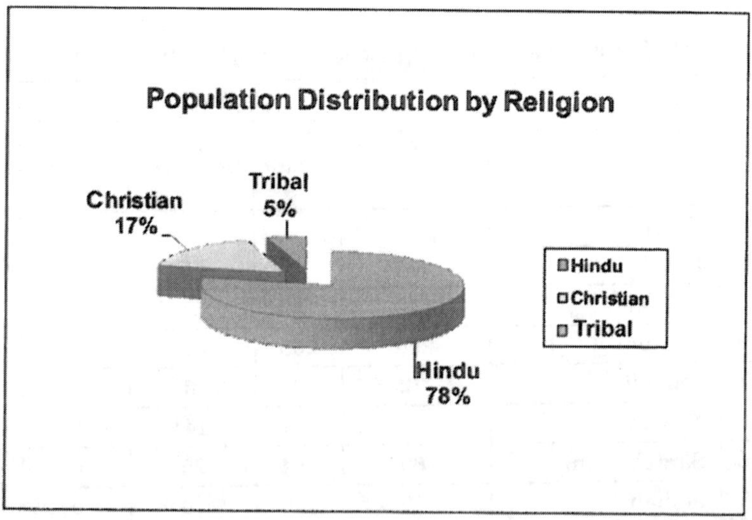

Chart 4.

And at the same time inter-linkages among all the groups living in that area persist in the sphere of social, economic, political, religious and cultural activities. But such inter-linkages are sometimes harmonious and conflict in nature which are determined by availability as well as sharing of resources in a given time and space. Similipal being a bio-geographical region hosts the habitats of various ethnic groups, spread in different villages within its boundary and also in its periphery. The people living in that area have their close link with the nature and their socio-cultural pattern is basically eco-based and eco-centred.

In order to find out the prevailing ethnoscape of the area the primary data on ethnic composition were gathered from nine buffer zone villages, which were simultaneously substantiated with information collected from different core, buffer and transitional zone villages of Jashipur and Bangriposi blocks. The data reveal that the area is predominantly an **adivasi** region. Though it is an adivasi dominated area some caste population are also sporadically distributed (table 9).

Table 9. Buffer Zone Villages

Block	*Gram Panchayat*	*Village*	*Community*	
			Tribe	Caste
Jashipur	Gudgudia	1. Badkasira	Kol/Ho, Bathudi, Santal/ Majhi	
		2. Barigaon	Kol/Ho, Kharia	Goud
		3. Bharrachua	Kol/Ho	
		4. Bilapagha	Kol/Ho, Bathudi,	Kudmi Mahato
		5. Chandikhaman	Kol/Ho	
		6. Gudgudia*	Kol/Ho, Kharia, Mahali, Bathudi	Goud, Teli, Bindhani Lohar, Kaibarta, Sadgop/ Gobra
		7. Khariadunguri	Kol/Ho	
		8. Khejuria/Khejuri	Kol/Ho, Kharia	Goud

Block	Gram Panchayat	Village	Community	
			Tribe	Caste
		9. Kuanribil	Kol/Ho	Goud
		10. Kumari *	Bathudi, Bhumij	Goud
		11. Kusumi	Kol/Ho, Bathudi, Bhumij	Bindani/ Lohar, Goud
		12. Nenjaghosra/ * Lanjighosra	Kol/Ho	
		13. Sankasira	Kol/Ho, Bathudi	
		14. Saharpat	Kol/Ho, Bhumij	Bindani/ Lohar, Goud
		15. Kondabil/Kundabil	Bathudi, Bhumij	Bindani/ Lohar, Goud
		16. Kabatghai*	Kol/Ho, Santal/Majhi	
		17. Jamunagarh*	Kol/Ho	
		18. Jenabil*	Kol/Ho, Bathudi	
		19. Bakua*	Kol/Ho	
	Barheipani	1. Asabani/ Asanbani	Kol/Ho, Munda	
		2. Barheipani	Kol/Ho	Goud, Bindhani/ Lohar,
		3. Bad-Uski	Kol/Ho	
		4. Jhatipani	Kol/Ho, Munda	
		5. Kolha	Kol/Ho	
		6. Nuniaguda	Kol/Ho	
		7. Rautola	Kol/Ho, Santal/Majhi	

Block	*Gram Panchayat*	*Village*	*Community*	
			Tribe	*Caste*
		8. San-Uski/ Sanasoki	Kol/Ho	
		9. Bandribasa	Kol/Ho, Munda	
		10. Barsia	Kol/Ho	Bindani/ Lohar
		11. Chakundakocha *	Kol/Ho	
		12. Haladia	Kol/Ho	
		13. Kuljhari/ * Kolajhari/ Koljhari	Kol/Ho	
		14. Adelkacha	Munda	
		15. Jojogoda	Munda	
		16. Phulbari/ Phulbadia	Munda	
		17. Chahala	Abandon	
		18. Brundaban	Abandon	
		19. Karkachia	Abandon	
	Astakunar	1. Astakunar	Kol/Ho, Kharia	Bindani/ Lohar
		2. Balarampur	Kol/Ho,	
		3. Burhabalanga	Kol/Ho, Munda	
		4. Garh-Similipal/ * Gad Similipal	Kol/Ho, Munda, Santal/Majhi	Kudmi Mahato, Goud
		5. Gopinathpur	Munda	
		6. Kukurbhuka	Kol/Ho, Munda	
		7. Lembhujharan	Kol/Ho, Munda	Bindani/ Lohar

Block	*Gram Panchayat*	*Village*	*Community*	
			Tribe	*Caste*
		8. San-Uski/ Sanasoki	Kol/Ho	
		9. Bandribasa	Kol/Ho, Munda	
		10. Barsia	Kol/Ho	Bindani/ Lohar
		11. Chakundakocha *	Kol/Ho	
		12. Haladia	Kol/Ho	
		13. Kuljhari/ * Kolajhari/ Koljhari	Kol/Ho	
		14. Adelkacha	Munda	
		15. Jojogoda	Munda	
		16. Phulbari/ Phulbadia	Munda	
		17. Chahala	Abandon	
		18. Brundaban	Abandon	
		19. Karkachia	Abandon	
	Astakunar	1. Astakunar	Kol/Ho, Kharia	Bindani/ Lohar
		2. Balarampur	Kol/Ho,	
		3. Burhabalanga	Kol/Ho, Munda	
		4. Garh-Similipal/ * Gad Similipal	Kol/Ho, Munda, Santal/Majhi	Kudmi Mahato, Goud
		5. Gopinathpur	Munda	
		6. Kukurbhuka	Kol/Ho, Munda	
		7. Lembhujharan	Kol/Ho, Munda	Bindani/ Lohar
		8. Makabari	Kol/Ho, Munda	Bindani/ Lohar

Block	Gram Panchayat	Village	Community	
			Tribe	Caste
		9. Nawana/Naana	Kol/Ho	Bindani/ Lohar
		10. Nikhirda	Kol/Ho, Munda	
		11. Saruda *	Munda	
		12. Jadadihi	Kol/Ho	
		13. Kiajhari	Kol/Ho	
		14. Daldali	Abandon	
		15. Dhudruchampa	Abandon	
	Paudagarh	1. Rajupal/Rajapal	Kol/Ho	
Bangiriposi/ Bangriposi	Sarispal	1. Basilakacha *	Santal/Majhi	
		2. Amdapani	Santal/Majhi	
		3. Chakidi (Chakidihipanipal)	Santal/Majhi, Kol/Ho, Kharia	
		4. Jamtolia/ Jamtalia	Santal/Majhi	
		5. Jerkeni/Jarkani	Santal/Majhi	
		6. Khadighati	Santal/Majhi	
		7. Kukurbhuka	Santal/Majhi, Kol/Ho,	
		8. Kusumtota	Santal/Majhi, Kol/Ho,	
	Brahmangaon	1. Charbandha	Santal/Majhi, Kol/Ho	
		2. Dantiakocha	Santal/Majhi, Kol/Ho	
		3. Alapani	Santal/Majhi, Kol/Ho	
		4. Barubeda/ Baribeda	Santal/Majhi, Kol/Ho	
		5. Bhoduakocha	Santal/Majhi, Kol/Ho	
		6. Phuljhari/ Fuljhara	Santal/Majhi, Kol/Ho	
		7. Purunapani	Santal/Majhi, Kol/Ho	

*Buffer zone study villages *Core zone villages

The people living in such area have had long history about their migration, settlement, occupation etc. though the scope of its detailed analysis is out of the purview of present exercise. In this context only a brief discussion on above noted aspects may throw some light about their relationship with the environment.

The ethnic composition of different villages clearly indicates that among all groups the Kolha/Ho and Santal/Majhi are almost widely distributed throughout Similipal region. The other tribal groups viz Munda, Bathudi, Bhumij, Khadia, Bhuniya etc. are also found to live, but their distribution is limited to some villages of Gudgudia, Barheipani and Astakunar, Dhalabani, Paudagarh gram panchayats of Jashipur blocks. The main concentration of the Ho population is in north, central and western part of Similipal while in its eastern and north eastern side is dominated mainly by the Santal population. The Mankhadia or more familiar as Makdiya (also called as Birhor) until recently were nomadic type, have been given settlement by the Govt. in different transitional villages. Like the Makdiya, the Khadia have also been settled in some villages.

Unlike the tribal population, concentration of caste groups barring the Kudmi Mahato (who are fighting for tribal status in recent time) does not prevail in the said area. The Goud, Lohar/Bindhani, and other castes are scatteredly found in different buffer and transitional villages, but they are less in number. The Kudmi Mahato are predominantly concentrated in north, north western and north eastern fringe of Similipal, mostly found in transitional villages bordering Keonjhar district of Orissa and Jharkhand state. Of all study villages only Gad Similipal is having such group. In transitional villages several other castes namely Kumbhar (potter), Dhoba (washerman), Patra/Tanti (weaver), Barik (barber), Karan (scribe) etc. are found to live with various tribal groups. The villages in transitional zone are, by and large, multiethnic in character.

The ethnic composition in whole of the Similipal area clearly pinpoints that the adivasis form the dominant group in the sphere of numerical strength and wider distribution. Majority of them were originated from same racial stock, popularly known as *Austro-Asiatic* origin and they have also some linguistic and cultural affinities among several groups. Rice is their staple food and they are very much addicted to rice-beer, known as *handia.* Addiction to rice-beer is regarded as part of their culture and even various caste populations are not free from this habit.

The tribal groups such as Kolha, Munda, Santal, Khadia, Mankhadia, Bhumij, Mahali are bilingual in nature. And many of them are accustomed to speak in Hindi and Bengali. The Bathudi have given up their own language and Oriya has become their mother tongue. The caste people barring the Kudmi Mahato and the Lohar are mainly the Oriya speakers. Of course, sometimes they are bilingual. However, the Kudmi Mahato still retain their own language. As Similipal falls under the political boundary of Orissa state, Oriya gets the status of lingua-franca of the people. Thus from the language speaking behaviour of the people indicates that the people of the area, whether they may be the tribe or the caste, are in the process of multilingual and heterogeneous cultural atmosphere where each ethnic specific language and culture remain docile for a longer period due to influences of the dominant state language like Oriya and regional languages such as Bengali and Hindi. And in course of time such language may be forged with the dominant language and culture as happened among the Bathudi and other groups. In fact, the geographical location of Similipal in such a place where the people are in the process of constant influences of three dominant regional languages i.e. Oriya, Hindi and Bengali. Because the place is encircled by three states - Orissa, Jharkand and West Bengal.

From historicity it is revealed that a mass of the existing people belonging to different ethnic groups living in the area once migrated to this part of the country from neighbouring Jharkhand (prior to November 2000 it was the part of Bihar), West Bengal and also from other parts of old Mayurbhanj State. That is why Similipal region has become a multilingual and multicultural zone and even after merger of the Mayurbhanj State with Orissa in 1949 such feature remains alive that still continues. One point may be mentioned that how the villages came into existence inside the forest that part of history remains obscure. But from people's view point it may be taken into consideration that the villages were set up during the feudal days as the inhabitants used to be engaged for different assignments like palanquin bearers, making forest path etc. when the feudal kings and their advisors entered into the forest for hunting wild games and holiday-making. Some remnants still bear the witness of such activities during that time in some locations like Chahala, Barheipani etc. which are now regarded as the place of interests to the tourists.

In the past interactional pattern among different ethnic groups were conspicuously observed in the sphere of hunting, commonly familiar in the area as *akhand sikhar* (a mass hunting of wild animals was practised by the tribals on an auspicious day, known as *Pana Sangkranti* day observed usually during April), sharing of labour during house-building and agricultural activities etc. During those days community cohesiveness was more rigid and individualistic attitude never got importance in community life. This type of relations clearly reflects that in an isolated geographical terrain like Similipal where the people of different groups used to follow a code of cooperation for sharing of available resources for their sustenance and their needs were minimum. But unlike the past particularly in post independence era, such situation does not prevail now. Interactional pattern among various communities has been transformed considerably. Now inter-personal and inter-community relations are primarily determined by economic yardsticks like individual landholding, manpower, holding of livestock population and other goods having certain economic value.

The 'in-migration' and 'out-migration' of insiders and outsiders into such forest location and elsewhere as found in Gudgudia and other villages of buffer and transitional areas provide a space for interactions among various communities. Such interactions are having wider implications in the arena of community, gender and socio-economic empowerment. The existence of various women self-help groups (constituted from the women of different communities namely Kolha, Bathudi, Bhumij, Lohar, Goud, Keot or Kaibarta, etc.) prevailed in different villages is the reflection of on-going process of one kind of gender empowerment. Functioning of self-help groups is an outcome of induced economic development in this backward area. For smooth running of SHGs ethnic sentiment of various communities gets special importance. Because community identity cannot be overlooked or ruled out as it plays a covert role in strengthening its (SHG) longevity.

Maintaining ethnic identity in a particular space is one of the vital components of ethnoscape. Hence every group in Similipal still maintains their own identity despite various onslaughts of cultural invasions. Modernity gradually penetrates in the life of the communities and this cannot be stopped. Similipal being a tourist spot attracts a large number of tourists every year from whom the forest dwelling communities borrow various cultural traits which are being penetrated in their social system. For instance, tribal youths irrespective of their ethnic background are now accustomed to use trousers and shirts which were hitherto unknown to them.

It is noticed that some villages are ethnically localised i.e. the village or hamlet is exclusively occupied by a particular ethnic group and some are overlapped where ethnic boundary is redundant specially observed in multiethnic villages. This has prevailed under various socio-economic and natural calamities conditions such as drought and famine, in search of livelihood resource, space crunch for making new habitation area that has been compounded by imposition of restrictions on further encroachment of forest lands by the forest authorities. Restrictions on encroachment of forest lands compelled various communities to share the space (which were unutilised and remained vacant or fallow), available in the village and also to adjust among themselves. In that case maintaining ethnic boundary has become sidelined; rather survival interests get priority to them.

Encompassing ethnoscape it is necessary to examine some other aspects namely livelihood condition primarily related to occupation, cultural pattern, inter-ethnic relations etc. These aspects are relevant and interwoven in finding the actual nature of ethnoscape prevailed in the area.

Land is the principal economic resource to the people of the region and agriculture is the mainstay of their economy which is substantiated by forest resources available in their immediate environment. That means, the people are sustained by both agriculture and forest resources which are complimentary to each other and significant in their life. The people living in the fringe area of Similipal are mainly marginal and small farmers and landless labour as far as landholding position is concerned. The landholding data were not available for critical analysis. Land records on some study villages available in revenue department were very clumsy and obscure that need for proper scrutiny for drawing a concrete inference about their actual landholding position. However, the overall living condition, socio-economic background and material possession of the people indicate that they are practically poverty-stricken people and whatever they produce are inadequate to sustain their life. In order to mitigate their problem for sustenance almost every family is engaged as wage labour/casual labour. Both male and female take part in agriculture, collection of forest produces, road works etc.

Paddy is the principal crop of the area grown mainly in rainfed condition. All communities barring Makdiya (Birhor) and the Khadia (who are mainly dependent on gathering of various minor forest produces like date palm leaves, *sabui* grass, sal leaves, bamboo etc. used for making of mats, rope, plates, baskets etc) maintain their livelihood through agriculture. But this does not mean that they do not exploit forest resources. Once they are in the midst of forest location as well as in the fringe of forest area it is obvious that certain forest resources such as fuel wood, building materials (wooden posts, grasses, leaves, bamboo etc.) are collected by them for generations and forest is the only source of these materials to them. Hence forests play a significant role in the life of all the communities living in SBR region and it cannot be segregated from their economy.

Socio-economic background

Unlike the foraging communities the people of different ethnic origin living in SBR are primarily settled in nature and their economy is based mainly on subsistence agriculture which is substantiated by forest resources. Both land and forest resources available to them are not adequate to maintain their livelihood. Because agricultural lands are mainly confined to valley area, which are being, cultivated for generations without any crop rotations that has retarded the yield of crops. And of late, with the declaration of biosphere by the state scope for further expansion of agricultural area has been restricted. The biosphere issue has not only stopped the expansion of agricultural land but have also curtailed their right for extraction of forest produces considerably as they used to enjoy this privilege conventionally in the past. Now their economy is practically in jeopardized condition.

To them two major resources - land and forests are most vital for their survival. Since they live in the forests for generations, obviously their socio-cultural and economic systems are largely influenced by the forest ecology. Their economy is operative in three ways - by gathering of minor forest produce, by shifting agriculture and by settled agriculture. Of these three systems first two were more effective in the past among various ethnic groups as they were in isolation from the purview of market economy because of inaccessibility to their terrain from the influence of market forces. Traditionally, barter system was most effective mechanism to share resources, especially related to various minor forest produces and some kind of agricultural produces grown by *podu* (shifting) cultivation. Till the second quarter of the 20th century no market was in existence inside the SBR forests and their exposure to market economy was too limited and they were not empowered to understand the importance of money value and its use. Later when the communication facilities were extended to that area by virtue of tourism promotion, implementation of various welfare programmes by the Govt. and the NGOs, functioning of educational institutions, self-help groups (SHGs), weekly markets in different parts of the SBR the people have gained some empowerment

Table 10 Occupational Category in Study Villages

Sl. No.	*Village*	*Community*	*Occupational Category*	
			Primary	*Secondary*
1	Gudgudia	Kol/Ho	Agriculture	1. Labour 2. Artisan 3. Medicineman 4. Business 5. Driver
		Kharia	Minor Forest Products Collection	Labour
		Mahali	Basketry	Labour
		Bathudi	Service	Agriculture
		Goud	1. Agriculture 2. Service 3. Labour	1. Fishing 2. Carpentry 3. Business (Tea Shop Owner) 4. Liquor Business
		Lohar/ Bindhani	1. Service 2. Blcksmith 3. Labour	Labour
		Teli	1. Agriculture 2. Business	1. Agriculture 2. Business 3. Service
		Kaibarta	Service	Agriculture
		Sadgop (Gobora Caste)	Service	
2	Nenjaghosra/ Lanjighosra	Kol/Ho	1. Agriculture 2. Labour 3. Cattle Herding	1. Labour 2. Tailoring 3. Liquor Business 4. Business

(contd...)

Sl. No.	Village	Community	Occupational Category	
			Primary	Secondary
3	Kumari	Bathudi	1. Agriculture 2. Service	1. Agriculture 2. Service 3. Labour 4. Business 5. Minor Forests Products Collection
		Bhumij	1. Agriculture 2. Labour 3. Service	1. Labour 2. Minor Forests Products Collection 3. Agriculture
		Goud	1. Agriculture 2. Labour	1. Agriculture 2. Labour
4	Kuljhari/ Koljhari	Kol/Ho	Agriculture	1. Labour 2. Livestock Business 3. Minor Forests Products Collection 4. Business
5	Chakundakocha	Kol/Ho	Agriculture	1. Labour 2. Minor Forests Products Collection
6	Garh-Similipal/ Gad Similipal	Kudmi Mahato	1. Agriculture 2. Service	1. Service 2. Labour 3. Business 4. Agriculture
		Kol/Ho	Agriculture	1. Labour 2. Carpentry
		Munda	Agriculture	Labour
		Santal/Majhi	Agriculture	Labour
		Goud	Agriculture	Labour
7	Saruda	Munda	1. Agriculture 2. Service	1. Labour 2. Agriculture
8	Basilakacha/ Basilakocha	Santal/ Majhi	Agriculture	Minor Forests Products Collection
9	Durdura Tanda	Makdiya/ Birhor	1. Labour 2. Rope Making 3. agriculture	Minor Forests Products Collection

Table 11. Live stock population in study villages

Sl. No.	*Village*	*Cow*	*Bullock*	*Buffalo*	*Goat*	*Poultry*	*Others*
1	Gudgudia	95	68	19	143	383	9
2	Nenjaghosra / Lanjighosra	31	37	4	113	146	3
3	Kumari	46	40	10	121	184	0
4	Kuljhari/ Koljhari	60	41	0	171	190	0
5	Chakunda-kocha	16	26	8	66	43	0
6	Garh Similipal / Gad Similipal	93	46	18	177	165	4
7	Saruda	63	4	2	104	199	0
8	Basilakacha / Basilakocha	30	15	0	31	54	0
9	Durdura Tanda	2	0	0	39	21	0
	Total	436	277	61	965	1385	16

and level of awareness in determining the value of their commodities. Now they are considerably exposed to market economy.

Both barter system and shifting cultivation are no longer operative in the area. Almost all the communities follow settled cultivation though some sort of gathering economy is still in practice among the Khadia and the Makdiya or Birhor. The existing occupational categories of different ethnic groups provide some insights of economic reality prevailed in such location. Agriculture is regarded as the primary occupation to almost all communities barring the Mahali, the Khadia and the Birhor (table 10). But labour gets the status of secondary occupation to all communities. This indicates that the people cannot sustain exclusively by agriculture because of uneconomic holdings. The land holding position in every village envisages that apparently majority of the families belong to the class of marginal and small farmers and the cereal crops (primarily paddy) they produce are inadequate to sustain their families. To augment their family income they have no alternate to engage themselves as wage labour (engaged in road works, agricultural activities etc). This is not only found inside the SBR villages but also found in its peripheral areas.

Apart from land and forest resources, livestock is also regarded as another economic resource that plays significant role in the life of the people. Every community is having plough and milch cattle, goat, poultry birds etc. and these are part and parcel of any agrarian community. Majority of the community in SBR are basically the peasants as reflected in their occupational pattern. In agrarian society livestock population is an asset and rearing of cattle, goat, poultry birds etc. are part of their economic life. In SBR a large number of cattle population are seen to graze within and outside the forest boundary. Rearing of livestock is a reality observed in every study village (table 11). From conservation point of view grazing of cattle in forests is harmful for regeneration of vegetation. But grazing activities are a prevailing phenomenon in this territory because of existence of villages.

Social background

Family Size and Type

The people living inside the forests are regarded as the forest dwelling community whose livelihood depends on available forest resources that are often remained inadequate to support large family as proved universally. The dearth of food security, healthcare, transport and communication facilities (that provide better opportunities for socio-

Table 12. Community Wise Average Family Size

	Community	*Number of Population*	*Number of Family*	*Average Family Size*
	Tribal Groups			
1	Kol/Ho	504	104	4.85
2	Kharia	142	36	3.94
3	Bathudi	205	39	5.26
4	Bhumij	72	13	5.54
5	Munda	141	24	5.88
6	Santal/Majhi	69	13	5.31
7	Mahali	10	3	3.33
8	Makdiya/Birhor	68	18	3.78
	Tribe Wise Average Family Size			4.74
	Caste Groups			
9	Goud	85	20	4.25
10	Kudmi Mahato	63	15	4.20
11	Lohar/ Bindhani	21	5	4.20
12	Teli	12	3	4.00
13	Kaibarta	4	1	4.00
14	Sadgop (Gobora Caste)	3	1	3.00
	Caste Wise Average Family Size			3.94

economic mobilities) etc. are directly or indirectly responsible for making the family size small. In other sense, nature influences to curb population growth to some extent. This phenomenon is quite applicable to study area where the people are in dearth of livelihood resources, healthcare and communication facilities.

The study (based on nine buffer zone villages) reveals that the average family size of the tribal and caste people living in that area is around 5.00 and 4.00 respectively (table 12, chart 5 and 6).

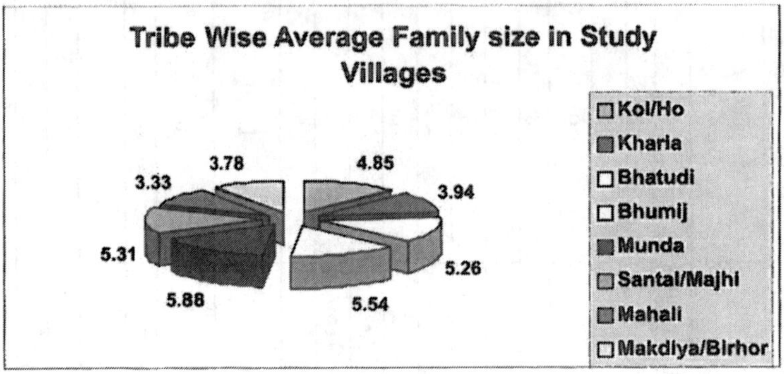

Chart 5

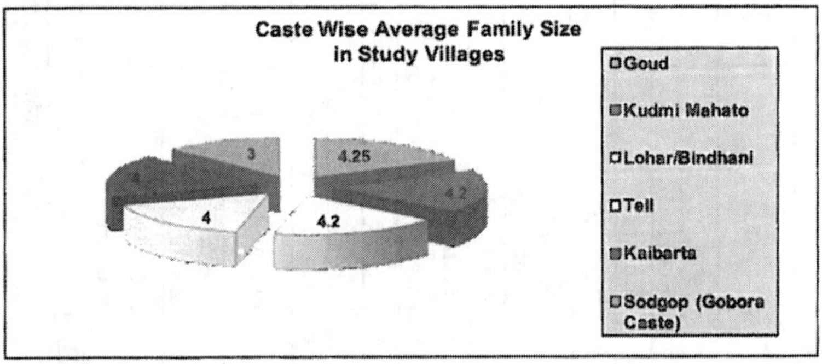

Chart 6

The population growth during the span of last 50 years (1961 - 2001) is not so significant in such location because of the aforementioned reasons. Even migration of population hailing from different ethnic background over the years did not influence much in such context.

Table 13 Population Distribution by Family Type in Study Villages

Sl. No.	*Village Name*	*Community*	*Family Type*			
			Nuclear	*Joint*	*Broken*	*Total*
1	Gudgudia	Kol/Ho	30	5	0	35
		Kharia	31	5	0	36
		Mahali	2	1	0	3
		Bathudi	1	0	0	1
		Goud	13	1	0	14
		Lohar/Bindhani	5	0	0	5
		Teli	1	2	0	3
		Kaibarta	1	0	0	1
		Sadgop (Gobora Caste)	1	0	0	1
		Total	85	14	0	99
2	Nenjaghosra/ Lanjighosra	Kol/Ho	18	1	2	21
3	Kumari	Bathudi	32	6	0	38
		Bhumij	9	4	0	13
		Goud	5	0	0	5
		Total	46	10	0	56
4	Kuljhari/Koljhari	Kol/Ho	16	1	1	18
5	Chakundakocha	Kol/Ho	9	2	0	11

(contd...)

Sl. No.	*Village Name*	*Community*	*Family Type*			
			Nuclear	*Joint*	*Broken*	*Total*
6	Garh-Similipal/Gad Similipal	Kudmi Mahato	9	5	1	15
		Kol/Ho	15	4	0	19
		Munda	1	0	0	1
		Santal/Majhi	1	0	0	1
		Goud	1	0	0	1
		Total	27	9	1	37
7	Saruda	Munda	14	8	1	23
8	Basilakacha/ Basilakocha	Santal/Majhi	8	4	0	12
9	Durdura Tanda	Makdiya/Birhor	13	3	2	18
Total			236	52	7	295

With regard to family type it is highlighted that three types of families are in existence among all ethnic groups of which nuclear category secures predominant position (80 %), followed by joint (18 %) and broken (2%) types (tables 13 and 14, charts 7, 8 and 9).

Table 14. Community Wise Population Distribution by Family Type

Community		*Family Type*			
Tribal Groups		*Nuclear*	*Joint*	*Broken*	*Total*
1	Kol/Ho	88	13	3	104
2	Kharia	31	5	0	36
3	Bhatudi	33	6	0	39
4	Bhumij	9	4	0	13
5	Munda	15	8	1	24
6	Santal/ Majhi	9	4	0	13
7	Mahali	2	1	0	3
8	Makdiya/ Birhor	13	3	2	18
Total		200	44	6	250
Caste Groups					
9	Goud	19	1	0	20
10	Kudmi Mahato	9	5	1	15
11	Lohar/ Bindhani	5	0	0	5
12	Teli	1	2	0	3
13	Kaibarta	1	0	0	1
14	Sadgop (Gobora Caste)	1	0	0	1
Total		36	8	1	45
Grand Total		236	52	7	295

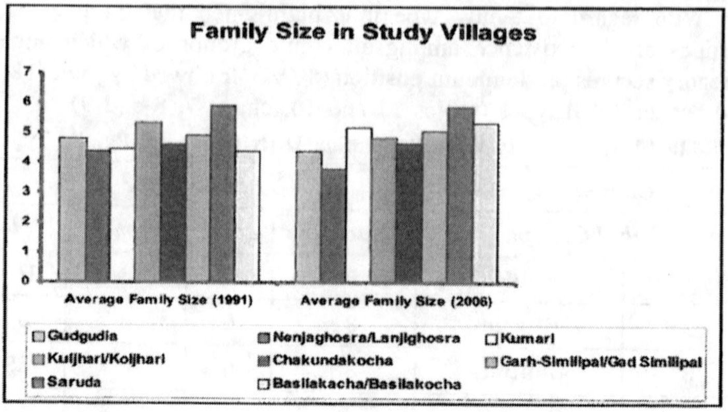

Chart 7

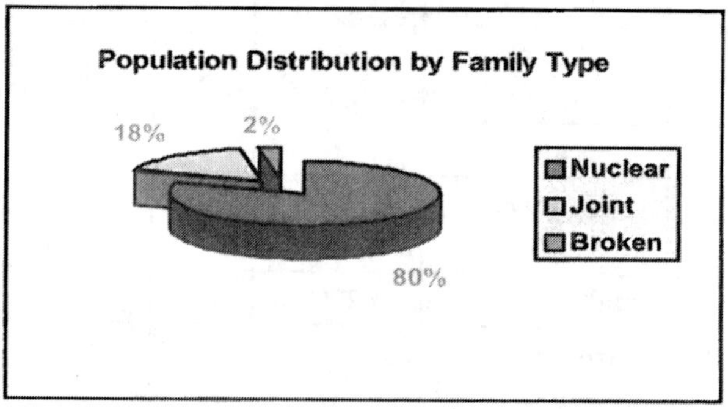

Chart 8

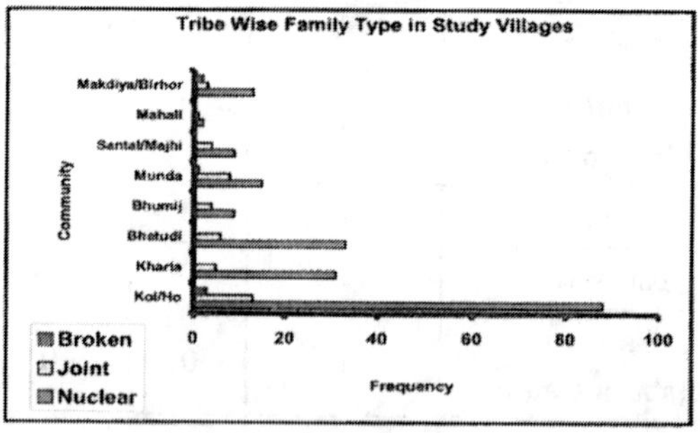

Chart 9

MARITAL STATUS

Monogamy is the rule among all communities. As per marital status the population is classified into four categories viz. unmarried, married, widowed and divorcee. It is significantly noticed that both unmarried and married categories are dominant among all groups and the rests are insignificant who are part of the system (tables 15, 16 and charts 10, 11).

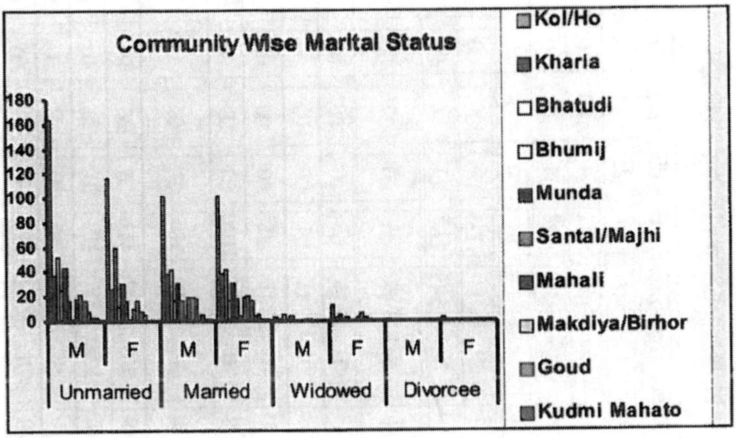

Chart 10

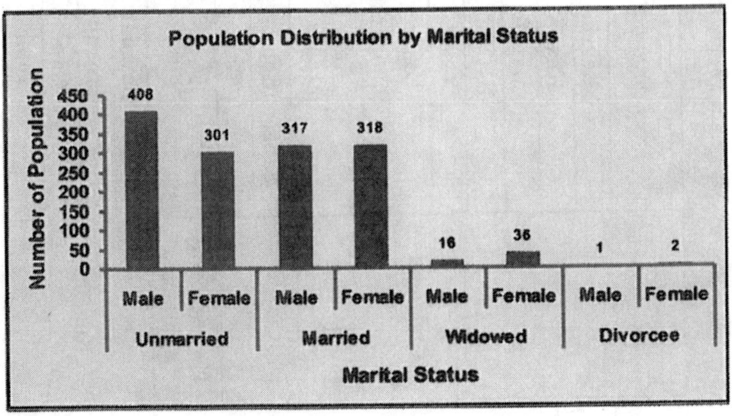

Chart 11

Table 15. Population Distribution by Marital Status in Study Villages

Sl. No.	Village Name	Community	Unmarried		Married		Widowed		Divorcee		Total		
			M	F	M	F	M	F	M	F	M	F	Total
1	Gudgudia	Kol/Ho	57	50	31	31	1	10	0	1	89	92	181
		Kharia	37	25	38	38	1	3	0	0	76	66	142
		Mahali	1	1	4	4	0	0	0	0	5	5	10
		Bathudi	3	1	1	1	0	0	0	0	4	2	6
		Goud	13	8	14	15	1	4	0	0	28	27	55
		Lohar/ Bindhani	7	5	4	4	0	1	0	0	11	10	21
		Teli	2	0	5	5	0	0	0	0	7	5	12
		Kaibarta	2	0	1	1	0	0	0	0	3	1	4
		Sadgop (Gobora Caste)	1	0	1	1	0	0	0	0	2	1	3
		Total	123	90	99	100	3	18	0	1	225	209	434
2	Nenjaghosra/ Lanjighosra	Kol/Ho	32	8	19	19	1	0	1	0	53	27	80
3	Kumari	Bathudi	49	58	41	41	5	5	0	0	95	104	199
		Bhumij	25	11	17	17	1	1	0	0	43	29	72
		Goud	5	6	4	4	0	2	0	0	9	12	21
		Total	79	75	62	62	6	8	0	0	147	145	292

(contd...)

Sl. No.	Village Name	Community	Unmarried		Married		Widowed		Divorcee		Total		
			M	*F*	*M*	*F*	*M*	*F*	*M*	*F*	*M*	*F*	*Total*
4	Kuljhari/ Koljhari	Kol/Ho	26	25	16	16	1	3	0	0	43	44	87
5	Chakunda-kocha	Kol/Ho	11	13	13	13	0	0	0	1	24	27	51
6	Garh-Similipal/ Gad Similipal	Kudmi Mahato	17	8	18	17	1	2	0	0	36	27	63
		Kol/Ho	38	21	23	23	0	0	0	0	61	44	105
		Munda	1	3	1	1	0	0	0	0	2	4	6
		Santal/ Majhi	2	1	1	1	0	0	0	0	3	2	5
		Goud	4	3	1	1	0	0	0	0	5	4	9
		Total	62	36	44	43	1	2	0	0	107	81	188
7	Saruda	Munda	42	28	29	29	4	3	0	0	75	60	135
8	Basilakacha/ Basilakocha	Santal/ Majhi	15	16	16	17	0	0	0	0	31	33	64
9	Durdura Tanda	Makdiya/ Birhor	18	10	19	19	0	2	0	0	37	31	68
Total			408	301	317	318	16	36	1	2	742	657	1399

Table 16. Community Wise Population Distribution by Marital Status in Study Villages

Community		Unmarried		Married		Widowed		Divorcee		Total		
		Male	*Female*	*Male*	*Female*	*Male*	*Female*	*Male*	*Female*	*Male*	*Female*	*Total*
Tribal Groups												
1	Kol/Ho	164	117	102	102	3	13	1	2	270	234	504
2	Kharia	37	25	38	38	1	3	0	0	76	66	142
3	Bathudi	52	59	42	42	5	5	0	0	99	106	205
4	Bhumij	25	11	17	17	1	1	0	0	43	29	72
5	Munda	43	31	30	30	4	3	0	0	77	64	141
6	Santal/ Majhi	17	17	17	18	0	0	0	0	34	35	69
7	Mahali	1	1	4	4	0	0	0	0	5	5	10
8	Makdiya/ Birhor	18	10	19	19	0	2	0	0	37	31	68
	Total	357	271	269	270	14	27	1	2	641	570	1211

(contd...)

	Community	Unmarried		Married		Widowed		Divorcee		Total		
		Male	*Female*	*Male*	*Female*	*Male*	*Female*	*Male*	*Female*	*Male*	*Female*	*Total*
Caste Groups												
9	Goud	22	17	19	20	1	6	0	0	42	43	85
10	Kudmi Mahato	17	8	18	17	1	2	0	0	36	27	63
11	Lohar/ Bindhani	7	5	4	4	0	1	0	0	11	10	21
12	Teli	2	0	5	5	0	0	0	0	7	5	12
13	Kaibarta	2	0	1	1	0	0	0	0	3	1	4
14	Sadgop (Gobora Caste)	1	0	1	1	0	0	0	0	2	1	3
	Total	51	30	48	48	2	9	0	0	101	87	188
	Grand Total	408	301	317	318	16	36	1	2	742	657	1399

EDUCATIONAL STATUS

The people living in villages are almost deprived of the basic education facilities. Majority of villages barring Gudgudia and other some villages in whole of buffer and core zones are devoid of school facilities especially primary schools. The lack of school facilities creates problems to the children in pursing basic education and those schools located in some villages are not easily accessible to the small children who are, by and large, remained in their homes. And the schools functioning in some villages are not up to the mark due to various reasons. That is why a large number of illiterate population is found to exist among almost all communities. Illiteracy is more prevalent among the female population. The tables (17 and 18) and charts (12, 13 and 14) bear the witness of existing educational reality prevailed in study area.

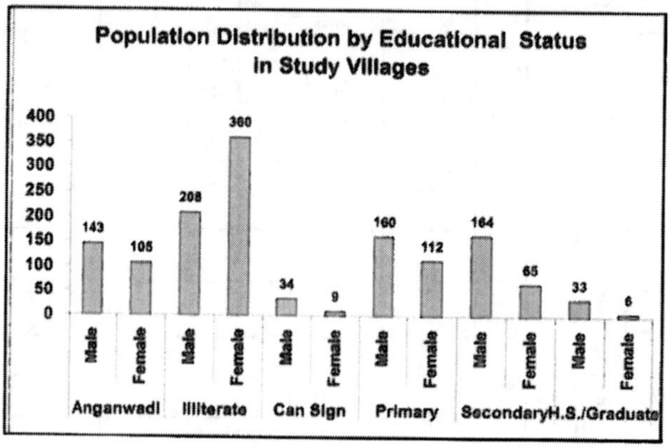

Chart 12

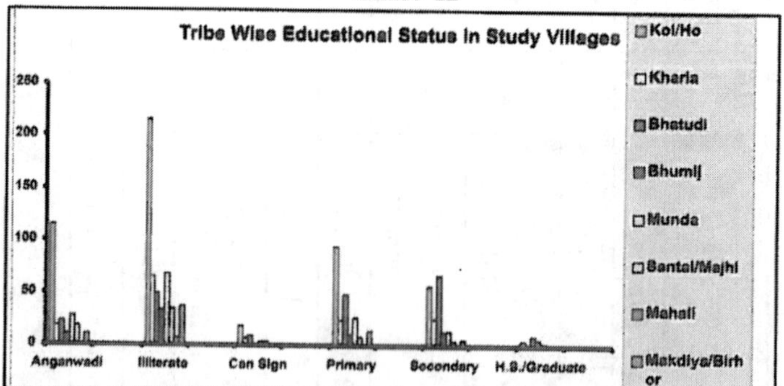

Chart 13

Table 17. Population Distribution by Educational Status in Study Villages

Sl. No.	*Village Name*	*Community*	*Literacy*												*Total*		
			Anganwadi		*Illiterate*		*Can Sign*		*Primary*		*Secondary*		*H.S./ Graduate*				
			M	*F*	*M*	*F*	*M*	*F*	*M*	*F*	*M*	*F*	*M*	*F*	*M*	*F*	*Total*
1	Gudgudia	Kol/Ho	16	17	23	46	3	1	21	19	23	9	3	0	89	92	181
		Kharia	11	8	24	42	6	1	12	12	22	3	1	0	76	66	142
		Mahali	0	1	3	4	0	0	1	0	1	0	0	0	5	5	10
		Bathudi	1	0	0	0	0	1	1	1	2	0	0	0	4	2	6
		Goud	6	3	4	15	0	0	8	3	10	6	0	0	28	27	55
		Lohar/ Bindhani	3	1	2	2	0	1	5	5	1	1	0	0	11	10	21
		Teli	0	0	0	1	0	0	0	1	2	1	5	2	7	5	12
		Kaibarta	0	0	0	0	0	0	2	0	0	1	1	0	3	1	4
		Sadgop (Gobora Caste)	1	0	0	0	0	0	0	0	0	1	1	0	2	1	3
		Total	38	30	56	110	9	4	50	41	61	22	11	2	225	209	434

(contd...)

Sl. No.	Village Name	Community	Literacy												Total		
			Anganwadi		Illiterate		Can Sign		Primary		Secondary		H.S./ Graduate				
			M	F	M	F	M	F	M	F	M	F	M	F	M	F	Total
2	Nenjaghosra/ Lanjighosra	Kol/Ho	15	3	19	21	1	0	12	2	6	1	0	0	53	27	80
3	Kumari	Bathudi	12	10	9	40	5	3	20	26	41	24	8	1	95	104	199
		Bhumij	8	2	15	18	0	0	8	2	6	7	6	0	43	29	72
		Goud	1	2	3	8	0	0	2	2	3	0	0	0	9	12	21
		Total	21	14	27	66	5	3	30	30	50	31	14	1	147	145	292
4	Kuljhari/Koljhari	Kol/Ho	10	15	20	23	0	0	9	5	3	1	1	0	43	44	87
5	Chakundakocha	Kol/Ho	8	9	10	17	0	0	5	1	1	0	0	0	24	27	51
6	Garh-Similipal/ Gad Similipal	Kudmi Mahato	1	3	7	9	1	0	11	8	11	5	5	2	36	27	63
		Kol/Ho	14	8	12	25	11	2	12	8	12	0	0	1	61	44	105
		Munda	0	2	1	2	1	0	0	0	0	0	0	0	2	4	6
		Santal/ Majhi	1	0	0	1	1	0	0	1	1	0	0	0	3	2	5
		Goud	0	3	5	1	0	0	0	0	0	0	0	0	5	4	9
		Total	16	16	25	38	14	2	23	17	24	5	5	3	107	81	188

(contd...)

Sl. No.	Village Name	Community	Literacy												Total		
			Anganwadi		Illiterate		Can Sign		Primary		Secondary		H.S. / Graduate				
			M	F	M	F	M	F	M	F	M	F	M	F	M	F	Total
7	Saruda	Munda	18	8	25	40	2	0	17	9	11	3	2	0	75	60	135
8	Basilakacha/ Basilakocha	Santal/ Majhi	10	7	10	24	2	0	5	2	4	0	0	0	31	33	64
9	Durdura Tanda	Makdiya/ Birhor	7	3	16	21	1	0	9	5	4	2	0	0	37	31	68
Total			143	105	208	360	34	9	160	112	164	65	33	6	742	657	1399

Table 18. Community Wise Population Distribution by Educational Status in Study Villages

	Community	Literacy												Total		
		Anganwadi		Illiterate		Can Sign		Primary		Secondary		H.S. / Graduate				
		M	F	M	F	M	F	M	F	M	F	M	F	M	F	Total
Tribal Groups																
1	Kol/Ho	63	52	84	132	15	3	59	35	45	11	4	1	270	234	504
2	Kharia	11	8	24	42	6	1	12	12	22	3	1	0	76	66	142
3	Bathudi	13	10	9	40	5	4	21	27	43	24	8	1	99	106	205
4	Bhumij	8	2	15	18	0	0	8	2	6	7	6	0	43	29	72
5	Munda	18	10	26	42	3	0	17	9	11	3	2	0	77	64	141
6	Santal/Majhi	11	7	10	25	3	0	5	3	5	0	0	0	34	35	69
7	Mahali	0	1	3	4	0	0	1	0	1	0	0	0	5	5	10
8	Makdiya/ Birhor	7	3	16	21	1	0	9	5	4	2	0	0	37	31	68
	Total	131	93	187	324	33	8	132	93	137	50	21	2	641	570	1211

(contd...)

	Community	Literacy												Total		
		Anganwadi		Illiterate		Can Sign		Primary		Secondary		H.S. / Graduate				
		M	F	M	F	M	F	M	F	M	F	M	F	M	F	Total
Caste Groups																
9	Goud	7	8	12	24	0	0	10	5	13	6	0	0	42	43	85
10	Kudmi Mahato	1	3	7	9	1	0	11	8	11	5	5	2	36	27	63
11	Lohar/ Bindhani	3	1	2	2	0	1	5	5	1	1	0	0	11	10	21
12	Teli	0	0	0	1	0	0	0	1	2	1	5	2	7	5	12
13	Kaibarta	0	0	0	0	0	0	2	0	0	1	1	0	3	1	4
14	Sadgop (Gobora Caste)	1	0	0	0	0	0	0	0	0	1	1	0	2	1	3
	Total	12	12	21	36	1	1	28	19	27	15	12	4	101	87	188
	Grand Total	143	105	208	360	34	9	160	112	164	65	33	6	742	657	1399

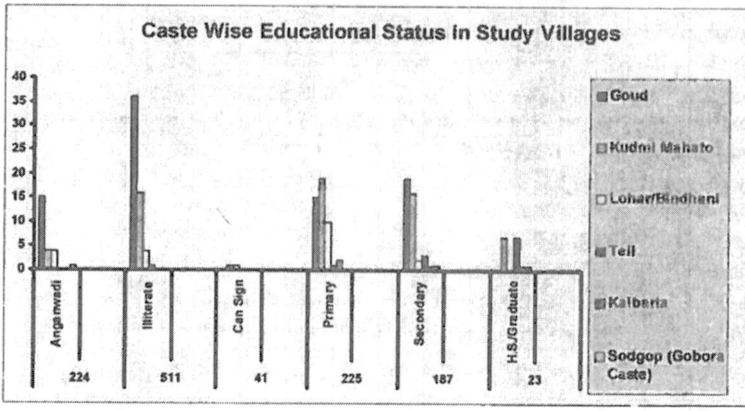

Chart 14

Socio-cultural background of the People

SBR is basically a multicultural zone where various communities live in harmony. Every community maintains their own identity. Despite community identity all communities are in the midst of constant interactions for their own survival. Interactions take place in the sphere of agricultural activities, socio-religious functions like marriage, obsequies, celebration of village deities, festivals etc., managing of natural calamities such as drought, flood, attack of wild animals and so on.

All communities follow monogamy. Sometimes polygyny is permitted under certain conditions. In case of marriage alliance community endogamy and clan exogamy are strictly maintained. Each group is having a good number of clans, known as *kilis* (see appendix I) which are exogamous and totemistic in nature. Adult marriage is the rule. Widow remarriage is allowed. Marriage by negotiation is the norm among all communities and follows patrilocal type of residence. Both nuclear and extended type of families exists. Patrilineal rule of inheritance is commonly prevailed i.e. sons inherit the parental properties. But succession goes to the eldest son only.

Every community used to be governed by their own system of polity which is almost disappeared now. Each village was commonly headed by a headman or Pradhan who used to be assisted by the village elders. That means a village or community council was operative for disposing various social offences. But this system of village governance has disappeared and this has been replaced by the present day Panchayat system formed on the basis of adult franchise which functions for a period of five years. They are now influenced by almost all mainstream political parties such as Congress (I), Jhrarkhand Party, Bharatiya Janata party (BJP), Socialist Unity Centre of India (SUCI) etc. of which SUCI has strong base among the people of SBR area. This party is having units in different SBR villages to organize the poverty stricken forest people for mitigating various grievances from the Govt. Apart from these mainstream parties, Maoists' menace is also gradually taking place in the area as reported by many local inhabitants. This is an emerging trend observed among the people of whole of the SBR region.

The people in SBR, by and large, are primarily the nature worshippers. Conservation of *jahiras* (sacred groves mainly consisted of sal trees, found in every village), worship of *Singbonga* (Sun God), *Badam Thakur* and other spirits bear the witness of their nature worship. Apart from this they worship various Hindu pantheons like Shiva, Krishna, Lakshmi, Jagannath, Manasa, Durga, Kali etc. They also celebrate major Hindu observances such as *ekadasi*, *janmastami*, *shivaratri* etc. Worshipping and celebrating of aforementioned Hindu gods and goddesses and observances are the indication of their assimilation process to the larger Hindu society that leads them to claim as Hindu by religion.

Conclusion

The Similipal Biosphere Reserve is one of the prominent biosphere reserves in India, located in the central part of Mayurbhanj district of Orissa. This biosphere reserve is rich in biodiversity in terms of human population, wildlife and vegetation. The present study is mainly confined to human population only distributed in different villages of core, buffer and transitional zones.

The study covers various aspects of population such as civic amenities, ethnoscape, community wise population structure, socio-cultural background etc. The field investigation was carried out in eight buffer zone and one transitional villages in order to ascertain the aforenoted aspects. The people are broadly divided into two groups - tribes and castes. However, SBR is basically a tribal-dominated area consisted of various ethnic groups namely Kol/Ho, Khadia, Bathudi, Santal/Majhi, Mahali, etc. The castes such as Goud, Teli, Lohar/Bindhani etc. are sporadically distributed in different villages. Both castes and tribes live in a harmonious way. The villages are of both uniethnic and multiethnic type in character. Majority of villages are devoid of basic amenities like school, dispensary, post office, potable water etc. Among all buffer zone villages Gudgudia gets the status of a prominent centre in whole of SBR in terms of its infrastructure, location and exposure.

Population distributed in core and buffer zone villages are not creating any significant pressure on its resources as reflected from population data available during the span of last 50 years (1961-2001). The life span beyond 60 years of age is less due to varied reasons such as lack of healthcare facilities, balance diet, diseases etc. It is noticed that in every ten years of duration on and average number of population growth in the area is around 200 individuals only. This indicates that the area experiences low population growth. Hence the available resources are still able to sustain the existing population in years to come.

The people living in such territory belong to Austro-Asiatic race and majority of them were migrated from the present day Jharkhand region as their social history bears certain witness of this premise. Moreover, geographical location of SBR and the racial, linguistic and cultural features of the people are similar to the people of Jharkhand region. People are, by and large, bilingual and Hindu by religion. It appears from ethnosscape that all parts of SBR are inhabited by the above mentioned communities whose livelihood primarily depends on agriculture and forest resources and they are practically treated as forest dwellers. Their economy, socio-cultural system is governed by nature i.e. they are eco-centric in nature.

Findings

- Existence of human population is a reality in any biosphere reserve like SBR that cannot be overlooked under any circumstances. They are regarded as forest dwellers.
- Available population data during the period of last 50 years (1961-2001) in its core and buffer zone villages do not create much alarming signal in the sphere of biodiversity conservation and maintaining the ecological balance of the area.
- Ethnoscape encompasses the nature of human surface and the livelihood pattern prevailed in the area.
- SBR is basically a tribal dominated region and the people belonging to different caste and tribal communities are in the midst of multicultural zone.

- Majority of the people are illiterate due to lack of educational facilities.
- Agriculture stands as the main economic pursuit of almost all communities living in and around its boundary. Apparently majority of the cultivators are, by and large, marginal and small farmers.
- Crops, mainly paddy grown by them are not adequate to sustain the whole population. To overcome such problem majority of them are engaged as wage labour in and outside the biosphere reserve.
- People are non-vegetarian in nature and rice is their staple food.
- Poor civic amenities such as road and transport, educational institutions, primary health centre etc. lead the people to remain as impoverished since generations. And recent biosphere reserve declaration has hindered all development activities that increases additional misery and agony of the people.
- Monogamy is the rule. Adult marriage and marriage by negotiation are customarily followed by all communities. People are mainly Hindu by religion. Animists and Christians are less in number. However almost all communities worship the nature as reflected in conservation of *jahiras* (sacred groves) where the abode of village deity, known as Gram Devi is located. Deity is ascribed on stone slabs or tree.
- People being the forest dwellers, dependency on forests is inevitable and their social system is largely governed by nature. Hence evacuating them from the biosphere reserve (as debated in administrative and political circles) under the pretext of biodiversity vis-à-vis wildlife conservation is a challenging task to all concerns.
- People are being conscious to retain their right on forests and other natural resources available in their immediate environment under the influence of various mainstream political parties such as SUCI, Cong (I), BJP and Jharkhand parties. Of late, the local youths are being politically motivated by the Maoists in order to emancipate from deprivation and frustration. Such political forces are in a subtle way creating a space for the people to establish their legitimate right to live in that area despite various constraints for managing survival needs. What situation will emerge on the issue of their
- Survival in future that would be judged only in time perspective.

Appendix I List of Kilis (Clan) recorded among different communities during study: Tribe and Caste wise list of clans/ *kilis*

Kol / Ho	*Kharia*	*Munda*	*Santal*	*Bhumij*	*Bathudi*
1. Dangil	1. Naiko	1. Tuti	1. Soren	1. Kariakuchi	1. Nakuar /Sinnokia
2. Simku	2. Siyar/ Seoro	2. Purti	2. Hembrom	2. Hansda	2. Madhai
3. Gagrai	3 Sankiyaro/ Sankaro	3 Odeya	3. Murmu	3. Narayanisamar	3. Khandai
4. Samar	4. Benedia	4 Soy	4. Marandi	4. Kuchi	4. Saluk/ Renesia
5. Pirua	5. Mirindia	5. Kandir	5. Hansda	5. Pachri	5. Khatua
6. Boipai	6. Phiriria	6. Champia	6. Tudu	6. Barde/ Barda	6. Patra
7. Bodhra/ Barda	7. Pradhan	7. Kalundia	7 Baske	7. Sandi	7. Matia
8. Kara/ Koda	8. Nag	8. Baghe		8. Kauri	8. Parasar
9. Chatar	9. Bindia/ Birindia	9. Mundey		9Samara/ Sauti/ Samar	9. Senapati
10. Karwa	10. Kusamghatia	10. Sanga			10. Ganesarri/ Ganesha/ Ganasara
11. Dara	11. Duarbandha/ Duorobandhi	11. Hansda			11. Andarjuria /Kancha
12. Jarikapat	12. Sanghatia	12. Hembrom			12. Nageswar
13. Jamuda	13. Nageswara	13. Horo			13. Jamadia
14 Suri	14. Patrabichha	14. Birua			14Dhanphul
15. Soy	15. Bhaluoriya	15. Sirka			15. Purti
16. Tariba	16. Paturia	16. Samad			16. Kulapadia
17. Alda	17. Suar	17. Tiu			17. Surya
18. Adwasoy	18. Bhalumia	18. Mundri			18. hanakula
19. Purti	19. Panibonia				19. Biswal
20. Tintwa	20. Kusmaria				20. Paturia
21. Bandra/ Bangra	21. Duarsaini				21 Matian
22 Ghoshra	22 Pichiria				22. Kunora
23. Birua/ Birwa					23. Bisar
24. Guia					24. Giri
25. Hansda					
26. Hembrom					
27 Laguri					
28 Gundua					
29 Sirka					
30 Khuntia					
31 Dorai/ Durai					
32 Patralamal					
33 Badujankho					
34 Khedai/ Kerai/ Khirai					
35 Pingua					
36 Bangsing					
37 Tiu					
38 Baghe/ Bagh					
39 Saluket					
40 Hambira					
41 Bhari					
42 Saoni					
43 Bajra					
44 Kalundia					
45 Chakki					
46 Bhengra					
47 Dubraj					
48 Gama/ Pingua					
49 Mouchel					
50.Jiwanbalmuchu					
51. Sainya					

Bathudi	*Birhor*	*Kudmi Mahato*	*Goud*	*Lohar Bindhani*
1. Nakuar /Sinnokia	1. Murmu	1. Bomshiar	1. Suya	1. Braman Baikar
2. Madhai	2. Hembram	2. Chilbindha	2. Chainsa	2. Mehendia
3. Khandai	3. Kumbha	3. Kankatiar/ Katiar/ Katial	3. Kainsa	3. Kachhima
4. Saluk /Renasia	4. Mahali	4. Hindua	4. Mantri	4. Kaithar
5. Khatua	5. Dadujal / Dadujat	5. Nangtuwar	5. Bagha	5. Sonari
6. Patra	6. Goud	6. Bir Banwar/ Banwar	6. Thela	
7. Matia	7. Jaksaria	7. Dongria	7. Barik	
8. Parasar	8. Kherwar	8. Bansria	8. Sindura	
9. Senapati	9. Chaule	9. Jalbandha	9. Kudia	
10. Ganesarri / Ganesha / Ganasara	10. Nagkuriya /Naikuriya	10. Makdia	10. Makud	
11. Andarjuria /Kancha	11. Bhagrati	11. Sankhuad		
12. Nageswar	12Tikriya /Galowar			
13. Jamadia				
14Dhanphul				
15. Purti				
16. Kulapadia				
17. Surya				
18. hanakula				
19. Biswal				
20. Paturia				
21. Matian				
22. Kunora				
23. Bisar				
24. Giri				

2

Watersheds Management in Similipal

Surajit Kumar Pal

Air, water and food are the basic to the survival of life. Of which air and water are free gifts of nature on which all living organisms/ species of the planet have been sustaining since the beginning of creation. But the recent years have witnessed a steady deterioration of the availability of fresh air and water in terms of their quantity and quality due to man made disasters which adversely affect the sustenance of life in earth. In course of time water has become socio-political concern of the entire humanity. Any discussion on water touches upon various socio-cultural, economic and ecological domains unlike the earlier view on water as an unlimited natural gift. This study intends to explore various aspects relating to the watersheds of Similipal Biosphere Reserve of Orissa in agreement with people, natural resources and pattern of consumption.

In practical sense, water is considered as elixir of life. Obviously there is an inherent relationship that exists between humans and water and therefore also between water and religion. Seventy percent of the globe is covered by water and also seventy percent of human body is constituted with water. And therefore it would not be wrong to claim that water is coterminous with life and it seems almost impossible to separate water, life and religion. Hence water bears a special significance in the sphere of various ritual activities of all human societies belonging to different faiths of the world. That is why, water is considered as an essential and sacred thing to all mankind.

Considering its importance in respect of survival of life, maintaining of earth's ecosystem, progress of human civilization etc. this precious natural resource requires to be conserved for judicious use and its reckless wastage should be eliminated under any circumstances. Today water crisis is a global problem emerged due to man's self-centric attitude towards nature, bypassing the survival and entity of other living things whose contributions in maintaining earth's ecosystem are immense and invaluable. Now the question is from where water comes?

In our common parlance it is believed that water comes from watershed, existed in the form of natural springs, streams, lakes, rivers etc., originated mainly from the hills, mountains, plateaus etc. covered by forests and many of them are connoted as wildlife sanctuaries, national parks and biosphere reserves where nature's bounty in the sphere of biodiversity, beauty of waterfalls, flowing rivers, landscape, vegetations etc. is overwhelmingly witnessed. In recent era debates on the issue of conservation of the watersheds/water resources have got paramount importance because it leads to influence the framing as well as other issues. As focus of study is watershed, it requires to be defined for a common understanding.

The term 'watershed' implies that an area of land from which precipitation flows to a single stream or water body. It is basically a geo-hydrological unit with a common drainage outlet, synonymous as catchment area and drainage basin of a larger area in size, demarcated by ridges or gullies. However in other sense, watershed is having three basic components namely land, forest/vegetation and water which are closely interwoven to influence its perennial character. Along with above three basic components other two components - people and animals are also taken into consideration as integral part of watershed. Watershed, drainage area or catchment area is usually interchangeably used meaning for the same. Progress of human civilization, people's well-being, and existing of other life forms are closely related to a watershed in some manner or other. In recent time it is recognised as one of the key and direct development units in rural development. Because any kind of development activities depends on steady supply of water. Thus protection, conservation and management of watersheds are needed in this era of water crisis.

Watershed is classified into two broad categories - natural and man-made. Springs, streams, rivers, wetlands etc. are treated as natural watersheds, while the man-made type includes wells, ponds, tanks, canals, reservoirs, dams etc. (in other sense these are treated as water harvesting structures). Usually every biosphere is primarily endowed with first category of watersheds located in different geo-climatic conditions. The second type has come into existence later particularly with the advancement of human activities pertaining to agriculture and industrial development since ancient time. The Similipal Biosphere Reserve under study is considered a rich natural watershed area. Its richness can only be ascertained by examining the status of three principal components - soil, forest and water that have direct role in influencing the local ecology, people's livelihood as well.

Under this backdrop the present study aims to highlight the existing status of essential components of watershed and its management prevailed in the concerned biosphere. The study also deals with people's wisdom and knowledge pertaining to traditional water harvesting system nurtured for generations that gets special weightage as they are inseparable from the realm of its conservation and management.

Study area, objectives and methodology

The Similipal Biosphere Reserve (SBR) is a densely forested hill range, known as the Similipal hill range, which forms a part of northern plateau, located in the central part of the erstwhile Mayurbhanj State, now Mayurbhanj district of Orissa, lying close to the eastern most end of the Eastern Ghats. It falls in the Mahanadi Biogeographical Region and within the biotic Province, Chotanagpur plateau; spreads over an area of 2750 sq. km. The name 'Similipal' has been derived from '*simul*' trees (silk cotton) with red flowers abundantly found in the area.

Of its total area the entire core zone (845.00 sq. km.) and the major part of the buffer zone (1905.00 sq. km.) are constituted together as the 'Similipal Tiger Reserve' (STR). Within the STR boundary altogether 65 villages (core - 4 and buffer - 61) are found to exist which are revenue category in nature. Its transitional area covers an area of 2575.00 sq. km, having 1200 villages are inhabited by a large number of tribal and non-tribal population. The SBR is also considered as the home of richest watershed giving rise to several perennial rivers and innumerable springs and streams on which the livelihood of the people of Mayurbhanj and other neighbouring districts of Orissa depend as well as it influences the weather condition of the eastern states particularly Orissa, West Bengal, Jharkhand and Bihar. Hence its watersheds, valleys and hill ranges contribute an ecological niche to the diverse kind of species.

For collection of empirical data fieldwork was conducted in the SBR during November - December 2006. The study was primarily confined to eight buffer zone villages (Gudgudia, Lanjighosra, Kumari, Kolhjari, Chakundkocha, Gad Similipal, Saruda and Basilakocha) and part of one transitional village (Durdura MakdiyaTanda) selected randomly that may fulfil some specific purposes of study. The villages are treated as unit of study selected on the basis of accessibility, size as well as feasibility to cover, having ethnic representation, availability of water, land and forest resources etc. which are considered the primordial components of any watershed study. Considering the nature of geographical location of different villages as well as time constraints to cover various aspects of the concerned topic both survey and anthropological methods were simultaneously employed for culling field information within a stipulated period through which the objective of study can be reflected.

The sample villages were selected from different directions of the SBR namely east, west, north - west, north - east barring the southern direction where core villages are located, kept outside the purview of the present study. Every village bears certain specific features as far as ethnic composition, occupation of the inhabitants, harnessing of various natural resources etc.

The people, by and large, belong to the tribal communities such as Kolha, Bathudi, Khadia, Santal, Munda, Makdiya (Mankhadia/Birhor). However several castes groups namely Goud (Magdha Goud), Teli, Lohar (or Bindhani), etc. are also distributed in both buffer and transitional villages. In both zones villages are either uniethnic or multiethnic in character. The multiethnic villages are predominantly demarcated by hamlets, locally known as '*sahi*' or '*sai*' named after the community name. For instance, '*Kolha*' or '*Khadia sahi*' etc.

In SBR majority of the villages are accessible by jeep because of extension of road network, though kachcha type, spread in its length and breadth. In fact some sort of road network existed prior to merger of the Mayurbhanj State with Orissa. During the feudal rule the king used to go for hunting in Similipal. For easy movement to the forest as well as to the villages' road was extended during those days. In post-independence era this has been further extended to almost all its parts and maintained by the forest authorities for proper conservation and management of forest resources, wildlife as well. There are several entry points to approach different villages located inside the SBR of which Jashipur (western side) and Lulung (south - eastern side) are treated as prominent approach points. Approach to different study villages barring Basilakocha (done through Bangriposi side) and Durdura (north-western side) through Jashipur and all of them are spread within a radius of 40 - 50 km from Nawana, the centre point of SBR. From Jashipur direction the average distance of each study village barring Basilakocha and Durdura (about 125 and 15 km respectively) is around 30 - 70 km.

The study bears certain objectives. They are:

To document water related issues in order to overcome the present dilemma on conservation of biodiversity, environment and ecology which are being depleted and degraded fast in the name of so-called development!!

To understand people's wisdom and knowledge in protecting, conserving and managing this resource for sustainable use.

To make database for further study that may be useful in undertaking welfare programmes for the people and also to be useful for modifying the policy of conservation, management of various watersheds prevailed in that area.

Every study village is situated nearby the perennial water source, having land resource and forest vegetation that provide the food security to the people. The people living in and around the biosphere are mainly depended on subsistence agriculture and majority of them are primarily the marginal farmers - cum - owner cultivators. In addition to agricultural activities they are also engaged as agricultural labourers or as wage labourers. Such agrarian status of the people reflects their real economic condition prevailed in the area. To sustain themselves in such situation every family member barring young children take part in agricultural activities, harnessing of available natural resources such as collection of fuel wood, house - building materials, grasses, ropes, wooden posts etc.

In dealing with various components of watersheds rainfall and temperature occupy first priority for examination as they play a significant role in maintaining its health and longevity and also climatic condition of that area.

Rainfall, Temperature and Humidity

Rainfall, temperature and humidity are directly linked with the perennial character of any watershed. The SBR receives adequate rainfall during monsoon, but due to its undulating character major part of rainwater goes wastage as runoff. The incessant rainfall makes the weather cool and also leads to thick vegetative cover in the area. The major rivers, innumerable streams, nala, falls etc. are originated mainly from the Similipal hill range which are considered not only the lifeline of the people of Similipal but also to the people of Mayurbhanj, Baleswar and Bhadrak districts.

It appears from rainfall data that the average annual rainfall of the district is around 1600 to 1700 mm. While in SBR it is around 2250.00 mm. based on 18 years (1988-2005) of rainfall data. The monsoon usually starts in June and gets weak from the middle of October. The intensity of precipitation in the area takes place during June to September and major annual precipitation occurs during this period. The average and percentage of precipitation in such period are around 1700.00 mm. and 76.00 per cent respectively. And this period experiences on and around 81 rainy days, while the total annual rainy days are 121 days. The history reveals that some pockets of the Similipal used to bear the witness of uninterrupted rainfall during the monsoon period, but in recent time this has become occasional. The reasons for such phenomenon are of manifold - such as deforestation, impact of climate change etc.

The climate of Similipal is moderate warm and humid in nature. The area experiences tolerable summer as the temperature hardly goes beyond 40°C. The night temperature in summer is cool and pleasant. The winter temperature falls to about 5°C to 8°C. Winter begins from November and becomes severe in December lowering the temperature to 5°C in many parts of the hill particularly at Meghasini, upper Barhakamuda, Tinadiha, Debasthali, and Nawana and these places also experiences frost during such period. The mean maximum and minimum temperature in summer ranges between 33.4°C to 37.4°C, while in winter it ranges between 7.2°C and 11.1°C. The air is quite humid with relative humidity always exceeding 70%. That is why the climate of the area always remains humid. This humid climate directly or indirectly influences the health of soil and water resources and also the forest vegetation.

Land and soil resource

Land and soil are the integral part of watershed management. Percolation of rainwater into earth largely depends on soil types of the area that contributes groundwater recharge and moisture holding capacity of soil. Geologically Similipal is the part of Gondwana land, underlained by an assemblage of volcano-sedimentary rocks and many of these rocks are metamorphosed. It has three concentric cups of metamorphic rocks, inter-bedded with submetamorphic layers. Metamorphic and sub-metamorphic rocks have contributed largely to its soil formation. The subterranean water holding capacity in these hills is consequently much higher than in general elsewhere and hence there is perennial flow of water all over these hills.

Soils of Similipal are acidic in nature, broadly grouped as lateritic, red and yellow, mixed red and black and sandy loam. Lateritic soils are found in plateau area, having the capability to support good tree growth. Heavy clay is also found in the wide flat basins. In deep layers it supports good growth of both plants and animals. Where the depth of the soil is thin it favours the growth of grasses. However this type of soil classification is mainly confined to scientific and academic circles, but not to the common people who are not acquainted to the above mentioned terms. Instead they have their own perception on land and soil types according to its productivity and water holding capacity, nurtured for generations.

Land and soil are commonly perceived by the people as *jami* and *mati*, respectively. Land is again broadly classified into two types - *ghara diha* or *kala diha* (homestead land) and *chasa jami* or *dhan jami* (cultivable land), based on topography of the area. While the cultivable land is categorised into four types - (a) *ool / beda / khal jami* or 'A' class land, (b) *daiam / dhipa jami* or 'B' class land, (c) *saiam / nali jami* or 'C' class land and (d) *guda / dahi jami* or 'D' class land. All these types of cultivable land are purely based on fertility and productivity conditions of soil.

Like the land, soils are of three types - *bali mati* (sandy soil), *ponkhua mati* (clay mixed alluvial soil) and *pakutia mati* (clayey soil), classified on the basis of texture, fertility and nature of water holding capacity. The sandy type is mainly found in 'D' class land, the second category or clay mixed alluvial is seen in both 'A' and 'C' class land, while the clayey type is seen in 'A' class land (chart 1).

Chart 1. Land and Soil Types in Similipal Biosphere Reserve

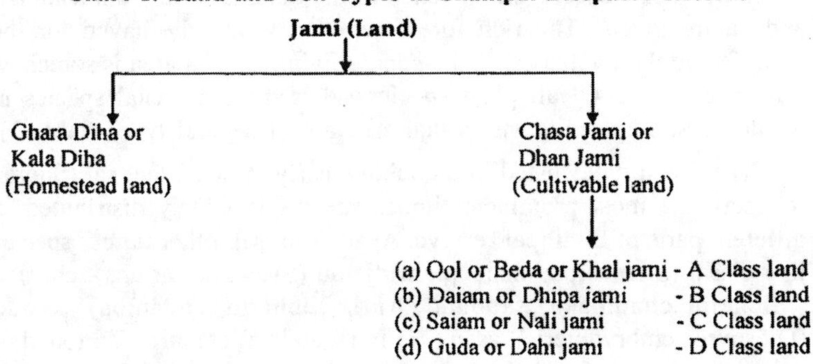

Mati (Soil)

Bali mati (Sandy soil)	Ponkhua mati (Clay mixed alluvial soil)	Pakutia mati (Clayey soil)
↓	↓	↓
Found in guda jami	Found in both beda & nali jami	Found in beda jami

Moisture / water holding capacity in soils varies from soil to soil. Moisture content in sandy soil is less as it does not hold more water because it is porous in character. While in clay mixed alluvial and clayey categories of soil water holding capacity is more due to fine texture, humus and rich biomass content. Hence the lands having these two types of soils are considered more ideal for crop production by the people of the area.

Farm activities mainly take place in flat and terrace type of lands seen in the valley area where the soil possesses good water holding capacity. The land and soil types narrated earlier vary from village to village (tables 1 and 2).

Forest resource

Similipal is famous for its rich bio resources of which plant vegetation bears special significance to the botanists, environmentalists and nature lovers. The rich forest vegetations are the haven for the wildlife and the livelihood of the people living in such area is somehow influenced by it. Of all plant species sal is the most vital species in whole of Similipal. Forests in that area are of several types (table 3).

The forest types listed in the table clearly indicate that sal (Shorea robusta), the most prominent timber species is widely distributed in different parts of Similipal reserve. Apart from sal, other timber species like piasal (pterocarpus marsupium), jamu (Syzygium arjuna), champa (Michelia champaka), kadam (Anthocephalus cadamba), kendu (Diospyros embryopteris), asan (Terminalia alata) etc. are also found to exist in such area. Scrub forest and grasslands are found in some small pockets which are important for the sustenance of wildlife. The semi-evergreen and deciduous vegetation along the perennial water sources

Table 1. Village Wise Land Types

Sl.No.	Village	Type of land		
		English term	Local term	Remarks
1.	Guidgudia	A class paddy land	Ool/beda/khal jami	Major part of the paddy area fall under this category
		B class paddy land	Daiam/dhipa jami	
		C class paddy land	Saiam/nali jami	
		D class paddy land	Guda/dahi jami	
2.	Lanjighosra	B class paddy land	Daiam jami	Covers more paddy area
		C class paddy land	Saiam jami	
		D class paddy land	Guda/dahi jami	
3.	Kumari	A class paddy land	Ool/beda jami	Both have covered more sizeable area
		B class paddy land	Daiam jami	
4.	Koljahari	C class paddy land	Saiam jami	
		D class paddy land	Guda/dahi jami	
5.	Chakundakocha	D class paddy land	Guda/dahi jami	
6.	Gad Similipal	C class paddy land	Saiam jami	
		D class paddy land	Guda/dahi jami	
7.	Saruda	C class paddy land	Saiam jami	
		D class paddy land	Guda/dahi jami	
8.	Basilakocha	C class paddy land	Saiam jami	
9.	Durdura	C class paddy land	Saiam jami	
		D class paddy land	Guda jami	This type of land of land are used for cultivation of vegetable, pulses and oil seeds

Table 2. Village wise Soil Types

Sl. No.	*Village*	*Soil type*		*Remarks*
		English term	*Local term*	
1.	Gudgudia	(a) Clay mixed alluvial	Ponkhua mati	
		(b) Sandy-cum-clayey	Bali mixed Pakutia mati	
		(c) Clayey	Pakutia mati	
		((d) Sandy	Bali mati	
2.	Lanjighosra	(a) Clayey	Pakutia mati	
		(b) Sandy loam	Bali mixed Pakutia mati	
		(c) Sandy	Bali mati	
3.	Kumari	(a) Clay mixed alluvial	Ponkhua mati	
		(b) Clayey	Pakutia	
4.	Koljhari	(a) Sandy loam	Bali mixed Pakutia mati	
		(b) Sandy	Bali mati	
5.	Chakundakocha	(a) Sandy loam	Bali mixed Pakutia mati	
		(b) Sandy	Bali mati	
6.	Gad Similipal	(a) Sandy loam	Bali mixed Pakutia mati	
		(b) Sandy	Bali mati	
7.	Saruda	(a) Sandy loam	Bali mixed Pakutia mati	
		(b) Sandy	Bali mati	
8.	Basilakocha	(a) Clayey	Pakutia mati	
		(b) Sandy	Bali mati	
9.	Durdura	(a) Sandy loam	Bali mixed Pakutia mati	
		(b) Sandy	Bali mati	

Table 3. Forest Types in Similipal Biosphere Reserve

Sl. No.	*Forest types*	*Area of concentration*
1.	Tropical semi - evergreen forest	Confined mainly to deep & damp valleys along the perennial streams and nalas, found in small pockets inside the Similipal hills
2.	Moist mixed deciduous forest	Found in patches all over the hills having favourable edaphic conditions, mainly drainage and moisture congenial for sustenance and growth of sal
3.	Moist peninsular sal	Found all over the area in Similipal hills upto an elevation of 800 metres above mean sea level
4.	Moist peninsular high level sal	Occurs primarily on the hill tops and plateaus above 800 metres altitude containing pure sal crop above 75 %
5.	Moist peninsular low level sal	Widely spread and occupies most of the hill slopes upto an elevation of 400 metres, characteristically of pure sal type spread over plain forest and slopes in the Similipal reserve
6.	Very moist peninsular sal	Occupies mainly in deep and damp valleys
7.	Dry peninsular sal forest	Confined to Satkosia, Jari, Kanapat, Tunguru, Sarali and few other places of Similipal; due to lack of perennial streams area is quite dry and the species are deciduous type
8.	Dry deciduous hill forests	Extends over eastern and the southern Similipal on steep and exposed slopes where poor quality of sal forest along with deciduous species is seen
9.	Grassland	Found upper most hill slopes and their upper summits over 900 m altitude above mean sea level

Source: *Database on Similipal Biosphere Reserve*, Orissa.

form the reparian zones. Forests of this area are practically considered an important economic asset to the people as this provides various livelihood resources to them.

Major forest resources having economic value are: (a) timber of different varieties, (b) silk / tasar, (c) honey, (d) sabai grass, (e) medicinal plants, (f) orchids, (g) other minor forest products that include sal seeds and leaves, kendu leaves, silk cotton, fruits, resins, various kinds of edible tubers and fruits etc. In the past, say over a century back these forest resources used to be abundantly available in Similipal reserve.

History reveals that with the establishment of railways and mining in neighbouring states and encouragement for setting up the human settlements in the forest in the late 19th and the early part of 20th Centuries in order to ensure cheap labour supply for extraction of timber resources, loss of timber and other forest wealth started to reduce. Later, particularly in post independence era availability of these resources has considerably reduced because of large scale felling of trees by the state and the private parties and other human activities. This has largely contributed the loss of forest cover and at the same time depleted the forest wealth in Similipal. Such historical facts can never be overlooked that bear certain relevance in portraying future ecological history of that region.

Despite such situation Similipal always gets special significance in maintaining biodiversity, perennial character of all major streams / nalas and rivers and food security of both human and animal population. The existing forest vegetation, consisting of 1170 plant species including orchids (as recorded by the SBR authorities), is in paramount importance, being safeguarded and conserved under regeneration and silvicultural activities in recent time for the cause of maintaining future entity of human and wildlife surface of the area.

In this context it is relevant to note that with the declaration of biosphere reserve forest vegetations in some parts of Similipal are getting dense gradually where wild animals can roam freely. This is a positive trend as far as restoration of denuded forest cover is concerned. Biosphere rules have restricted some sort of illegal tree felling and poaching of wild animals, though not absolutely, done by the timber mafia and poachers which until recently was unabated in nature. The existing forest cover still provides an ideal home for over 350 species of fauna (as per the SBR record), consisting of mammals (41 type), birds (264), reptiles (32), amphibians (12) etc. This reflects the real biodiversity scenario prevailed in Similipal. Thus the richness of Similipal is largely influenced by its geographical location, climate, rainfall and available water resources.

Water Resource

Water is a key factor in any wildlife habitat and its importance is significant when the tract falls in a seasonal rainfed area. Similipal hill range is considered the richest watershed area, endowed with a large number of springs, nalas / streams and rivers that form the principal water sources to wild life and human population, distributed in different parts of its boundary. Major perennial rivers of Mayurbhanj district have originated from the Similipal hill ranges which are also fed by innumerable nalas /streams (table 4). The area receives good rainfall during monsoon and major portion of rainfall flow along the nalas / streams, rivers as runoff that often causes land degradation by severe soil erosion, adding to the farmers' agony.

The study highlights that every study village is commonly located nearby some perennial water sources like nalas or streams and rivers (table 5) and these are treated as lifeline of the residents of all villages. The names of these nalas / streams are eco-specific i.e. the names familiar to the people are associated either with any animal or plant or with any place. For instance, Gad Similipal nala bears the name of a place i.e. the Gad Similipal Village. Similarly Barheipani nala is associated with the name of an animal like *baraha* (wild boar). Thus the names of various nalas / streams are basically indigenous type perceived and nurtured by the people for generations which in reality bear eco-sense that embedded in their cognitive level.

In this context it is noted that the history of human settlements in different phases of our civilization has had always a close link with water sources, whether natural or man-made, which are considered the pre-condition of establishing any human habitation in any parts of the world. The present study based on different villages of Similipal is no exception to this fact. Since the inception of their migration to this territory the people started to establish their habitation nearby the nalas / stream in such a way that they can sustain their livelihood under any circumstances. And in course of time they were familiar with local topography, different water sources, hills, vegetations etc. that influence their livelihood in one or other way. People's close association with all these natural things for maintaining their survival interests led to get them (natural things) specific nomenclatures which are inseparable from their social system.

Table 4. River System in Mayurbhanj District / Similipal Biosphere Reserve

Sl. No.	*Name of the river*	*Origin*	*Tributaries*
1.	Budhabalanga/ Balanga / Balangi	Similipal hill range, falls at Barheipani that flows easterly direction and gets the shape of a river and finally falls into Bay of Bengal	(a) Palpala (b) Chipat
2.	Gangahara	Similipal hill range that joins the Budhabalanga at Phullarighat in Balasore district	
3.	Sone / Sono	Similipal hill range	(a) Sanjo
4.	Jamira / Jambhira	Similipal hill range, flows towards Balasore district and falls into Bay of Bengal	(a) Barua nala (b) Mahanti nala (c) Gulpha nala
5.	Khadkai	Similipal hill range, flows north westerly direction; it forms the boundary between Singhbhum and Mayurbhanj districts and meets the Surbarnarekha river	
6.	Salandi	Southern slope of the Meghasini peak and flows to the south for some distance and then flows easterly direction till its confluence with Boula nala and traverses through Balasore district and meets the Damra river	
7.	Deo	Similipal hill range, flows towards the west in Pachpir subdivison	(a) Budhi nala
8.	Khairi-Bhandan	Similipal hill range, actually Khairi and Bhandan are two rivers that flow to the west in Pachpir subdivision and meet together near Jasipur. The combined stream has become Khairi-Bhandan, flows to the west and passes through Raruan. Near Khiching it meets the Kantakhair and finally meets the Baitarani river which flows on the border of the district	

Table 5. Status of Natural Water Sources in Study Villages

Sl. No.	Village	Nala / Stream / River	Origin	Status
1.	Gudgudia	(a) Amjhari / Chirugoda nala, flows westerly & meets in Khairi,	Amjhari pahad, located quite adjacent to the village	Gets dry in summer months
		(b) Serengda nala, flows southerly & then westerly which ultimately meets the Khairi,	-do-	-do-
		(c) Sendiri nala, flows westerly and also meets the Khairi,	-do-	Perennial
		(d) Judia nala, flows westerly and meets the Khairi	-do-	Gets dry in summer months
		(e) Khairi river, flows south westerly direction	Similipal hill range	Perennial
2.	Lanjighosra	(a) Sendiri nala, flows southerly & then turn westerly	-do-	-do-
		(b) Khairi river, located 0.5 km. away from the village	-do-	-do-

(contd...)

Sl. No.	*Village*	*Nala / Stream / River*	*Origin*	*Status*
3.	Kumari	(a) Sanda nala, flows Southerly and falls in Khairi	Asuraghara pahad	Perennial
		(b) Khairi river, draws south-eastern boundary of Kumari	Similipal hill range	-do-
4.	Koljhari	(a) Barehipani nala flows in eastern boundary of the village,	Similipal hill range	-do-
		(b) Pakhri nala, flows in north-south direction of the village,	Chandan pahad	-do-
		(c) Bhaduakad nala, flows in eastern boundary of the village	Bhaduakad pahad	-do-
5.	Chakunda kocha	(a) Barehipani nala, draws the southern boundary of the village, flows east-west direction	Similipal hill range	Perennial
		(b) Daukocha nala,	Daukocha pahad, located in eastern side of the village	Gets dry in summer
		(c) Misin nala	Chandan pahad?	-do-

(contd...)

Sl. No.	*Village*	*Nala / Stream / River*	*Origin*	*Status*
6.	Gad Similipal	(a) Gad Similipal nala, flows from eastern boundary of the village, (b) Astakunar nala, flows in northern boundary of the village,	Bakua/Bukua pahad, located around 8 km. away from the village Similipal hill range	Perennial -do-
7.	Saruda	(a) Budhabalanga river, flows in north-south boundary of the village, (b) Saruda or Horinggada nala, flows in east-west boundary of Saruda	Bodorgoda pahad, about 8 km away from the village and draws its southern boundary Chetani pahad, about 3 km. away from eastern side of Saruda	-do- -do-
8.	Basilakocha	(a) Budhabalanga river, flows in south-western side of the village (b) Khadkai river, draws boundary line between Basilakocha and Amdapani village, flows in southern side of the village and ultimately meets the Budhabalanga river after certain distance	Similipal hill range -do-	-do- -do-
9.	Durdura	Bhandan nala, located in western side of Durdura		Gets dry in summer

Table 6. Status of Additional Water Sources available in Study Villages

Sl. No.	*Village*	*Type of water sources*				*Status*
		Dug well	*Ponds / tanks*	*Tube well*	*Total*	
1.	Gudgudia	-	1*	5+	6	* Community type mainly meant for pisciculture; + all are community type of which one is out of order
2.	Lanjighosra	4*	1+	1@	6	* For drinking & domestic use and for little bit vegetable cultivation during winter; one is community type and the rests are individual; + Individual; @ Community type
3.	Kumari	-	-	3*	3	* All are community type
4.	Koljhari	1*	-	1+	2	* Community type and but abandoned due to full of filthy substances; + Community type
5.	Chakundakocha	-	-	-	-	Residents are exclusively depended on natural nala / streams (see table)

(contd...)

(contd...)

Sl. No.	Village	Type of water sources				Status
		Dug well	Ponds/ tanks	Tube well	Total	
6.	Gad Similipal	3*	2+	1@	6	* Community type, out of which two are abandoned; + one is community type meant for both agriculture and pisciculture and the other one is individual meant for agriculture only; @ Community type
7.	Saruda	1*	-	1+	2	* Community type, but abandoned; + Community type
8.	Basilakocha	-	-	1*	1	* Community type
9.	Durdura (Makdiya Tanda)	2*	1+	1@	4	* Community type, but abandoned; + Community type meant for both agriculture and pisciculture;@ Community type, but out of order
	Total	11	5	14	30	

Apart from natural water sources some additional sources of water, which are man-made in character, are also found to exist in different villages (table 6). Dug wells, ponds etc. are often categorised as water harvesting structures (WHS) through which rainwater percolates considerably in earth that influences the recharge of groundwater level and at the same time water impounded in these structures are useful in mitigating water scarcity during summer months.

The Table 6 listed additional sources of water have certain positive dimension in the sphere of rainwater harvesting, mitigating water scarcity during natural calamities, crop cultivation etc. But due to poor maintenance harnessing of full potential of all these sources remains untapped and many of them are abandoned by the people. Some dug wells are owned by some individuals and some are community type. The community wells and tube wells, usually made by the Public Health Engineering Dept (PHED) under welfare schemes, are inadequate as per demand of the inhabitants. Majority of dug wells as seen in different villages are full of filthy substances which are ideal for breeding of mosquitoes and other microbes and the water available in such wells are unsafe for human consumption.

Ponds are, by and large, community type excavated in some buffer zone villages for the purpose of agriculture and pisciculture. However such defined purpose in many cases remains unfulfilled for varied reasons. Moreover, command areas of all existing ponds have not been demarcated and major part of lands adjacent to such structure always remains uncultivated due to lack of ensured water sharing among the beneficiaries, inputs for cultivation as well. Thus, despite good rainfall people living in the area are deprived of getting adequate water for safe drinking, domestic and agricultural purposes.

In this context it is mentioned that as part of wildlife management the SBR authorities have recently constructed a few water harvesting structures (WHS) in some perennial nalas/streams spread in some locations (table 7) mainly for facilitating water to the wildlife during dry season. These WHS are normally devoid of any diversion channel through which some portion of flowing water can be channelised to the farm lands for crop cultivation. This causes a kind of deprivation among the people whose demand for irrigation has been either sidelined or ignored by the authorities on the ground of biosphere rules and regulations where survival interests of the forest dwellers (who are usually marginalised in nature) often get less weightage.

Table 7. Water Harvesting Structures in Similipal Biosphere Reserve

Sl. No.	*Name of the range*	*Name of the beat*	*Name of the site*	*Type of Structure*	*Remarks*
1.	Pithabata	Bhajam I	Bhajam	Masonry Anicut	In good condition
2.	Nawana	(a) Joranda	Joranda watch tower in ring word Pansia	-do-	-do-
		(b) Nigirda		Earthen dam	-do-
		(c) Bakua	Athatdeuli	-do-	-do-
3.	Chahala	(a) Chahala	Chahala	Masonry Anicut	-do-
		(b) Kairakacha	Kairakacha	Earthen dam	Needs renovation
		(c) Brundaban	Brundaban	Masonry Anicut	In good condition
		(d) Karkachia	Karkachia	-do-	Needs renovation
4.	Upper Barakamuda	(a) Meghasini	Meghasini	Earthen tank	-do-
		(b) Debasthali	Debasthali	Masonry	-do-

Source: Conservator of Forests and Field Director, Similipal Biosphere and Tiger Reserve, Baripada.

As conventionally perceived by the authorities that under biosphere management two major components - conservation of wildlife and vegetations get more priority than the human survival; but totality of any biosphere like Similipal depends on the coexistence of wildlife, vegetations and the human population whose presence in such location is a reality that cannot be undermined under any circumstances. Rather, avenues for socio-economic development of existing population should be created as part of its (biosphere) overall management. And for that purpose a comprehensive plan for harnessing of available sources of water need to be adopted at this juncture which is indispensable for increasing agricultural production as it provides the subsistence of the larger masses.

The existing agricultural pattern in the area, primarily rainfed in nature, needs to have a look where productivity of different crop varieties has remained static or comparatively less than other areas due to varied reasons. The local agricultural practice cannot be segregated from watershed management that to be mentioned in the following paragraphs for understanding the overall agrarian scenario prevailed in such area.

Agriculture and cropping pattern

As per agro-climatic classification Similipal falls under the zone of north central plateau where climate is hot and moist sub-humid in character. For such climatic condition the area is suitable for cultivation of various types of crops that give rise to a kind of cropping pattern being customarily followed by the people for generations.

Cropping pattern is one of the important dimensions closely interwoven with watershed development. Because water requirement of an area especially rural belt or any other isolated place is measured by the agricultural practice prevailed in such place. On the basis of climatic condition and available soil and water resources the people grow various crops for several years which are climatically acclimatised to that area and get the shape of cropping pattern. Soil types and availability of water resources lead the people to devise the nature and technology of production which in course of time becomes the part of their cultural system.

Similipal being the forest area has limited scope of agricultural practice particularly within its boundary. Agricultural lands are mainly found in valley area and no scope for further expansion of agricultural area exists due to biosphere rules. The same land is being cultivated by the people for generations. Agriculture takes place mainly during *barsa*

Table 8. Cropping Pattern in the Similipal Biosphere Reserve

Type of crops	*Crop name*	*Local term*	*Sowing season*	*Harvesting season*	*Remarks*
I. Barsha Fasal (Kharif Crops):					
Dana fasal (Cereals)	Paddy	Dhan	June-July	Nov.-Dec.	
	Maize	Maka	May-June	Aug.-Sept.	
Masala fasal (Spices)	Turmeric	Halud	May-June	Dec.-Jan.	
	Ginger	Ada	May-June	Dec.-Jan.	
Taila fasal (Oil seeds)	Niger	Rasi	June-July	Jan.-Feb.	
Dali fasal (Pulses)	Black gram	Biri	June-July	Sept.-Oct.	
Pariba fasal (Vegetables)	Pumpkin	Baita	May-June	Aug.-Sept.	
	Gourd		May-June	Aug.-Sept.	
	A kind of gourd	Kakharu	May-June	Sept.-Oct.	
	Brinjal	Baigun	May-June	Aug.-Sept.	
	Chilli	Lanka	May-June	Aug.-Sept.	
	Leafy vegetables*	Shag	May-June	July-Aug.	* Includes, khada shag
II. Rabi Fasal (Rabi crops):					
Dana fasal (Cereals)	Paddy	Dhan	Jan. - Feb.	April - May	
	Wheat	Gam/Gaham	Jan. - Feb.	April - May	
Masala fasal (Spices)	Garlic	Rasun	Nov. - Dec.	Feb. - March	

(contd...)

Type of crops	*Crop name*	*Local term*	*Sowing season*	*Harvesting season*	*Remarks*
	Coriander	Dhania	Nov. - Dec.	Feb. - Mar.	
Taila fasal (Oil seeds)	Mustard	Sarisha	Oct. - Nov.	Feb. - Mar.	
Pariba fasal (Vegetables)	Potato	Alu	Oct. - Nov.	Feb. - Mar.	
	Tomato	Bilati	Oct. - Nov.	Feb. - Mar.	
	Cabbage	Bandha kabi	Oct. - Nov.	Feb. - Mar.	
	Cauliflower	Fula kabi	Oct. - Nov.	Feb. - Mar.	
	Brinjal	Baigun	Oct. - Nov.	Feb. - Mar.	
	Onion	Pneyaj	Oct. - Nov.	Feb. - Mar.	
	Carrot	Gajar	Oct. - Nov.	Feb. - Mar.	
	Sugar beet	Beet	Oct. - Nov.	Feb. - Mar.	
	Chilli	Lanka	Oct. - Nov.	Feb. - Mar.	
	Radish	Muda	Oct. - Nov.	Jan. - Feb.	
	Leafy vegetables*	Shag	Oct. - Nov.	Jan. - Feb.	* Includes khusala shag, palang shag etc.

(contd...)

Type of crops	*Crop name*	*Local term*	*Sowing season*	*Harvesting season*	*Remarks*
III. Horticultural/ fruit crops+:					
	Mango	Amba			Fruit bearing takes place after a few years of planting
	Jack fruit	Panasha			Fruit bearing takes places after a few years of sowing/ planting
	Banana	Kadali			Fruit bearing gets after 2-3 years of planting the sapling
	Papaya	Pipa			Fruit bearing gets after one season of sowing

+ Crops listed under such category do not belong to a particular season

Table 9. Crop Varieties Actually Grown in Study Villages

Sl. No.	*Village*	*Crop varieties during kharif season*				
		Cereals	*Pulses*	*Oil seeds*	*Spices*	*Vegetables*
1.	Gudgudia	Paddy, Maize	Black gram	Rasi	-	Pumpkin, Gourd, Chilli, Brinjal
2.	Lanjighosra	Paddy, Maize	-	-	-	Pumpkin
3.	Kumari	Paddy, Maize	-	-	-	Pumpkin, Gourd,
4.	Koljhari	Paddy, Maize	-	-	-	Pumpkin, Gourd,
5.	Chakunda kocha	Paddy	-	-	-	Pumpkin
6.	Gad Similipal	Paddy,	Black gram	-	-	Pumpkin, Gourd
7.	Saruda	Paddy, Maize	Black gram	-	Ginger	Pumpkin, Gourd
8.	Basilakocha	Paddy, Maize	-	-	-	Pumpkin, Gourd
9.	Durdura	Paddy*	Black gram*	Niger*	-	Pumpkin, Gourd

*All these crops are mainly grown by the Santal and other communities of the village, but not by the Makdiya; only one or two Makdiya persons grow pumpkin, kakharu or gourd in their courtyard for family consumption.

+Recently introduced by two families.

(contd...)

Sl. No.	*Village*	*Crop varieties during rabi season*				
		Cereals	*Pulses*	*Oil seeds*	*Spices*	*Vegetables*
1.	Gudgudia	-	-	Mustard	Coriander, Garlic	Tomato, cabbage, cauliflower, radish, sugar beet, carrort, chilli, onion, ool kabi, brinjal, leafy vegetables*
2.	Lanjighosra	-	-	Mustard	-	Tomato, cabbage, cauliflower, radish, pinach
3.	Kumari	Paddy, wheat	-	Mustard	Coriander	Potato, tomato, radish, cabbage, ool kabi, brinjal, cauliflower, onion
4.	Koljhari	Paddy	-	-	-	Cabbage, cauliflower, radish
5.	Chakunda kocha	-	-	-	-	-
6.	Gad Similipal	-	-	Mustard	-	Cabbage, cauliflower, brinjal, chilli, radish
7.	Saruda	Paddy	-	Mustard	-	Cabbage, ool kabi, cauliflower, brinjal, chilli, radish, sweet potato
8.	Basilakocha	-	-	-	-	-
9.	Durdura	-	-	Mustard	Coriander	Cabbage, ool kabi, cauliflower, tomato, chilli, radish, brinjal, leafy vegetables*

*khusala shag, plang shag (pinach) etc.

kala (rainy season) and the crops grown in such condition are known as *barsa* or *kharif fasal*. However, in recent time during rabi season some varieties are being grown in some villages. The crops produced in both seasons are primarily meant for subsistence only, categorised into cereals, pulses, oil seeds, vegetables and fruits/horticultural crops (table 8). All these categories of crops are constituted together to take the shape of cropping pattern of that area.

The crop varieties mentioned in the above table are not grown uniformly in all SBR villages. But the pattern itself provides a broader outline of agricultural practice prevailed in such area though production of all crop varieties does not necessarily applicable or followed to each and every village. Practically it depends on availability of adequate inputs, resources, labour, know-how, feasibility etc. Variation of crop cultivation is markedly observed in study villages due to aforestated reasons (table 9).

Among all crops *dhan* (paddy) is the principal crop in the area grown in every village during kharif season. In recent time some common winter vegetables namely *bilati* (tomato), *bandha kabi* (cabbage), *ful kabi* (cauliflower) and *ool kabi* (tuber like), *lanka* (chilli) etc, are being grown in some villages (see table 9), during rabi season which were until recently not familiar to the people of this territory. Along with these crops some other varieties like mustard, coriander etc. are also grown which do not require much water as the moisture content in soil is able to sustain the crop growth to some extent. Cultivation of summer paddy or *kharatia dhan* is getting importance in such season particularly where some sort of irrigation facilities are available.

It is reported that about two decades back no rabi cultivation in the villages of Similipal forests was prevailed as the people were ignorant about the know-how of such cultivation due to their isolation as well as less exposure from the contact of neighbouring peasants. Moreover, in the past the livelihood requirements of the people were limited and whatever they used to produce in kharif season, were relatively adequate for maintaining their sustenance. But later with promulgation of biosphere reserve many restrictions came into force for harvesting forest resources (which are economically useful for their livelihood) and practice of agriculture beyond the existing village limit is banned that creates problems in producing more food crops for growing population.

To sustain themselves with limited land resources they have adopted the strategy of survival. Under certain compulsion and finding no other alternative means of livelihood a few families have paid attention to rabi cultivation and this is being done in some small patches of land located normally in nala/stream side from where watering of crops can be made. Of the total families only 19.00 percent families are involved in such cultivation (table 10). It is reflected in the table that rabi cultivation in such area is still in incipient stage. This situation needs to be changed by promoting intensive agriculture in dry season through which people's socio-economic condition can be improved as well as their dependency on forest resources can also be minimised.

Table 10. Rabi Crop Cultivators in Study Villages

Sl. No.	*Village*	*Total no. of families/ households*	*No. of families/ households involved in cultivation of rabi crops*	*Per-centage*
1	Gudgudia	99	16	16.16
2	Lanjighosra	21	3	14.29
3	Kumari	56	17	30.36
4	Koljhari	18	5	27.78
5	Chakun-dakocha	11	-	-
6	Gad Similipal	37	5	13.51
7	Saruda	23	7	30.43
8	Basilakocha	12	-	-
9	Durdura	18	3	16.67
	Total	295	56	18.98

Apart from vegetables in some villages particularly in Kumari, Saharpat and Kandibil summer varieties of paddy are grown as the cultivators apply their own device (by making narrow earthen channel, locally known as *maria*) to channelise river or stream water for such cultivation. As reported that the cultivators belonging to Gudgudia, Kumari, Saharpat and Kandibil, who initiated to introduce rabi cultivation in their villages, may be considered the front runners of agricultural production among all communities living in SBR.

Traditional water harvesting system

To sustain in harsh climate the people make various devices for harvesting rainwater and groundwater used for drinking, agriculture and other domestic purposes. The devices they adopt for such activities are eco-friendly and rational in nature. In Similipal some traditional water harvesting systems are still in existence, nurtured by the people for generations that may have certain relevance even today. The devices they use for water harvesting are as follows:

In the past every village of Similipal forests was having own system of water harvesting for collection of drinking water. A kind of well, locally familiar as ***chua***, about one or one and half feet deep and 2 ft. diameter hole, used to be dug by the people nearby the nala or stream or river from where groundwater outflow automatically to such well and the water is harvested for drinking and other domestic purposes. The people used to apply their own wisdom or common sense where such type of well is to be dug that could be perennially alive for water harvesting. In a village two to three such type of wells were in existence and all these were meant for community use and managed by the villagers. But this system of water harvesting has become obsolete now. The remnants of this system are still visible in Gudgudia, Gad Similipal and some other villages. Now the chua system has been replaced by tube wells installed in many villages for mitigating drinking water problem of the people.

Apart from chua, the other system of groundwater harvesting is in practice visible in Koljhari and some other remote villages. The system is that a small hollow bamboo pipe is fixed in the mouth of a small hole located in hill slope from where groundwater comes out, there an earthen pitcher or any metal container is placed down below the pipe so that the container automatically gets filled with water after sometime. Water harvested by such device is the usual domain of womenfolk used mainly for drinking and other domestic purposes such as cooking, cleaning of utensils etc. In fact, this system prevails where fetching of drinking water from other distant places is troublesome and strenuous job to the children and womenfolk and the other sources of water particularly the tube wells or dug wells are not available or if available, majority of them remain out of order due to lack of proper maintenance.

For irrigating agricultural field a device, commonly knows as ***maria***, a type of earthen channel, is made from source (particularly in the hill slope from where water flows) to the field where groundwater or surface water flows can reach to field directly. Such channel lies almost parallelly in the hill slope. This practice is mainly followed in case of irrigating the terrace field. In some cases earthen made diversion channels, also knows as maria, are made from the side of check dam to the field for bringing surface water from nala or stream flows nearby the crop field. This system is adopted usually in order to save the crops from dry spells or erratic monsoon behaviour and also to irrigate crops grown during rabi season. The channel is made in such a way that water should not go wastage or gets seepage in ground. The length, breadth and depth of the channel depend on location, distance of the crop field and nature of water flow in stream or nala. Sometimes it is 200' - 300'ft. or more in length, 2' ft. width and 1' to 1.5' ft. depth.

The other system of harvesting groundwater is called as *dungi.* This method is followed when the crop land is located in other side of a stream or nala where water harvesting facilities are not available or not possible to lift water from the nala or stream whose bed is too deep. Water (usually outflows from a hole located in the hill or forest) is channelised from hill slope through hollow bamboo pipes with some strong bamboo or wooden posts from one side to the other side of the stream. Under such system water is carried directly by some hollow bamboo pipes (each of its length is about 30' to 40 ft which are joined together to make a channel) from the source of water to the field. This dungi system is also applied for harvesting surface water from nala or stream for irrigating the agricultural field.

To harvest surface water from nala / stream, some bamboo pipes are tied to make a channel through which water can reach to the field automatically. This provision is followed where making of earthen channel is not possible (due to hard rock or for porous soil) and the location of crop field is in the other bank of the stream. Sometimes, in order to irrigate a sizeable land area several bamboo pipes are placed horizontally about 10 to 15 ft. intervals (which served as artificial channel through which more water can be harvested at a time) on diversion channel coming from check dam as observed in Saruda.

Terrace cultivation prevails in many villages particularly where the crop fields are located in hill slopes. In such case conservation of rainwater and improvement of water regime in hill slopes get priority. By making earth embankment as contour bunding the people layout small terrace fields on the hill slopes and also on table lands and grow paddy. By means of water channels, they regulate water flow from natural hill streams or rainwater impounded during rain and irrigates the terrace fields. As flow irrigation is not possible in the undulating topography, the runoff water is stopped or impounded in chambers of square or rectangular shape (used as crop field) made from earth ridges at different levels and allowed to soak deep into the soil. This results in availability of adequate moisture at root levels, which helps in better productivity of crops.

The above narrated techniques and devices of the traditional water harvesting system used by the people, which are being disappeared, still bear the scientific relevance in the present context especially in the sense of conservation outlook. Since ancient time various types of traditional water harvesting system are prevailed in different parts of India. For instance, the *kunds* (tanks), *johads* (small check dams) of Rajasthan, *bodis* (ponds) of Vidarbha region of Maharastra, *ahars* (rectangular catchment basin with three sided embankments) of Bihar, *katas* (water harvesting structures) and *bandhas* of orissa, community wells of Gujarat and over two lakh irrigation tanks spread all over the country bear the witness of traditional water harvesting.

There is a widespread impression that these systems have become derelict and obsolete - physically and institutionally. Tube wells, bore wells and piped public water systems drawing water from rivers and dams / reservoirs have indeed distracted attention from traditional system, overshadowed them and in many cases, even put them out of use. The progressive reduction in the area irrigated by ponds, tanks and other minor surface sources reported by official statistics would seem to corroborate the impression of widespread and progressive decay.

Govt. policy (framed by the technocrats, bureaucrats and ministers) prone to large dams and reservoirs has accelerated the process of decay of traditional systems. The process of decline has reached in such level that the traditional water conserving systems are regarded as outdated and unsuitable for mitigating water crisis in the arena of agricultural production and industrial development. Such perception from the part of the government vis-à-vis policy makers is debatable and not rational also. Rather, devices and techniques used traditionally by the people are complimentary to large dams and reservoirs as proved in many places of our country. Their systems do not bring any adverse effects on nature. In the era of water crisis, traditional water harvesting systems which are being decayed across the country need to be revitalised and encouraged in every micro level that could be more beneficial to the people and also useful for sustainable development.

Terrain like Similipal biosphere requires to have special treatment as far as water resource management is concerned where construction of large dams, reservoirs and canals is prohibited and undesirable also. Instead, small water harvesting structures like wells, ponds, tanks etc. should be promoted under watershed programme in each village and at the same time traditional systems need to be rejuvenated for the sake of increasing productivity and also for maintaining ecological balance of the area.

Water Requirement

Conservation and management of water resources stand closely on nature of needs and consumption practice related to drinking, crop production and industrial activities. It is evidenced that since the beginning of human civilization this prime resource has been ruthlessly plundered by man for his own use leaving aside other living species that creates a threat to the entire life system of the earth. To face this situation and to maintain certain balance a standardised pattern of average water requirement has been formulated by the scientists in respect of drinking, agricultural activities and industrial purposes. Water requirement mainly depends on water demand, increased manifold in recent time due to different facets of economic development in both rural and urban areas. Water demand is of several types viz. domestic, agricultural, industrial, institutional and commercial, public uses etc. It varies from place to place or location to location, depending upon the water availability conditions.

Similipal being a forest area (or rural belt) where agriculture gets the status of principal economic pursuit of the people is the centre of our discussion as far as water requirement for agricultural practice is concerned. The existing rainfall data indicate that water availability in the area is apparently adequate, but this aspect needs to be scientifically measured by the experts through which actual water scenario may be ascertained.

In agriculture measuring of water requirement of a crop depends upon the duration of crop, water holding capacity of soil, root zone and critical stages of crop growth. At the same time it is closely associated with the physiographic nature of the terrain, land and soil types. Soil types of Similipal barring lateritic (already discussed elsewhere), are apparently having good water holding capacity that leads the people to grow paddy (a water intensive crop) and other crops. However, the Dept. of Agriculture has prescribed a standardised pattern for water

Table 11. Water Requirement for Different Crops as per the Base Period (in mm.)

Crop varieties	*Base period (i.e. No. of days)+*	*Range of total delta (i.e. depth of water requirement /acre as per the base period)*	*No. of watering / irrigation*	*Crop varieties*	*Base period (i.e. No. of days)*	*Range of total delta (i.e. depth of water requirement /acre as per the base period)*	*No. of watering/irrigation*
Cereals:							
Rice	100/ 120 -130/ 135 -145	900-1200	12-15	Millet*	115-120	450-550	5-6
Wheat	110 -120	400-450	3-4	Bajra*	110-120	450-550	5-6
Maize	100 -110	500-600	6-7	Ragi*	90-110	500-600	5-6
Jowar*	110 -120	500-600	5-6				
Pulses:							
Biri (Urad)	75 -90	200-250	2-3	Red gram*	160	400	5-6
Green gram*	75	150	2-3	Horse gram*	90	150	2-3
Peas*	90 -120	150-200	3-4	Gram*	100	150	2-3
Oil seeds:							
Mustard	90 -100	300	3-4	Safflower*	100	250	3-4
Rasi (Niger)	100 -110	300?	3-4	Sunflower*	90-100	400	3-4
Castor*	130	500	4-5	Groundnut*	110-120	480-550	6-8

(contd...)

Crop varieties	*Base period (i.e. No. of days)+*	*Range of total delta (i.e. depth of water requirement /acre as per the base period)*	*No. of watering / irrigation*	*Crop varieties*	*Base period (i.e. No. of days)*	*Range of total delta (i.e. depth of water requirement /acre as per the base period)*	*No. of watering/irrigation*
Fibre crops:							
Cotton*	130 -150	650-700	8-10				
Jute*	120 -130	500-550	4-6				
Vegetables:							
Potato	100 -110	500-650	6-8	Pea*	90-120	150-200	3-4
Brinjal	120 -130	500-600	6-8	Bhendi*	90-120	500-600	5-6
Chilli	120 -130	700-750	10-12	Gourd	90-100		
Onion	105 -120	300-350	4-5	Beet	70-80		
Tomato	100 -110	500-550	5-6	Bean*	90-120	150-200?	3-4
Cabbage	125 -130	600-650	6-8	Pumpkin	90-100		
Cauliflower	120 -130	600-650	6-8	Radish	70-80		
Cash crop:							
Sugarcane*	170 -180	1200-1500	8-10				

*Not grown by the farmers of SBR

+Base period of different crops depends on soil fertility, types and climatic conditions of an area or a region.

Source: Agronica, 2005. Directorate of Agriculture, Govt. of Orissa.

requirement (formulated by the agricultural scientists and water experts) for cultivating various crops applicable throughout the state irrespective of agro-climatic zones (table 11).

From the table 11 one can get a broad idea about the crop wise water requirement when the crop cultivation is done under irrigation method. The term delta is nothing but the depth of water provided to the crop from the date of sowing to the date of harvest (i.e. base period). And both delta and base period vary from crop to crop and it amounts to increase or decrease of numbering of watering (as shown in the table). Watering of crop also depends on soil condition where it is grown. Thus water requirement for crop is = total delta x area irrigated.

Similipal biosphere is primarily a monocrop area where paddy gets the status of staple crop in the area grown under rainfed condition. Of late, in some villages summer variety of paddy is being grown by some farmers who can manage to harvest surface water from perennial nalas or stream flowing nearby their field. Table shows, of all crops paddy and sugarcane require more water from sowing to harvest. But cultivation of sugarcane and other varieties of cash crops is yet to be familiar to the cultivators of such area. It is because their age-old agricultural practice does not have the tradition to cultivate different crops other than paddy and some varieties of vegetables. Moreover, their ignorance and lack of exposure to adopt know-how for cultivation of new crops are responsible for such situation. In the past, the area had the tradition of *podu* or shifting cultivation. Later this became obsolete and was completely replaced by settled cultivation. Now this type of cultivation has become the main agricultural practice of that area.

Practice of irrigated agriculture in Similipal is incipient stage which is confined to some cultivators only. It is observed that during rabi season the arable land in villages remains uncultivated due to lack of assured irrigation facilities though the area is blessed with innumerable perennial nalas, streams and rivers. In spite of adequate water resource (its major percentage usually goes waste) the people experience water scarcity as far as crop production is concerned in dry season. To overcome this situation assured irrigation facilities need to be provided under watershed programme. The people are eager to introduce multiple crops of both cereal and non-cereal crops (which are having certain commercial value) in their arable land that will bring some prosperity in their life. But this has not been materialised due to aforesaid reasons.

In development paradigm water is considered one of the most vital components of human progress and its demand in various sectors of present economic development has increased manifold. In comparison to other areas people's demand for water in Similipal is still limited. Water collected from available nalas or streams and rivers is used for drinking and some essential domestic purposes like cooking, cleaning of utensils, smearing of floor and courtyard etc. Bathing is normally done in rivers or streams and these sources also meet the requirement for cattle population. This indicates that without natural water sources sustenance of the people in such location will be more miserable.

To them, agricultural pursuit is more vital than any other economic activities as it provides major subsistence to them. Kharif production is now inadequate to sustain the whole population throughout the year. This requires more attention so that major percentage of cultivable land can be brought under irrigation facilities during rabi season through which additional production could be made. Procurement of any modern irrigation inputs like pump set or diesel generator used for lifting stream or river water a large sum of finance is involved which is not affordable to them. This can only be done by the biosphere authorities or any other government or non-governmental donor-cum-development agencies. Measuring of water consumption in Similipal is primarily be confined to crop cultivation than any other purpose. And it is not wrong to state that existing water sources are enough to meet future requirement for such purpose also.

Conservation and Management of Watersheds

Conservation and management of natural resources get more importance in the sphere of future entity of the mankind. The issue is being debated among the environmentalists, social and political activists, policy makers etc. The forest dwelling communities have their own perception how to tap and conserve natural water resources for sustainable use. But the people's wisdom and knowledge as perceived and nurtured by them for generations are being disappeared due to influence of modern economic forces and lack of state patronage. However traditional watershed management is still in existence among many tribal and non-tribal communities in India. The devices they use for harvesting rain and groundwater for drinking and agricultural practices are mainly eco-centred and eco-friendly which are not responsible for damaging the carrying capacity of our mother earth. The system they follow is more significant to the practitioners of modern management and technologies.

Of late, watershed management stresses on conservation and rational use of land, soil and water resources for optimum production with minimum hazard to natural resources. However, this objective is somewhat redundant to the forest dwelling communities whose sense of wisdom and knowledge bears similar parity, though not codified by them in the framework of modern scientific management. In fact, the acumen they possess is eco-culturally based, meant for sustainable use of all kinds of natural resources available in their immediate locale or environment not only for their own sustenance but also for maintaining the entity of other living organisms.

Conservation of water is closely associated with the people's religious belief because water always gets special treatment in spiritual activities of the mankind. Hence, in spiritual sense, water treated as sacred thing and its conservation cannot be isolated from religion as well. This phenomenon is quite applicable to the people of Similipal whose religious belief is centre around the hills, rivers, trees, stone slabs etc. where abode of their gods and goddesses is located. Presence of *jahiras* (sacred grove) in every village located nearby the nala/stream or river or hillside bears the testimony of such belief system. The vegetations in *jahira*, normally sal groves are never cut down by them. Conserving *jahira,* where village deity lives, is the task of all village persons where no profane activities are allowed under any circumstances. Protecting the sal or other tree groves in each jahira is the symbol of their wisdom to conserve the natural resource.

The hills namely the *Chandan pahad, Aamjhari pahad, Bhaduakad pahad, Badargoda pahad, Atharadeuli pahad, Daukocha pahad, Bakua pahad, Sarua pahad* etc. are deeply rooted in their cultural system which bear the witness of people's eco-cultural sense. From conservation point of view forests available in these hills get more importance as they protect soil resource from erosion during monsoon and also absorb rain water that enhance water regime in catchment areas of different streams and rivers flowing down from these hills. Thus for maintaining sustenance the people do not harm the forest cover available there. This sense of conservation outlook remains operative in their cognitive level also.

From the above discussion it is clearly understood that people's traditional wisdom and knowledge related to resource management are very much relevant today that cannot outrightly be sidelined or ignored. Rather, it needs to be incorporated as an additional input in the frame of holistic approach of watershed management. And community participation in this task should be ensured for long run maintenance, benefit as well.

Watershed Development

In recent time watershed development is regarded as one of the key factors for rural development. It is more applicable in arid, semi-arid, rainfed and drought prone parts of the country. To mitigate water scarcity in the sphere of agriculture, rearing of livestock, drinking and domestic use various watershed development programmes are coming up in aforestated regions as part of rural development. Under this programme afforestation, water harvesting structures (like ponds, tanks, wells etc.) mainly for rainwater and for groundwater recharge are being constructed in villages in order to employment generation for the hapless farmers and landless communities (whose economy is based on agriculture and wage labour).

Resource conservation, generation and utilisation are treated as the basic principles of watershed development. And for that purpose certain measures are adopted. They are - utilising the land based on its capacity, protecting the topsoil, minimising sediment inflow to streams/nalas, rivers, ponds, tanks and other water bodies, conservation of rainwater, safe diversion of surface runoff to storage structures, water harvesting for irrigation, construction of check dams for increasing groundwater recharge, improving socio-economic status of the people by increasing cropping intensity, etc.

Similipal is basically a rainfed region of Orissa which receives adequate rainfall and major part of rainwater goes wastage as runoff during monsoon that leads to severe soil erosion and land degradation. But to curb this situation any comprehensive watershed programme is yet to be seen in the area. Dependency on agrarian economy need to have greater attention to protect and conserve land and soil, water and forest resources for improving people's socio-economic condition on the one hand and on the other it is essential to maintain ecological balance of the area. Because water regime and climatic condition of Mayurbhanj, Balasore, Bhadrak districts of Orissa and other neighbouring eastern states viz. Jharkhand, West Bengal and Bihar are largely influenced by these resources.

Considering its importance it is essential to adopt soil conservation measures that not only reduce the sediment inflow during monsoon but also enhance the productivity of the land resources. Sedimentation takes place from number of sources. They are - (a) erosion from agricultural, forest and waste lands, (b) stream channel erosion including bank cutting and bed degradation, (c) erosion caused by incessant rain and floods,

(d) Gullying by concentrated runoff, (e) erosion from deforestation (meant for making roads, human habitations, institutional areas etc), (f) mass movement of soil from land slides, slumps and soil crops, (g) mining and dump left as wastes, etc.

Of all these sources of sedimentation, majority are applicable to Similipal as many of its principal rivers and their tributaries, different perennial streams and nalas are seen to have severe siltations. Bank cutting and bed degradation are quite common in streams flowing adjacent to Saruda and other villages. Sediment inflow causes severe siltation and also affects natural course of the rivers and streams. Hence soil conservation measures need be adopted at the first instance in order to protect topsoil which is important for crop growth and productivity of soil. That is why soil conservation measures are considered as an integral part of any watershed development.

In SBR no comprehensive plan for soil conservation measures are yet to be seen. Due to topsoil erosion moisture holding capacity of soil has significantly reduced over the year that causes crop production. This has largely contributed with the process of denudation of forests in upper catchments that probably reduces quantum of rainfall retention in the soils of the catchment and has surface runoff. Both have adverse effects on productivity. It is known from available evidences that in most states there has been little increase in productivity of rainfed lands over last two and half decades. Similipal biosphere is no exception to this phenomenon. To overcome this problem and to enhance agricultural productivity in such area it is required to adopt a comprehensive watershed management programme and in this regard community participation needs to be ensured who can play a significant role in such task. In other sense, watershed programme may be taken as a mission of sustainable development through which soil, water and other natural resources of that area can be conserved.

The watershed approach actually benefits the farmer through improved soil health, better drainage and more efficient use of rainwater with the possibility of excess water being stored in suitable structures for use during scarcity periods. By watershed development food security of the people is largely ensured. Because it greatly enhances recharge of groundwater level and also improves soil health in desired level that leads the people to grow more crops and to pursue other economic pursuits. Watershed management also protects the downstream farmlands and human habitations from floods, reduces siltation of dams and various other irrigation structures and protects natural resources.

The existing check dams or water harvesting structures (only a few have come up recently in some areas of SBR) made in some streams did not bring any fruitful result as far as soil conservation, irrigation of farm lands are concerned. In real sense, basic principles of watershed development has been overlooked that neither reduces the topsoil erosion nor fulfils the irrigation demands of the people. Moreover, check dams are devoid of any demarcation of command areas which are essential part of proper utilisation of harvested water for crop cultivation.

The main objective and rationale of watershed development is ensured that as much as possible of the local precipitation is harnessed and retain for use in the watershed. This can be done by restoring vegetative cover in the upper slopes to reduce the runoff; soil conservation measures to increase the quantum of precipitation absorbed and retained in the soil; and creation of small ponds to store part of the precipitation and to facilitate larger recharge into the local aquifers.

It is noted that how much more water can be retained in watershed depends on numerous factors. The scope for increasing the quantum of moisture retained depends on the existing soil cover status (in terms of depth and structure) and the extent to which soil conservation measures can improve these parameters. According to soil experts that soil fertility is mainly found in about 6-7 cm. depth of the topsoil. Below this, is the topsoil which holds the water. Hence, the primary task of watershed development is to increase water retention capacity of soil and its potential increase can be done if surplus monsoon rains (normally go wastage) are stored by means of excavating farm ponds, tanks, etc. and conservation forest cover in runoff (or upper slope) and intermediate zones of watershed.

In addition to that the terrace fields, where agriculture takes place, should have proper bunds on all sides that will arrest loose soil erosion which is easily washed away along with rainwater during monsoon or during irrigation and settles elsewhere. This is strikingly visible in Gad Simlipal, Saruda, Chandikhaman etc. where terrace fields are located. In recent past the SBR authorities had entrusted the work of watershed development to the Sambandh, a non-governmental development agency in some buffer zone villages (Gudgudia, Lanjighosra, Kumari and Chandikhaman). The name of the programme was Khairi Watershed. Under this programme some soil conservation measures were taken up by constructing some concrete bunds and gully treatment on some nalas coming from nearby the hills, but the said work fails to bring any desired result. And after sometime the agency ceased the work. In fact, this was a piece-meal venture that never made any remarkable impact in terms of resource generation, soil conservation and improvement of people's socio-economic life. This experience pinpoints that instead of peace-meal venture, all the existing schemes affecting land, water and forests should be merged into a single unified watershed programme where the scope of people's participation from planning to execution levels is ensured.

Movement for Water

Movement for water has got momentum among the people of SBR in recent years. Their assertion is to have assured irrigation water during dry season, but this has not yet been materialised. The grievances on several issues including water they have made over the years lead them to come under the influences of some mainstream political parties, particularly the SUCI (Socialist Unity Centre of India). This party is having strong base almost in every village of Similipal area. In the recent past a number of tussles between the people and the forest authorities took place on various issues of which demand for water gets special significance. As discussed before that optimal use of available water for agricultural activities still remains unutilised, which normally goes wastage due to lack of adequate water harvesting structures.

The dearth of water (though adequate in the area) compels the people to remain idle particularly during rabi/dry season. Arresting of nala or river water for irrigating larger area in dry season is not in the hands of the people and also not possible to them. It links with massive financial commitment that could not be affordable to these underprivileged and marginalised masses whose survival, rather, is at stake. To overcome such problem they have made number of pleas to the SBR authorities, but their pleas did not make any headway in resolving the issue. It is reported that "if runoff water is stored in proper structures agricultural scenario in the area will be of different status that will bring some sort of economic prosperity in their life and at the same time seasonal migration in search of employment will also considerably be curbed". This reflects how the people are eager to have this prime resource that may be useful in creating an opportunity to make them self-supported in all respects.

In order to establish their right on water and other resources required for survival they are becoming politically empowered and in such task SUCI's role is undoubtedly immense. Since the inception of their movement for water the local M. L. A. belonging to this party took up the matter with the state authorities. But the matter still remains unresolved. It is relevant to note that during a span of last two decades a good number incident on water issue took place of which some are notable to mention that bear certain politico-historical implications in the arena of people's movement in recent time (table 12).

Table 12. Incidents on Water Issue

Sl. No.	*Year*	*Place*	*Cause of incident*	*Remarks*
1.	Early 1990s	Kukurbhuka, Astakunar G.P.	Making cross bandh on Mirgikacha nala	Still pending
2.	Early 1990s	Gitilpidi, Astakunar G.P.	Making cross bandh on Marang Gagda nala	-do-
3.	Early 1990s	Gad Similipal, Astakunar G.P.	Making cross bandh on Gad Similipal nala	Recently a cross bandh has come up

The listed incidents on such issue are still remembered by the people of Similipal wanted to have assured irrigation for cultivation of crops during dry months. With active initiative of Sri Ananda Birua, the then Sarpanch of Astakunar G.P. the people of that area tried to make diversion channel by making cross bandha on the above noted perennial nalas for irrigation. The water of Mirgikacha nala was meant for irrigating Kukurbhuka village, the second was meant for Gitilpidi village and the third one was for Gad Similipal and Benipur villages. They made the channels, but were dismantled by the SBR authorities on the plea of biosphere rules where no human activities are allowed that may threat to wildlife as well. The first two cases are still unresolved while cross bandh on Gad Similipal stream has come up recently with government assistance.

Their movement was not confined to Astakunar G.P only. It was spread in all other G.Ps like Barheipani, Gudgudia, podagarh etc. In fact, it became a pan Similipal movement that started date back to mid 1970s or early 1980s when the people came in contact with some political parties. Of late, under ITDP scheme diversion channels on some perennial nalas have come up in some buffer zone villages (Saharpat, Kusumi and Bilapogha of Gudgudia G.P.) in order to extend irrigation facilities during dry season.

In this context it is noted that about a decade back or more so the people of afore mentioned G.Ps pressed their demand to the district authorities for a comprehensive plan to make irrigation dams in Khairi river and other perennial streams. But this issue still remains pending in both policy and political levels. Even in mid 1980s a multipurpose water scheme in Khairi river named as Uttaras Water Scheme at Uttaras village of Dhalabani G.P. of Similipal was mooted and taken up by the present sitting M.L.A. But this has not yet been executed as biosphere rules are impediment to such project in and around its periphery.

Thus movement for water has not in end now; it is basically an on-going process that has got the special status in political circle. Their movement was absolutely not in vain. Its impact has not only brought some positive changes in their mindset but has also created some space in policy level. Now question comes how long the people will remain silent to resolve their grievances? Despite adequate resources available in such territory maintaining livelihood is a great challenge to them. Depriving them from basic livelihood resources may bring some undesirable and adverse situations in years to come that would affect the overall management of biosphere. The place is one of the backward areas of Orissa where some ultra political parties are making a base for their activities and the local residents are not remaining silent, rather getting influenced by them. And at this juncture it is difficult to foresee what new situation will emerge in near future that can only be judged in time perspective.

Summary and Conclusion

In the preceding pages an attempt has been made to highlight the existing status and nature of conservation and management of some micro watersheds prevailed in some buffer zone villages of Similipal Biosphere Reserve. The study covers the interrelationship of its basic components namely land and soil, water and forest resources. Perenniality of any watershed depends on the better health of these resources and nature of its conservation and use. In the arena of conserving these resources human role is essential which is also considered as an integral part of watershed study. That's why the present study includes all these components in order to finding the existing status and their relationships.

The SBR is predominantly a rich watershed area which gives rise to a good number of perennial rivers and innumerable streams and nalas. This territory is having a large number of human settlements, inhabited mostly by the adibasis (Kolha/Kol, Khadia, Santal, Bathudi, Munda etc.), who have been residing for generations. Almost each and every village is blessed with perennial water sources consisted of small nalas and streams. The catchment areas of these streams and nalas are located mainly in hill slopes and below of it agricultural fields are found. These natural sources of water are considered as lifeline of the people.

In dealing with watershed traditional water harvesting system, agricultural pattern have been given more stress as people's livelihood cannot be isolated from this paradigm. In the present context watershed development does not mean only the conservation of biodiversity and wildlife. As the area is having the human population their well-being needs to be integrated as part of its management and under no circumstances their wisdom and knowledge pertaining to water harvesting cannot be undermined. Traditional water harvesting systems are by no means dead or unsuitable in the era of water crisis.

The devices they use for harnessing water from natural sources are quite eco-friendly, though this cannot be comparable with mega structures like dams, reservoirs etc. The mega water projects often bear far reaching implications on local environment and ecosystem of the area. Traditional water harvesting system is complimentary to the big dams and surface reservoirs. Place like Similipal, where ecology gets fragile day by day due to loss of forest cover and severe soil erosion, needs to have more small water harvesting structures like ponds, tanks, wells and small check dams that could be more useful to recharge groundwater level on the one hand and on the other crop productivity would also be increased by its (harvested water) full potential use during dry season. In order to improve the socio-economic status of people in such location micro watershed programmes should be promoted in a large scale through which soil and water resources can be arrested in a fruitful manner.

It is incorrect to suggest, large and small water storages of surface and groundwater as mutually exclusive alternatives, but maintaining the balance between them is considered a vital issue. Much greater attention to local harvesting is essential for raising the productivity of rainfed lands, for reducing the dependence on larger reservoirs, for exploiting the overall water potential, and for more efficient management of water in large system. The traditional water harvesting systems in Similipal are in a diminishing stage. There is much to be learnt from both engineering techniques and institutional arrangements of traditional systems. It requires state intervention for the cause of the people and also for future conservation of the concerned biosphere.

However, many of the traditional forms and practices need change in the light of changing demographic, socio-political and economic contexts. Many new techniques of engineering, construction and management are available. We should not, in the name of preserving traditions, deny ourselves the benefit of using these new techniques. And at the same time we must be conscious of dangers of blind and indiscriminate adoption of the modern techniques. The latter should be used selectively, modified and adapted in specific contexts wherever it would be more suited.

Under present economic set up people of Similipal cannot remain isolated from mainstream economy. Being the bonafide citizens of India they need to have emancipation from century old backwardness. For this purpose, a comprehensive programme integrating all aspects of watershed development need to be adopted at the earliest through which proper conservation of natural resources can be made. And in this context egoistic attitude of state authorities is required to be changed in the interest of biosphere conservation in tuned with the welfare of the local people.

Findings and Suggestions

- SBR is basically a rainfed region which receives adequate rainfall, having rich biodiversity.
- Watersheds studied in some of its buffer zone villages are apparently considered as micro type and perennial in character.
- Depletion of forest cover over the years causes severe soil erosion in runoff zones of different watershed area. No protective measures are yet to be seen to arrest soil erosion as well as to maintain soil productivity.
- Due to topsoil erosion moisture holding capacity of soil has significantly reduced over the years that have adverse effects on productivity of land resources.

- Runoff water flows in different perennial streams / nalas and rivers gets wastage throughout the year and barring some piece-meal type of works no proper measures are yet to be seen to harvest such runoff water for the purpose of agriculture and other gainful economic activities of the people.
- Name of hills, forests, nalas / streams, rivers etc. is eco-specific, culturally bound and place specific, closely associated with people's livelihood.
- Sacred tree groves, locally perceived as *jahiras* found in every village are protected by the people that stand as the symbol of people's wisdom of nature conservation.
- All existing water harvesting structures especially ponds, wells etc. lack proper demarcation of its command area which is important for rational use of impounded rainwater for domestic and other purposes.
- People are eager to have adequate water for agricultural production in dry season.
- Traditional water harvesting system (still prevailed in some villages) is in diminishing stage due to lack of state patronage as well. People's wisdom and knowledge on water harvesting still have relevance that needs to be respected for the sake of sustainable development.
- Absence of comprehensive watershed development programme in village level leads to water scarcity during dry season and is also responsible for some sort of seasonal migration.
- Existence of human settlements in such biosphere reserve is a reality, predominantly inhabited by the adivasis. Conservation of land /soil, water and forest resources cannot be possible without people's cooperation and involvement.
- Catchment areas of all nalas / streams should immediately be surveyed by the hydrogeologists and other allied experts in order to assess the actual situation of soil and forest cover that play significant role in recharging water level.
- Detailed landuse survey needs to be undertaken at the earliest through which land-man relationship could properly be assessed.
- There is a need to conserve rainwater, which flows out as runoff, whereas this should be harnessed by constructing water harvesting structures in every village. Identifying hydrological units and

developing them in an integrated manner should be the new strategy for water management.

- The village should be identified as the basic unit of watershed. All villages should also be brought under a comprehensive watershed programme for promoting sustainable utilisation of the natural resources by the community involved while encouraging restoration of the ecological balance.

3

Religion and Worldview of the Buffer Zone Villages

Arup Ray

The anthropological approach on biosphere reserve is a new orientation in respect of biodiversity and cultural diversity. Relationship between biodiversity and cultural diversity is a new area of study and requires attention. Throughout the world the conservation of pristine biodiversity is the prime agenda. Global warming, green house gases, pollution are major concerns of entire humanity and that is why the need of conservation of biodiversity gets momentum. The vast forest tract comprises not only diversified flora and fauna but also human being is to be considered as an integral part of the ecosystem.

The objective of the biosphere reserve reveals certain conditions in respect of core and buffer from administrative point of view; particularly a precondition has been imposed in respect of core zone where total ban on human intervention is demanded. On contrary, a school of thought strongly pleaded the theory of integral co-existence of tribal population and forest. This school of thought has raised a relevant question who is encroacher and who is destroyer? They argued that in India 65% of the forest cover is situated in 187 tribal dominated districts. Out of 50 dense forest cover districts, 49 are tribal dominated. Indian Parliament passed the bill "The Scheduled Tribes and other Traditional Forest Dwellers (Recognition of Forest Rights) Bill, 2006" has addressed the debate on "tigers vs. tribals". Despite that the issue is not only restricted on the ongoing debate between conservationist and human activists. The protective legislation is not the only way to solve the entire problem. The understanding of the entire situation needs a scientific outlook to address the problem and evaluate other factors, which are neglected or equally responsible for destruction of forest and wild life.

The present approach on religion and worldview is a scientific tool to understand the society and culture of the buffer zone villages of 'Similipal Biosphere Reserve' by three ways viz. 'Folk Taxonomy', 'Folk Concept' and 'Folk Awareness'. This study will help to evaluate the reciprocity between forest vis-a-vis nature and human elements which are also a part of the eco-system. Nevertheless, this approach may open many other invisible components, which are intricate in respect of a linkage between biodiversity and cultural diversity.

The relationship between biodiversity and cultural diversity could be understood in the backdrop of inclusion of human elements inside the biodiversity. The primary objective of set up of 'Biosphere Reserve' is to maintain and preserve pristine gene pool. The importance on development of human resource has not been denied. Inspite of that the sole objectivity for setting up of 'Biosphere Reserve' is inclined towards plants and animals.

The importance of the conservation in the present circumstance is immense. This importance is not denied by anybody, but the contradiction is arising on the question of the role of the people residing in the 'Biosphere Reserve'. The main drawback is that the role of the people residing in 'Biosphere Reserve' has not been answered in the Indian context particularly in terms of historicity. The anthropological orientation in the context of 'Biosphere Reserve' starts from that line of thought process. The role of the people residing inside the 'Biosphere Reserve' particularly in the buffer zone is not clearly indicated. However, so many restrictions are imposed on them. The sole objectivity of a 'Biosphere Reserve' is defeated while people's role in respect of conservation is not properly demarcated. This creates the problem in the adjustment of the people residing in the 'Biosphere Reserve' specifically in the buffer zone.

Under the prevailing circumstance, the approach on worldview is important. This approach will help to understand historical relationship between the people residing inside the 'Biosphere Reserve' and pristine biodiversity. The understanding of the linkage between biodiversity and cultural diversity is essential in respect of determination of their role in conservation process. The level of understanding in terms of conservation of biodiversity varies. Unless and until this level of understanding is ascertained, the local people cannot be motivated. The next discourse on the objective of study will help to understand properly.

There are two approaches in Cultural Anthropology concerning society and ecology. The first one is "Formal Structural Approach" which emphasis on ideal aspects of society, social system or activity which is thought to be contributing to the maintenance of 'Social Structure' (Radcliff Brown, 1948). The second one is "Ecological Approach" which emphasis on adaptation of the subsistence technology with the environmental ecology and the on going social system (Steward, 1955).

The two principal approaches have the limitation to project a comprehensive dimension. In both the approaches, the society is conceived of 'Structural – Functional entity with interaction to ecology'. Where as little or no attention is given to the ideas and sentiments of the people towards their society (Adhikary, 1984: 5). Beside these two approaches, many Cultural Anthropologists have given attention and effort to characterize the whole culture in logical abstract forms. They derived as many as concepts and technical terms viz. 'Ethos' (Bateson, 1936); 'Theme' (Opler, 1945); 'Pattern' (Benedict, 1949); 'Assumptions' or 'Postulates' (Hoebel, 1958). However, they have not used the term 'Worldview' to explain the underlying expression of the culture.

The term Worldview was first used by Sol Tax (1941) in his book "Worldview and social relations in Guatemala". He defines worldview as 'the mental apprehension of reality', which includes total knowledge and beliefs about nature and man. Later, Redfield used the term 'Worldview' in 1952-53 in the article "The Primitive Worldview" (published in 1962).

In nutshell the main contention of Redfield is as follows:

- Every human being differentiates himself/herself from other human being by classifying two categories viz. 'Intimate and Similar' as well 'Distant and Different'.
- Secondly, an effort was attempted to classify and categorize of the non-human entity or nature and space.
- Thirdly, the study looks into the crisis management or unresolved entities as supernatural by two ways i.e. religion by virtue of god and occult beliefs or practices.

Similarly, Mendelson (1968) conceives worldview as a combination of three systems – cognitive, attitude and action systems. Geeztz (1968) states worldview is a peoples 'picture of the way things their concept of nature, of self, of society'. Johns (1972) reviews different approaches to worldview and says that at the level of common denominator it involves beliefs. Douglas (1973) in her book "Natural Symbols" equates cosmology with worldview and says that each cosmology and each worldview 'has its hidden implications'.

Kearney (1975) defines worldview from a new dimension. He categorically differentiates the cognitive Anthropological approaches and worldview. He stressed that the worldview is different from the principles of cognitive Anthropology (which includes 'ethno science' and 'ethno semantics') in three ways. Firstly, it is considered to view culture by giving prominence to non-verbal behaviour. Secondly, not entirely concentrating on folk taxonomy or classification rather incorporates other ideas and concepts. Thirdly, giving much concentration on underlying implicit concepts on causality, time or human nature etc. derives the conclusion. Thus, his dimension was based on functional worldview rather static one. Ashim Kumar Adhikary (1984) in his book 'Society and Worldview of the Birhor' explore the completely meaningful world of the Birhor people and comprehend their mental state through the relation between the confronter and confronted forms as an important aspect.

After evaluating the points of above discussion, it can be concluded that the worldview of the people residing inside a biosphere reserve is regulated by two ways. The first one is the traditional beliefs and perceptions and secondly, the present circumstances and phenomena which may compel them to react differently to the changed social setting. Worldview is to be considered as psychic inertia for driving the life. Thus, the underlying truth prevails in the form of worldview in their cultural spectrum may give the direction to decide the adaptive strategy in new set up. The main reason is that the culture is a dynamics, which reincarnate in changing situation. The worldview is nothing but the inherent doctrine of the people who decides the direction of life and enable them to have choice.

Objective of the Study

The following objectives are considered in respect of this study.

- To understand the role of 'Nature' in respect of cultural doctrine of the communities inhabited in the buffer zone of Similipal Biosphere Reserve (SBR). Nature plays a vital role in determine of the cultural doctrine.

- To study the cultural ethos of the people inhabiting in the buffer zone villages of SBR.
- To understand the socio-cultural matrix of the communities inhabited in SBR and to ascertain the process of cultural assimilation in comparison to core cultural area in the states of Jharkhand and West Bengal.
- The understanding of the socio-economic integration of the communities in regard to historicity and political process is important in respect of worldview.

Focal Theme

Besides objective of the study, the focal theme is also important. This will enable to pin point the entire objective of the approach. The four basic focal themes are as follows.

- The three components viz. man, nature and supernatural power are detrimental to ascertain of worldview.
- To ascertain the relationship between man and supernatural power in the villages of SBR.
- To ascertain the relationship between man and nature in the villages of SBR.
- To ascertain the role of political economy and historicity in the villages of Similipal Biosphere Reserve.

Study Area

The Similipal biosphere Reserve (SBR) is one prominent biosphere reserve in India, declared in 1994 by Govt. of India, located in the central part of erstwhile Mayurbhanj state, now a district of Orissa. Prior to declaration as a biosphere reserve some of its area became a Tiger Reserve (1973), a Sanctuary (1978) and also a National park (1980). These three are constituted to form this biosphere, now functioned as biosphere as well as tiger reserve. The SBR, practically a hilly region, familiar as the Similipal Hill Range/Similipal Hills, covering an area of 5569 sq. km. with innumerable crests, valleys, perennial streams, nalas, and rivers, surrounded by dense forests having rich biodiversity, wildlife and diverse ethnic groups dominated mainly by the 'adivasis' (tribal). Like any other biosphere reserve the SBR has also three demarcated zonations - core (845.75 sq. km.), buffer (1904 sq. km.) and transitional (2575 sq. km). All the zones/areas are having human habitations. The core area has 4 villages, 61 villages in buffer area and 1200 villages are in transitional area. The total human population in and around the SBR is about five lakh and majority of them are directly and indirectly depended on it for sustenance.

The SBR region witnesses a subtropical climate with three distinct seasons - summer, rainy and winter. The climate is characterized by mild summer of three months duration from March to June, with temperature of less than 40°C; well distributed rainfall during south-west monsoon, conspicuously observed during June to September and intensity of rainfall comes down from October. Winter starts from November to February and the temperature during this period varies 50°C to 80°C. In winter frost occurs in upper Barhakamuda valley and

Table 1. Ethnic Compositions in Study Villages

Block	*Gram Panchayat*	*Village*	*Community*	
			Tribe	*Caste*
Jash-ipur	Gudgudia	Gudgudia	Kol/Ho Kharia Mahali Bathudi	Goud Lohar/ Bindhani Teli Kaibarta Sadgop (Gobora caste)
		Nenjaghosra/ Lanjighosra	Kol/Ho	-------
		Kumari	Bathudi Bhumij	Goud
	Barheipani	Kuljhari/ Koljhari	Kol/Ho	-------
		Chakundakocha	Kol/Ho	-------
	Astakunar	Garh-Similipal/ Gad Similipal	Kol/Ho Munda Santal/ Majhi	Kudmi Mahato Goud
		Saruda	Munda	-------
	Paudagarh	Durdura Tanda (MankdiyaTanda)	Makdiya/ Birhor	-------
Bangri-poshi	Sarispal	Basilakacha/ Basilakocha	Santal/Majhi	------

other adjoining areas of South Similipal and the Nawana valley of central Similipal also receives frost. The annual average rainfall is about 2200 mm. that influences the perennial character of watersheds available in the area.

The study area i.e. buffer zone villages are, by and large, located in hilly and plain terrains, populated mainly by the tribal namely Kol/ Ho, Kharia, Bathudi, Mahali, Munda, Santal/Majhi, Makdiya/Birhor. However some caste groups such as Teli, Goud/Maghdha Goud, Lohar/ Bindhani, Kudmi Mahato, etc. are also found to live with the tribal in some villages who are numerically less in number than the tribal. Among all these castes the Goud and the Kudmi Mahato are numerically more than other castes as seen in Gudgudia and Gad Similipal village. The features of village wise ethnic composition are given below for ready reference.

Cultural Domain

The Similipal Biosphere Reserve villages consist of many communities. Broadly, in common perception they are categorized into tribal and caste groups. Most of these populations had been migrated from the Singhbhum district of Jharkhand state. Few communities are migrated from Dhalbhum and Manbhum area of West Bengal adjacent to Jharkhand state. Each community had not been migrated at one point of time. Even, in each village communities had not been migrated at one point of time. Mainly the migration of the communities had been taken place at the time of rule of Mayurbhanj kings under their patronage. During the British rule, the migration was due to the Santal mutiny even after independence migration continued from the Jharkhand state (the then Bihar). Some of the caste groups for instance Mahato have been migrated from West Bengal. The routes and reasons may be different as well. The communities who have migrated in recent past may recall their memory to some extent. As for example, the Kudmi Mahato have come from the Purulia district of the West Bengal state to Mayurbhanj. Due to the scarcity of food they have migrated to the places in and around Jashipur and later a small section of them had migrated to the forest. It is said that Goud (a caste group in SBR) were the palanquin bearers who usually came to the forest along with the kings. Few of them decided to stay at forest. Three caste groups viz. Teli, Kaibarta and Sadgop are recent migrants who had come to the villages for the shake of business or service.

Tribal Groups

Kol/Ho

The Ho are the major tribal population in Mayurbhanj district. The Ho of Orissa migrated from the Kolha region, which, is now in Singhbhum district of Bihar. The Ho language belongs to the Austro-Asiatic group of languages. The mythological story of the origin of the Ho is believed that *Sing Bonga* created a male and a female child under a ring stone with an intention to create the human kind through them. They lived as brother and sister under the guidance of *Sing Bonga. Sing Bonga* was intolerable due to both of them were unable to perpetuate the human race. One day *Sing Bonga* advised them to be familiar with the forest trees and shrubs. After four days as per the advice the rice was fermented and produced rice beer (*Ili*) and consumed by both of them, as a result passion generated in them and they created population on the earth. The Ho claims themselves as direct decedents. Each Ho village has a *Jaherathan* or the place for their village deity usually a big sal grove. The Ho tribe is divided into 90 exogamous, totemic clans or *Killi.* At present, they could not remember the totemic origin. A male member of the Ho retains his clan identity throughout his life; a female member has to change her clan after the marriage. In case of remarriage of a widow, the parental clan is adhered to rather the clan of her deceased husband. Monogamy is the general rule but polygamy is also found. Adult marriage is practiced and mates are acquired through negotiation and mutual consent. The following clans or *Killis* are found in the studied villages of the Similipal Biosphere Reserve: *Adwasoy, Alda, Badujankho, Bajra, Bagh/Baghe, Bandra/Bhandra/Sonia Bandra (Birhor), Bangra/Bhengra, Bangsing, Barda, Bari/Bhari, Birua/Birwa, Bodhra, Boipai, Chakki, Chatar, Dangil, Darai/Dorai/Durai, Deogam, Dubraj, Gagrai, Ghoshra, Gundua, Guia, Hambira, Hansda, Hansda, Hembram, Hara, Jamuda, Jarikapat, Jiwanbalmuchu, Kalondia/ Kalundia, Kara, Karwa, Kerai/Khirai, Khedai, Khuntia, Kora/Koda, Korwa, Laguri, Mouchel, Patralamai, Pingua/Gama Pingua, Pirua, Purti, Saluket, Samar/Samad, Sainya, Saoni, Simku/Sinku, Singkuta, Sirka, Soy, Suri, Taki, Tariba, Tintwa, Tiriya/Tiria, Tiu, Tutti (Munda).*

They follow the patrilocal rule of residence. Widow remarriage is permissible. Both nuclear and extended families are in the community. All the sons inherit parental property equally, while the eldest son succeeds to the position of authority. They are marginal and small farmer, they work as agricultural labourers. Rice is the principal cereal of them.

Traditionally the Ho polity is centered on *Pradhan* who is considered as the representative of the king of the Mayurbhanj. He was the head of the village judiciary supported by the village elders. He had to collect land tax, road tax, and water tax, fruit tax on behalf of the king twice in a year. The post of the *Pradhan* was hereditary and the eldest son has the first right of becoming *Pradhan.*

The *Sing Bonga* or the Sun God is the supreme deity worshipped by them; he is worshipped first in every festive occasion. *Basuki Thakurani* is the earth goddess. Two stones decorated with vermilion inside the *Jahira* represent *Gram Siri* and his wife *Kiasuni Thakurani. Marang Bonga* is the deity of *Killi.* Different *Killis* have different *Marang Bongas. Buru Bonga* is the forest deity who is worshipped before communal hunting. *Buddhipat* is the goddess of rainfall. *Dessauli Bonga* is the goddess of diseases. The sprits of the dead ancestors are believed to be benevolent. *Baram Thakur* and *Gram Devi* (*Jahira*) are also considered as prime supernatural power. Besides *Sing Bonga* they believe in many other gods and goddesses who are controlling the different sphere of life. The deities may be classified into several categories according to their domains. The classification is as follows:

Table 2. Deity and Domain of Kol/Ho

Sl. No.	*Major Domain*	*Name of the Deity*	*Remarks*
1.	Economy	Basimata	Control Agriculture
2.	Nature	Nagira	Control water
3.	-Do-	Pahar Bonga	Mountain God
4.	-Do-	Sing Bonga	Sun God and Sky God
5.	-Do-	Chandu Bonga	Moon Goddess
6.	-Do-	Baktu Bonga	Thunder
7.	Social	Guru Bonga	Fertility Cult
8.	-Do-	Maran Bonga	Protect Domestic Animals.

The Ho believes in a number of benevolent and maleficent spirits who reside in and around the hills, forests, agricultural lands and their village premises. Thus, *Bonga* is the generic term, which is used to signify power and spirit. Santals and Munda also used the same term (Bonga) in the region. The real meaning of *Bonga* is a power, a force. Every individual, animal, plant, stream, river, tank, rock, tree, forest field and mountain possess the power, which is called as *Bonga*.

The religious head of the Ho is known as *Dehuri/Deuri*, who is a mediator between *Hor* (man) and *Bongas*. He works as a priest and officiates in all the village or community level worships. The sources of diseases, illness, drought and any other calamity etc. appear to him through dream. He is not able to know about the role of a *Bonga* directly in his day-today life. The medicine man of the Ho society is called as *Raulia*.

The Ho observes the influence of Hindu religion. A kind of cactus tree *Sidhi Daru* is considered as abode of *Dharma Devata* (*Shiv*) or *Shibu Dharam*, which is usually found in courtyard of a house. They also consider the sacred grove (*Jahira*) or *Gram Devi* is the village deity who protects from all evils. The abode of the sacred grove is usually in a corner of the village and in proximity of the forest. A tree symbolizes each sacred grove. In a Ho village, it is either *Behera* tree or *Panidharam* tree. The sacred grove is worshiped in *Bachanda* month (March) when the trees of the forest are full of flowers. The priest or *Dehuri* at that time sacrifice the fowl as part of ritual practice.

There are some common festivals in the villages of buffer zones of Similipal Biosphere Reserve. However, few communities observe some festivals with special prominence. The Ho observes the following festivals with due importance.

Table 3. Festivals of Kol/Ho

Sl. No.	*Name of the festivals*	*Month (Kol/Ho Calander)*	*Remarks*
1.	Jamna Parab	Indi	Harvesting Paddy
2.	Magho Parab	Push	New Year
3.	Ba Parab	Chait	Flowering in 'Sal' Tree
4.	Hero Parab	Goma	Rakha Bandhan

Kharia

Originally the Kharia live in the hilly and forest clad area of Orissa, West Bengal and Jharkhand states. Kharia life is inextricately related to the forest ecosystem. Usually they live in small villages situated in accessible forest tracts. They mainly depend on minor forest produce; in different season they collect different forest produce like-honey, resin, *Paro*, *Sal* seed, *Sal* leaf, firewood etc. The Anthropological literatures described the Kharia as food gatherers and hunters. In Mayurbhanj district of Orissa they are now settled population mainly dependent on the wages earned from agricultural labourers. In SBR villages, the Kharia are also dependent on collection of minor forest produce to a great extent. In SBR villages particularly in buffer zone, we could not find any village exclusively inhabited by Kharia population. In all the cases, they live along with other tribal and caste communities. Kharia are divided into exogamous clans. The following clans or *Killis* are found in the studied villages of the Similipal Biosphere Reserve: *Benedia, Bhalumia, Bhaluoriya, Bindia, Duarbandha, Duarsaini, Duorobandhi, Kusamghatia, Kusmaria, Kusumghatia, Mirindia, Nag, Nageswara, Naiko, Panibonia, Patrabichha, Paturia, Phiriria, Pichiria, Pradhan, Sanghatia, Sankaro, Sankiyaro, Seoro, Siyar, Suar*.

Marriage with the elder sister of the wife is strictly prohibited, but younger sister of the wife may be married. The widowed wife of the elder brother may also remarriage. The bride price is a common practice among them. Monogamy is widely prevalent but the polygamy is not tabooed. Marriage by exchange and cross cousin marriage are also prevalent among them.

The Kharia believe that the causes of most of the diseases are the sprits and deities, though certain spirits and deities are not manevolent. On the basis of the cause of the ailments they divide the spirits and deities as manevolent and benevolent. As a result of close proximity with the forest they are familiar with the trees and the plants, which may be used in regard to treatment of various diseases. Even, we have observed in buffer area villages that other caste and tribal communities are dependent on Kharia for the treatment of the ailments. Kharia like other tribal communities inhabited in SBR have a separate burial ground for cremation of the dead bodies. In the Similipal forest, we have noticed that there are two kinds of burial grounds viz. first one is the burial ground possessed by the tribal and other caste communities affiliated by Christianity and another one possessed by the non-Christian tribal and non-tribal communities. In case of that burial ground each dead body is buried along with a stone erected above the ground. On contrary, the Christian burial ground the "cross" made up of wood is erected above the ground. Kharia believes *Athraduar* or *Devasthali* as an abode of *Badam Thakur* who is the prime god. *Athraduar* is a place, which is situated, in the high altitude of deep forest. They consider the sacred grove (*Jahira*) or *Gram Devi* is the village deity who protects from all evils.

The Kharia also worshipped the *Karam* tree by the help of *Dehuri* (own community priest). The dead ancestors of the family are worshipped within the hut on two occasions namely – *Rakhi Purnima* (full moon) and last day of the month of *Paus*. The sun deity or *Dharam* is also worshipped in the same day. The Kharia recognised two hill deities viz. *Pat Devata* and *Baram*. The *Bhagut* is the tiger deity worshipped by the Kharia.

Bathudi

They believe that they are one of the original settlers of Bhanjabhum, which consists of the Mayurbhanj and Keonjhar districts of Orissa and the Singbhum district of Jharkhand. The Bathudi are divided into number of totemic clans. Monogamy is the general practice but polygamy is permitted in special cases. Bathudi practices clan exogamy. The specialty is that in each clan there is an association of *Bansa*, for example *Saluka* clan is associated with *Naradpadia-Bansa* and so on like *Kok-Kushmadia, Dahuk-Kapandia, Nag-madhai, Panipath-Jahariapal* etc. The following clans or *Killis* are found in the studied villages of the Similipal Biosphere Reserve: *Andarjuria/Kancha, Andhajodia, Biswal, Dhanakula, Dhanfula/Dhanphul, Ganasara/Senapati, Jamadia, Khandai, Khatua, Kulapadia, Kunora, Madhai, Matia, Matian, Nageswar/Madhai, Nakaro/Nakuar/Sinnokia, Parasar/Senapati, Patra, Paturia, Purti, Raut/ Senapati, Saluka/Ransia* and *Surya*.

The residence is patrilocal. Widow and divorce remarriage is allowed. The sons inherit the parental property equally and the eldest son succeeds his father. They either bury or cremate their dead bodies. They worshipped various gods viz *Shiva, Baramdeota, Gramdeota, Paolathakur* and the *Sitala* goddess. They observe the following festivals viz. Rashpurnima, Sarolpuja, Asharipuja, Dhuliapuja, Sitalapuja, Makarsankranti, Badnapuja and Shivachaturdasi etc. Their sacred specialist is called *Dehari*. The Bathudi are mostly Agriculturalist and work as labourers. They have their own traditional council controlled by the hereditary head *Pradhan*.

Bathudi believes *Athraduar* as an abode of all gods and goddesses. *Athraduar* is a place, which is situated, in the high altitude of deep forest. They consider the sacred grove (*Jahira*) or Gram Devi is the village deity who protects from all evils.

Bhumij

The Bhumij are mostly agriculturists and occasionally hunt and trap birds and animals in the forest. Some time they work as labourers. The Bhumij are divided into several exogamous clans. Community endogamy and monogamy are the general marriage rules. Marriages are commonly arranged by negotiation and bride price is customarily paid. Divorce, widower and widow remarriage are permitted. Nuclear family is common but joint extended families also exist. Inheritance and succession followed the male line.

The Bhumij are divided into several *Gotro* or *Killi*, which are totemic. The names of the *Killis* are as follows: *Kharisa, Rui* (A kind of fish), *Koyali or Koyala* (a kind of bird), *Uru* (a kind of black insect), *Hasda, Barda, Sandi, Pacri, Hembrom, Donda, Nag* (Cobra snake), *Birjilu* (a kind of stag), *Kamal* (lotus flower), *Tao* (a kind of bird), *Salkapar*, *Saaori* (palanquin). In our field data, we get the above-mentioned underlined *Gotro* or *Killi*. Besides that, we also get following *Killis* viz. *Kariakuch, Narayanisama, Kuchi, Kauri, Sauti/Samar, Sandi and Samara* though this list is not exclusive. In Similipal Biosphere Reserve, villages the Bhumij are not aware of totemic inclination in respect of the name of *Killis*. The following clans or *Killis* are found in the studied villages of the Similipal Biosphere Reserve: *Barda, Barde, Hansda, Kariakuchi, Kauri, Kuchi, Narayanisamar, Pachri, Patri, Samara, Sandi, Sauti/Samar*.

The Bhumij prefer Hinduism and simultaneously continue to propitiate some traditional deities. According to T. C. Das (1931:36), the Bhumij religion was essentially a tribal religion though it was reported that some minor Hindu goddesses' viz. *Manasa, Tulsi* etc. were included in their pantheon as influence of Hinduism. On contrary, our data reveals that they are totally switched over to Hinduism. Apart from that the major departure is that they worshipped *Dharam* (the prime god of Bathudi), *Singh Bonga* (the prime god of Santal) and *Jahira* (sacred Grove) etc. to show commonness in respect of the communities inhabited in Similipal Biosphere Reserve. The Bhumij belongs to Austro-Asiatic family of languages. Oriya and Hindi are spoken for inter group communication (K.S. Singh; 1998: 322-324).

Munda

According to S.C. Roy (1912) the Munda were originally inhabitants of northwest India, who latter moved to Chotanagpur and settled there. The Munda are primarily agriculturalist tribe. In some cases, they are also dependent on forest products. Monogamy is the common rule. They practise bride price. The Munda are patrilocal. Remarriage is permitted for widows, widowers and divorcees. Munda are divided into exogamous clan locally called *Killi*. However, *Killis* are totemic but now a day almost they have forgotten the totemic affiliation. Risley had given a list of Munda "secpts" which are as follows: *Bhengra* (horse), *Bodra* (ancestor did not wash their mouth after eating), *Parti* (crocodile), *Hansa* (swan), *Kandru* (fish), *Aind* (a kind of eel), *Jirhul* (a kind of flower), *Mundri* (quail), *Nimak* (salt), *Amba* (mango), *Balum* (salt), *Gondli* (a kind of grain). Some of the clan names bear the name of food articles. They eat those food articles, which they have social sanction. The following clans or *Killis* are found in the studied villages of the Similipal Biosphere Reserve: *Baghe, Balmuch, Birua, Chaki, Champia, Hansda, Hembram, Hinduar, Horo, Kalundia, Kandir, Marangdi, Mundey, Mundri, Mundu, Odeya, Purti, Samad, Sanga, Sirka, Soy, Tiu, Tuti.*

Parha is a traditional Panchyet comprises of about a dozen villages were all belongs to one *Killi.* Sometimes *Parha* does not coincide with the *Killi* group. It is reported that they had a dormitory system called *Gitiora.* The unmarried boys and girls had separate dormitories where they learned folk tails, customery laws, agricultural methods, hunting etc. Now a day the dormitory system has been abolished. *Sing Bonga* is the supreme God. The other deities of the village level are *Hatu Bongako, Desauli, Jaherpuri,* and *Chandi Bonga.* The ancestral spirits is called *Ora Bongako.* This sprit is worshipped to avoid the calamities. *Marang Buru* is the deity of rainfall and epidemic. *Ikir Bonga* is the god of Tank, Well and other water bodies. In Similipal Biosphere Reserve, we found the Munda who are also affiliated with Christianity, in that case though they recognize the traditional faiths but apparently, they are totally influenced by the local Churches.

The Munda celebrates the following religious festivals *Mage Parab* (Celebrated in the full moon of Poush month, spirits of the deceased ancestors are worshipped), *Phagu Parab* (Related with Holi festival), *Karam Parab* (Worshipped of forest trees), *Sarhul Parab* (Festival of flower), *Aouba Parab* (Before sowing of the first crop) and *Sohrai Parab* (Cattle worship).

Santal

The significant aspect is that the Santal are considered as single largest homogenous community mainly concentrated in Bihar, Jharkhand, West Bengal and Orissa. They are also called as *Majhi*.

The Santal are divided into twelve patrilineal totemic clans (*Pari*). The following clans or *Killis* are found in the studied villages of the Similipal Biosphere Reserve: *Baske, Hansda, Hembram, Marangdi, Murmu, Soren/Dandi Soren* and *Tudu*. Settled agriculture is the main occupation. They also work as industrial labourers and daily labours. Clan exogamy and community endogamy is practiced. Monogamy is the general rule. *Marang Buru* is the supreme deity in addition to that; they have clan and family deities or spirits (*Bonga*). Each Santal village has a grove of *Sal* tree, known as *Saran*, which is worshipped during various rituals.

Mahali

Mahali are mainly concentrated in the Jharkhand (Ranchi, Santal parganas, Singhbhum, Hajaribag, Dhanbad etc.), West Bengal and Orissa. Mahali is a basket making community. Basket making is their primary occupation, while agriculture and daily-wage labour are subsidiary one. They followed the clan exogamy and community endogamy as marriage rules. Marriages take place through negotiation or mutual consent. Monogamy is the general rule, polygamy is allowed in case the first wife is barren. Divorce and remarriage are allowed.

According to Risley (1891) Mahali are divided into five sub castes viz. *Bansphor-Mahli* (Bamboo basket maker), *Patar-Mahali* (Basket maker and cultivator), *Sulunkhi-Mahali* (Cultivators and labourers) *Tanti –Mahali* (Palanquin bearers) and *Munda-Mahali*. Mahalis are divided into exogamous totemic clans such as *Chardhagia*, *Charber* (a tree), *Dhilki*, *Dumria* or *Dungri* (wild pig) *Goondli* (grain forbidden) *Hansda, Hembrom*, *Khangar*, *Hermon*, *Khariar*, *Kathergach* (jack tree), *Kerketta* (a kind of bird), *Marri*, *Marumar* (Nilgao), *Topowar* (a bird), *Tirki* (bull) etc. The following clans or *Killis* are found in the studied villages of the Similipal Biosphere Reserve: *Tudu, Baske, Hembram, Kulara,* and *Marandi*. *Surji Devi* (Sun goddess) is the chief deity of them in the month of *Bhado* (August –September) they perform *Manasa Puja*, which is also called *Chhata Puja*. *Manasa* is the snake Goddess. In the month of *Ashwin* (September –October) they perform *Durga Puja* in each house.

Makdiya/Birhor

The Makdiya/Birhor are the primitive tribal group in Orissa according to the latest list of the PTG (Primitive Tribal Group) of the country. They mainly concentrated in the state of Jharkhand and West Bengal. They are an Austric-speaking people. The name of Birhor is supposed to have been derived from 'Bir' means Jungle and 'Hor' means man i.e. the man who lives in the jungle. They generally live in their traditional small settlements of conical type of huts (*Kumbha*) prepared out of grass and creepers and the settlement is known as (*Tanda*).

Traditionally the Makdiya/Birhor pursue few occupations viz. (1) rope making by using forest creeper, *Topa* (a kind of small basket made of barks foe extracting oil from oil seeds), (2) hunting of wild games particularly monkeys. In Similipal Biosphere Reserve after resettlement of the Makdiya/Birhor population, various developmental schemes have been introduced to change their dependence on the forest. These schemes are as follows: rope-making project, goatery, plantation of cashew seedlings etc. However, the results of the implementation of these schemes are not enthusiastic. Most of the goats are either died or sold. The ITDA has provided raw jute fibers and machines for preparing and twisting the rope. Nevertheless, the Birhor is not capable enough to use the machine, as they have no training and the machines are kept in idle. To stop the dependency on forest training has been imparted to the Makdiya/Birhor for the preparation of *Sal* leaf plate. Two machines have been supplied, it was reported that the machines were kept in idle because the Makdiya/Birhor are reluctant to prepare plate instead of bowl from the *Sal* leaf. The entire cashew seedling supplied by the ITDA dried up due to lack of cashew rearing knowledge.

Roy, S.C. (1925) mentions about 37 clans or *Killi* among the Makdiya/Birhor. Adhikary (1984) found 13 clans in Orissa. According to him the clans do not operate as a territorial entity and the clan solidarity is almost lost. Even in marriage the clan affiliation has not been considered prominently. In Similipal Biosphere Reserve we have observed the same kind of situation in the Makdiya/Birhor resettlement. The following clans or *Killis* are found in the studied villages of the Similipal Biosphere Reserve: *Bhagrati, Chaule, Dadujal, Golowar, Goud, Hembram, Jaksaria, Kherwar, Kumbha, Mahali, Murmu, Nagkuriya/Naikuriya, Tikriya/Galowar*.

Sing Bonga or sun god is considered as supreme deity. His wife is *Chandu Bonga* or moon. *Sing Bonga* is worshipped in the month of

Paus-Magh (January – February). Though *Sing Bonga* is the creator but *Dhatrimai* is a female deity and considered as creator of the natural resources on the earth. She is worshipped in the month of *Agrahayan* (November-December). There are two other important deities living in east and west, they are *Luguharon* in the east and his wife *Burhimai* in the west. The Birhor identified two categories of ancestral spirits namely *Hapram* and *Churgin*. The *Hapram* is again classified into *Bura Burahi* and *Chowrasi*.

Caste Groups

Goud

Goud is one of the earliest settlers of the Singhbhum district of Jharkhand. They migrated to Dhalbhum and other parts of West Bengal and Orissa in search of better land and mostly settled in the river basins. Goud is divided into four endogamous groups namely *Krishna Goud*, *Nand Goud*, *Mogda Goud* and *Jaharua Goud*. They have a number of lineages (*Bansha*) locally known as *Killi*, which are totemic. Fifteen identified *Killis* are *Nag* (serpent), *Kachchim*, *Nagesh*, *Sal* (a fish), *Jaim-Jodia*, *Jamjodia* or *Jamjadia*, *Chata* (a bird), *Bar-Behara*, *Bhaisa*, *Hathi*, *Hanumat*, *Kutar*, *Kudranosa*, *Palai* and *Rout* (K.S. Singh; 1998: 1092-1093). The following clans or *Killis* are found in the studied villages of the Similipal Biosphere Reserve: *Bada/Chadiya (Gond)*, *Chainsa*, *Mantri*, *Sindura*, *Suya*. They practice lineage (*Killi*) exogamy.

The primary occupation of the Goud is agriculture. The Jatiseba exercises social control of the Goud. *Patra-Bisoi* and *Mahakur* are the highest authorities. They worshipped *Hatikhera*, *Garam*, *Kudra*, *Munha*, *Barpahar*, *Panchbahini*, *Satbahini* and *Dharam*.They participate in *Shivagajan* dance, *Manasa*, *Tusu*, *Karam*, *Lakshmi*, *Mongla*, *Durga* festivals. Their sacred specialized are *Ojha*, *Naya* etc.

Kudmi Mahato

Kudmi Mahato is distributed in north and south divisions of Chhotanagpur, Santal Parganas of Jharkhand, Mayurbhanj, Keonjhar, Sundargarh, Balasore and Cuttack districts of Orissa, Purulia, Bankura, Midnapur, Burdwan, Birbhum and Nadia of West Bengal and tea gardens of Assam. They have more than eighty different totemic clans or *bansa*, some of which are as follows: *Baranag* (king Cobra), *Hindowar* (wild grass), *Keshria* (wild grass), *Khaidowar* (buffalo), *Katiar* (silk Cacoon), *Banoar* (monkey), *Baghowar* (tiger), *Hansadaq* (duck), *Phecha* (owl), *Chuha* (squirrel), *Kurum* (tortoise), *Musa* (rat) and *Chanch Mutra* (spider). The following clans or *Killis* are found in the studied villages of the Similipal Biosphere Reserve: *Bansria, Banwar, Bir Banwar, Bomshiar, Chilbindha, Dongria, Hindua, Jalbandha, Kankatiar, Katial, Katiar, Makdia, Nengtuwar, Sankhuad.* In Orissa they have the surnames viz. *Mahanto, Singh* and *Singdeo.* They speak in Bhanjabhumia or Kurmi dialect at home, Oriya and Bengali with outsiders. The Kudmi Mahato are mainly agrarian community.

Lohar/Bindhani

In Orissa they are also referred to as Lohura. They are mainly concentrated in the Mayurbhanj district of Orissa. There are number of *gotras* among them such as *Kaithwas, Hanswar, Soken*, etc. based on certain totemic objects. Members of the *Kaithwas* do not take gourd, while members of *Hanswar* do not kill duck. The following clans or *Killis* are found in the studied villages of the Similipal Biosphere Reserve: *Braman Baikar, Mehendiar, Kachhima, Kaithar, Sonari.* Cross cousin marriage, junior sororate and junior levirate are permissible in their community. The traditional occupation is ironwork. Some of them also own land, while some cultivate land belonging to the Santal, Kol and Bhumij. They are Hindu by faith and worshipped *Vishwakarma* as their tutelary deity.

Teli

In Orissa, The Teli have two sub groups namely *Ekadasi* and *Haldia.* The community is divided into several *gotras* and *Killis* viz. *Nag, Kashyap, Chandan* etc., which regulate marriage alliance and indicate ancestry. The *Nagashya* clan or *Killi* is found in the studied villages of the Similipal Biosphere Reserve. They are aware of the Varna system and perceived their position as Vaisya. Monogamy is the rule. Junior sororate is permissible in the community. Land is the main economic source. Both landholdings and landless families are found in the community. The traditional occupation of the Teli is business. They are Hindu and worship their family, village and regional deities like *Ganesh, Manasa, Kali, Thakurani* and *Gamota.* They have sacred specialists of their own who perform rites connected with lifecycle rituals. They participate in traditional fair and festivals like *Rathayatra, Sivarratri* and *Dolyatra* etc.

Kaibarta

The meaning of the word Kaibarta is the people who are depending upon water (*Ke-* water and *Barta-* depending upon). They generally practise endogamy. The main occupation of them is fishing and agriculture. They are mainly followers of *Mahapurusia Dharma* (*Vaishnavite* sect of Hinduvism). The clan or *Killi* are found in the studied villages of the Similipal Biosphere Reserve: *Nagasa*. They do not have any family, clan or village deities. They celebrate *Janmastami* and *Tithi* of *Sankardev, Madhabdev*, and *Haridev* (death anniversary of all these saints).

Sadgop (Gobora Caste)

They are also known as *Chasa* or *Chasi* in West Bengal. *Sarkar, Roy, Majumdar, Biswas, Mallick, Khan, Chowdhuri* and *Mondal* are their titles. The name Sadgop is derived from two Sanskrit words *Sad* or *Sat* and *Gope*. *Sad* or *Sat* means good and *Gope* means milkman. They believe that they are the descendent of Lord Krishna. However, they are now abandoning their pastoral pursuit. Paterson (1910) mentioned that the original settlement of Sadgop was at *Gopbhum*. They are divided into two sub groups viz. *Kulin* or *Kongar Sadgop* and *Moulik Sadgop*. Risley (1891) states that they are divided into two sub castes viz. *Purba Kuliya* (east *Bhagirathi* river) and *Paschim Kuliya* (east *Bhagirathi* river). They are mainly distributed in the districts of Burdwan, Hoogly, Birbhum, Bankura, Midnapur, 24 Parganas and Murshidabad of west Bengal and Singbhum of Jharkhand. Monogamy is the rule. Sadgop have some exogamous *gotras* or *Killis* viz. *Sandilya, Kashyap, Bharadwaj, Moudgulya* and *Alambayan*. The *Kashyap* clan or *Killi* is found in the studied villages of the Similipal Biosphere Reserve. They recognize their place as *Vaiysha* in the *Varna* system. They are mainly landholding community and agriculture is the primary occupation, whereas business, service, agricultural labours etc. are the secondary occupations. They follow Hinduism and worship *Hari, Lakshmi* as family deities. *Dharmaraj, Manasa, Shiva* etc. are their village deities. They believe that Gaya and Vrindaban are the pilgrim centres. The Brahman attends their different ritual performances. *Durga Puja, Kali Puja, Dharmarajer Gajan* etc. are the major festivals. They attend the fairs related to *Dharmarajer Gajan, Pirer Mela* etc. They make *Alpana* (floor paintings) at the time of their rituals.

Religion

The category religious affiliation of the communities inhabited in the buffer zone villages of SBR is as follows:

(1) Hindu

(2) Christian

(3) Tribal

The distribution of population on the basis of the above-mentioned category is given in next tables and charts.

Table 4. Population Distribution by Religion in Study Villages

Sl. No.	*Village Name*	*Community*	*Population*	*Type of Religion*		
				Hindu	*Christian*	*Tribal*
1.	Gudgudia	Kol/Ho	181	181	0	0
		Kharia	142	142	0	0
		Mahali	10	10	0	0
		Bathudi	6	6	0	0
		Goud	55	55	0	0
		Lohar/ Bindhani	21	21	0	0
		Teli	12	12	0	0
		Kaibarta	4	4	0	0
		Sodgop (Gobora Caste)	3	3	0	0
		Total	**434**	**434**	**0**	**0**
2.	Nenjaghosra/ Lanjighosra	Kol/Ho	80	80	0	0
3.	Kumari	Bathudi	199	199	0	0
		Bhumij	72	72	0	0
		Goud	21	21	0	0
		Total	**292**	**292**	**0**	**0**
4.	Kuljhari/ Koljhari	Kol/Ho	87	87	0	0
5.	Chakund akocha	Kol/Ho	51	51	0	0

(contd...)

Sl. No.	*Village Name*	*Community*	*Population*	*Type of Religion*		
				Hindu	*Christian*	*Tribal*
6.	Garh-Similipal/ Gad Similipal	Kudmi Mahato	63	63	0	0
		Kol/Ho	105	30	75	0
		Munda	6	0	6	0
		Santal/ Majhi	5	0	5	0
		Goud	9	9	0	0
		Total	**188**	**0**	**0**	**0**
7.	Saruda	Munda	135	0	135	0
8.	Basilakacha/ Basilakocha	Santal/ Majhi	64	44	20	0
9.	Durdura Tanda	Makdiya/ Birhor	68	0	0	68
Total			**1399**	**1090**	**241**	**68**

Table 5. Community Wise Population Distribution by Religion

Community		*Population*	*Type of Religion*		
			Hindu	*Christian*	*Tribal*
Tribal Groups					
1	Kol/Ho	504	429	75	0
2	Kharia	142	142	0	0
3	Bathudi	205	205	0	0
4	Bhumij	72	72	0	0
5	Munda	141	0	141	0
6	Santal/ Majhi	69	44	25	0
7	Mahali	10	10	0	0
8	Makdiya/ Birhor	68	0	0	68
Total		**1211**	**902**	**241**	**68**

Community		*Population*	*Type of Religion*		
			Hindu	*Christian*	*Tribal*
Caste Groups					
9	Goud	85	85	0	0
10	Kudmi Mahato	63	63	0	0
11	Lohar/Bindhani	21	21	0	0
12	Teli	12	12	0	0
13	Kaibarta	4	4	0	0
14	Sodgop (Gobora Caste)	3	3	0	0
Total		**188**	**188**	**0**	**0**
Grand Total		**1399**	**1090**	**241**	**68**

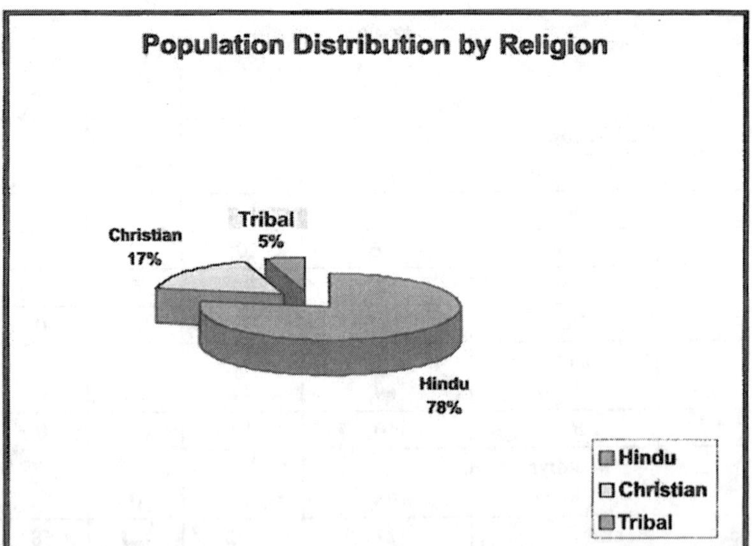

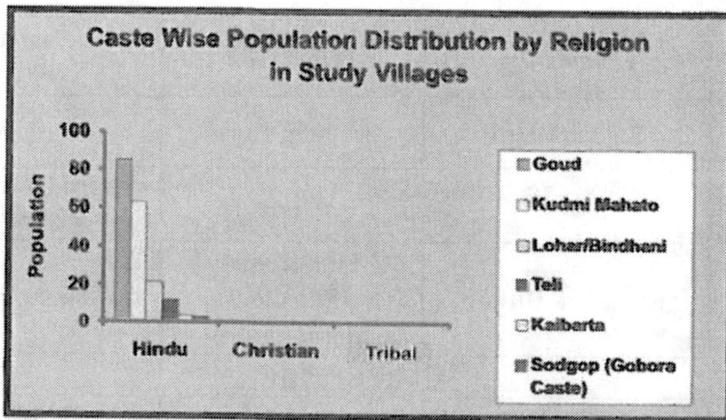

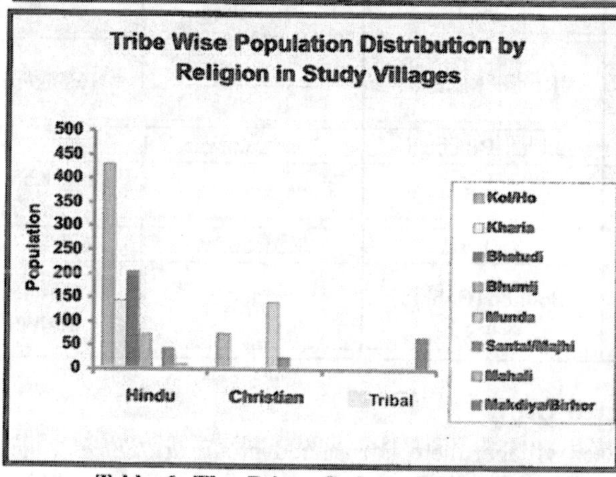

Table 6. The Prime Gods and Goddesses

Sl. No.	*Community*	*Prime God*	*Prime Goddess*
Tribal Groups			
1.	Kol/Ho	Baram Thakur and Singbonga	Gram devi
2.	Kharia	Badam Thakur Pat Devata	Bada durga
3.	Bathudi	Shiva/ Badam Thakur	Badla Muni
4.	Bhumij	Dharam Marang Buru	Manasa Tulsi
5.	Munda	Marang Buru	--------------

Sl. No.	*Community*	*Prime God*	*Prime Goddess*
6.	Santal/Majhi	Marang Buru	--------------
7.	Mahali	--------------	Surji Devi (Sun goddess)
8.	Makdiya/ Birhor	Sing Bonga (Sun God) Marang Buru	Chandu Bonga
	Caste Groups		
1.	Goud	Srikrishna	Parbati
2.	Kudmi Mahato	----------------	Kichakeshwari
3.	Lohar/Bindhani	Vishwakarma	----------------
4.	Teli	----------------	Kali Thakurani
5.	Kaibarta	Madhabdev	----------------
6.	Sodgop (Gobora Caste)	Shiva	Durga puja Kali puja

The Sacred Grove

In each village, there is a sacred grove or *Jahira*, which is also called as 'Gram Devi'. Irrespective of tribal or caste affiliation, each

Table 7. Sacred Grove or Jahira and Associated Trees (Village and Community wise)

Sl. No.	*Village Name*	*Community*	*Associated Trees*	*Remarks*
1.	Gudgudia	Kol/Ho, Kharia, Mahali, Bathudi, Goud, Lohar/Bindhan, Teli, Kaibarta, Sodgop (Gobora Caste)	Sal/Siris/ Assat/ Naha/ Pauri/	Each Community Worships as 'Gram Devi' or village deity.

Sl. No.	*Village Name*	*Community*	*Associated Trees*	*Remarks*
2.	Nenjaghosra/ Lanjighosra	Kol/Ho	-------------	'Jahira' of Gudgudia Village.
3.	Kumari	Bathudi, Bhumij, Goud	Sal, Sal, Simli, Amm, Jam, Kusum, Mahul, Jari, Champa, Bat	Each Community Worships as 'Gram Devi' or village deity.
4.	Kuljhari/ Koljhari	Kol/Ho	Behara Tree	The worship takes place during March (when the flowers are blossoming) by sacrifice poultry bird.
5.	Chakund akocha	Kol/Ho	Panidharam Tree	------------------------------
6.	Garh-Similipal/ Gad Similipal	Kudmi Mahato, Kol/Ho, Munda, Santal/Majhi, Goud	Sal/Asan	Jahira' is situated inside the dense forest.
7.	Saruda	Munda	Sal	-Do-
8.	Basilakacha/ Basilakocha	Santal	Sal	-------

Community considers the sacred grove as a deity who protects them from all evils. Thus, sacred grove in SBR is to be considered as a symbolic sacred entity, which is a binding force of commonness. Even the tribal groups who are Christian also respect the sacred grove. The following table will give an idea on the sacred grove and associated trees.

Nature and Supernatural Power

Irrespective of the community, affiliation people of SBR believe that hills are the abode of gods and goddesses. Therefore, they consider the hills as sacred natural object. The following table will give an idea on the associated beliefs about that.

Table 8. Sacred Natural Objects

Sl. No.	*Comm-unity*	*Sacred Hills*		*Other Sacred Natural Objects*	
		Male	*Female*	*Male*	*Female*
Tribal Groups					
1.	Kol/Ho	Anjhari, Barda, Bondar, Gansar, Samu, Athradeori, Jandi Badu, Chatun Badu	--------	Chandu Bonga (Moon), Baktu Bonga (Thunder), Sidh Daru (Cactus Tree)- abode of 'Shiv' god	Basimata (Earth), Nagira (water).
2.	Bhumij	Badar, Chatam Ram, Megasan, Lili Giri, Kaktipad, Ram Raja, Galsar, Duar Seni, Devathali, Dudra Champa, Chaula	Athradari, Buri Pahar	--------	-----
Caste Groups					
3.	Goud	Athraduar, Amjhar, Sanai Bhandar, Gansar Pahar, Daridara, Some Pahar	Danwa Dungri,	Barda Jungle	--------

Festivals

There are many festivals, which are celebrated by the communities inhabited in SBR. Festival is a binding force in society. In a particular festival, all communities take part despite differentiation. In a cultural milieu, festival has significance. The five elements of nature i.e. earth, air, fire, water and space (*Panchabhut*) which constitute biosphere are worshiped in various festivals.

As the communities inhabited in SBR are the settled agriculturists, most of the festivals are associated to their agricultural activities. Apart from their domestic traditional rituals, they also worship different *Bongas* communally either in their clan or village levels in different festivals. The traditional festivals are common in nature where the socio-religious as well as the recreational activities are performed simultaneously. The following table will give an idea on the cohesiveness in the society through festivals.

Table 9. Festivals

Sl. No.	*Festival*	*Month*	*Remarks*
1.	Ba Parab	Chait	Flower Blossoming in 'Sal' Tree
2.	Bandhana Parab	Kartik	At the time of 'Dewali' for Domestic Cattle
3.	Durga Puja	Ashin	Worship of mother goddess 'Durga'
4.	Goma Parab	Shavan	Raksha Bandhan
5.	Hero/Goma	Bhador	Raksha Bandhan
6.	Jamna Parab	Indi	Harvesting Paddy
7.	Kali Puja	Kartik	Worship of mother goddess 'Kali'
8.	Karma Parab	Fagun	Nature Worship
9.	Magho Parab	Push	New Year
10.	Makar Parab	Push/Magho	'Makar Sankranti'
11.	Noyakhai Parab	Bhado	Rain
12.	Rajo Parab	Ashar	Rain
13.	Shiv Ratri	Magha/ Fagun	Worship of 'Shiva' god
14.	Ura Parab	Baishak	New year

Worldview and Folk Taxonomy

Folk taxonomy is a tool to express the worldview. In a simple and homogenous society, the world is less categorized and segmented. In that respect there are as many as parametres to classify the universe. The nature of classification is called as paradigm. Each community has a unique perception in a particular aspect, which is called the cognition of the society. The each paradigm is a reflection of that cognition level of that society. In the context of biosphere, reserve the attempt has been made to derive the folk taxonomy on three categories viz. natural objects, cultural objects and socio-political objects. Again, each category has numerous objects. Thus, in each case few selected objects are chosen and illustrated. Few examples on each category are given in the next discourse.

Table 10. Folk Classification of Forests

Sl. No.	*Folk Name of the forests*
1.	Saura
2.	Abru
3.	Barda
4.	Amjhari
5.	Gutu
6.	Singh Buru
7.	Jamdila
8.	Birdaban
9.	Ranga Pahar
10.	Chandan
11.	Dankocha

The communities inhabited inside the forest differentiate the entire forest tract in cognitive ways. In general, people could not categories the forest, for them the entire forest is uniform or slightly different where the density is clearly prominent. The cognitive mode of perception in terms of classification of forest is something different. The folk name of the forests is related with certain principles for instance certain mountains, hills, hill tops are locally named for direction and identification. Few names are closely associated with soil types which inculcate with variation of colours. There are so many ways to identify and differentiation of the forest. These modes of identification and differentiation are made for practical purposes particularly to determine the direction.

Table 11. Folk Taxonomy on Domestic Animals

Sl. No.	*Category*	*Tribal Groups*				
		Kol/Ho	*Bathudi*	*Bhumij*	*Munda*	*Santal/ Majhi*
1.	Cow	Gunda	Gai	Gundi	Uri	Dargri
2.	Ox	Haria	Sanda	Sandari	Hara	Nahel
3.	Bull	-------	Balad	Hara	-------	-------
4.	Calf	Mine	-------	-------	-------	-------
5.	Male Calf	-------	Andia Bachuri	-------	-------	-------
6.	Female Calf	-------	Mai Bachuri	-------	-------	-------
7.	Buffalo	-------	-------	-------	Kada	Kada
8.	Goat	Meram	-------	-------	Meram	Meram
9.	Male Goat	Bodani	Boda	Bodasing	Bakra	Bayar
10.	Female Goat	Engal Meram	Mai Cheri	Merman	Inga Meram	Inga Meram
11.	Castrated Goat	Khasini	Khasi	-------	Khasi Meram	-------
12.	Kid	-------	Cheri Chua	-------	-------	-------
13.	Cock	-------	Gonga Kukra	Sing Sandi	Sandi	Sandi
14.	Hen	Chim	Mai Kukra	Sininga	Sim Inga	Sim
15.	Chicken	-------	-------	Siman	-------	-------
16.	Duck	Hansili	-------	-------	Gede	Gede
17.	Male Duck	-------	Andia Hansa	-------	-------	-------
18.	Female Duck	-------	Mai Hansa	-------	-------	-------

Each community or a cluster of communities have different connotation in respect of classify and identify of the domestic animals. The specific culture has a special form of classification in respect of age and sex. The classification is not a mare categorization and identification. It signifies many aspects of cultural domain, which in certain points of time coterminates with economic importance. This type of relationship with the domestic animals differs from place to place, not only that it differs in community level within a particular place. The particular type of classification is called the paradigm. Each paradigm is linked with the cultural connotation. Sometimes it depends on historicity. If we look on the above table, it reveals that each tribal group has different and elaborate terms on cow, ox and hen. This is because of the fact that all the tribal groups are practicing agriculture and hen is significant symbol in respect of social pride by virtue of cockfight.

Table 12. Folk Taxonomy on Natural Objects

Sl. No.	*Natural Objects (Folk Name)*	*Meaning*
1.	Marang Buru	Tall Hill
2.	Huring Buru	Small Hill
3.	Gutu	Hillock
4.	Marang Diri	Boulder
5.	Sirme/Sirna Chilmil	Sky
6.	Rinil	Cloud
7.	Goma/Da	Rain
8.	Hai	Wind
9.	Daha/Barkend	Water
10.	Gada	River
11.	Dig	Direction
12.	Ote	Earth
13.	Dhuru	Forest

Like the forest, the different natural objects bear different folk names irrespective of community affiliation. The categorization is distinctive and shows how the world of nature is classified. Though this categorization is exhaustive but reveals the diversified world where each segment of the world is characterized as with specific entity. This type of classification may be designated as common paradigm.

The folk taxonomy on colours is a classification of the colours by each group which they have perceived through ages and inherited traditional mode of way. The tribal groups have classified the colours in different ways. On contrary, the caste groups have a common categorization on colours irrespective of caste affiliation. The

Folk Taxonomy on Colours-Tribal Groups

Kol/Ho	Kharia	Bathudi	Bhumij	Munda	Santal/Majhi	Makdiya/Birhor
Khairati	Khairati	Khairia	Hendoti	Fika Rang	Ara	Sindur
Liltia	Hendotil	Beguni	Hendotil	Baiguni	Lil	Buda Lil
Sabujatia	Sabujatia	Sabujia	Liltia	Hariar	Lil	Shyam
Liltia	Hendoti	Nil	Hendoti	Nil	Lil	Lil
Sasanti	Sasanti	Haldia	Sasang	Sasan	Sasang Ara	Sasang
Arapia	Arapia	Lali	Arapia	Ara	Ara	Ranga

(contd...)

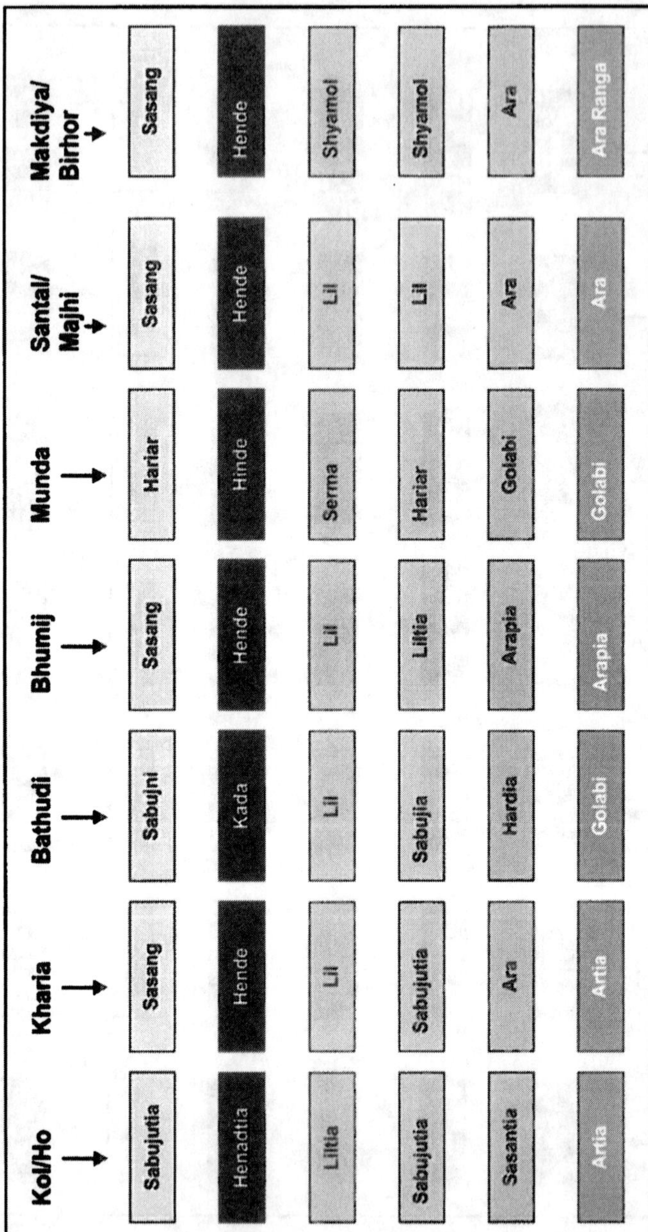
Kol/Ho
Sabujutia
Henadtia
Liltia
Sabujutia
Sasantia
Artia
Kharia
Sasang
Hende
Lil
Sabujutia
Ara
Artia
Bathudi
Sabujni
Kada
Lil
Sabujia
Hardia
Golabi
Bhumij
Sasang
Hende
Lil
Liltia
Arapia
Arapia
Munda
Hariar
Hinde
Serma
Hariar
Golabi
Golabi
Santal/Majhi
Sasang
Hende
Lil
Lil
Ara
Ara
Makdiya/Birhor
Sasang
Hende
Shyamol
Shyamol
Ara
Ara Ranga

interpretations in respect of differentiation of the colours in different ways signify classification of the entire world and society.

Folk Perception or Concept

Folk perception is another tool to ascertain worldview. In this presentation the folk perception is divided into two divisions' viz. folk concept and folk awareness. Folk concept is the indigenous idea to explain certain natural phenomenon. Folk awareness is another dimension to evaluate the level of perception in respect of social, economic and political process. Few examples are given to understand the both aspects.

Cloud in west direction indicates a heavy rainfall. Cloud in north direction means no rainfall. Cloud in southern means slight rainfall. Cloud in east direction indicates rain will run away by the effect of wind.

Makdiya/Birhor claims the descent from mythological story. They believe that they were the habitants of *Lanka* kingdom when *Ravana* was the king. *Hanuman* had destroyed their homes and that is why they are wandering here and there. Due to that hostility, they kill the monkeys and eat.

Folk Awareness

According to the people of the buffer zone of SBR, there are two administrative categorizations viz. core and revenue. They know the name of three political parties' viz. S.U.C.I., Congress and B.J.P. and one non-governmental organization *Samandh*. They know the hierarchy of the forest administration. They know the rights, entitlements and restrictions. They are well aware about their causes of plights and distinguish the remedial measures.

Conclusion

The religious beliefs and practices of each community irrespective of caste and tribal affiliation represent uniqueness in respect of cultural assimilation. The tribal groups have distinct cultural traits in their core cultural area of Jharkhand and West Bengal. On contrary, while they have migrated to Similipal Biosphere Reserve at different points of time, they engulf the cultural ethos of Orissa as well as left few own traits. Through the inter linkage with the fellow tribal and non-tribal groups in a new ecological set up each community share some commonness by the process of assimilation. The kaleidoscope (folk taxonomy, folk perception or conception and folk awareness) will help to reflect the same features while the worldview is too considered as product of these above-mentioned aspects looked into.

Before considering the relevance of this kind of study, it will be useful to analyze above discussed field data in a systematic manner. First, the religion and festival are two important components in cultural domain. If we look into these aspects, a marginal difference is observed irrespective of caste and tribal groups. The reason is that the people are living in a specific eco set up and facing the similar kinds of problems. Within forest, different groups require sharing a common platform. At the same time they express solidarity to overcome, the emerging constrains in new set up. The commonness compel them to come closer and sharing common traits by accept others religious traits as well as and discard few own cultural traits. If we analyze the religious traits of the SBR buffer zone communities as a whole, we get commonness i.e. concentric boundaries of the supernatural domain. The domain of the *Jahira* (sacred grove) is expanded up to village level. Every community believes that the abode of the all gods and goddesses is *Athradeul* which is a high hill top situated at core area of the Similipal Biosphere Reserve i.e. the domain of the other deities are spread over to the entire reserve area.

The second important components of worldview (consists of three aspects viz. Folk Taxonomy, Folk Concept and Folk Awareness) need to be analyzed in respect of above discussion. The analysis on folk taxonomy of colours reveals the fact that the tribal communities of the Similipal Biosphere Reserve are expressing different connotation on single colour. Even the classification of the colours is not uniformly dividing the whole spectrum. This indicates the pattern of cultural set up which is manifested through worldview of this particular tribal community. In true sense, specific culture has particular type of set of the components in a given ecological set up. These sets may consist of similar components but the arrangements and classifications will be different in case of each category. This type of derivation could be found out through componential analysis. One kind of that type of approach is attempted by classifying domestic animals. Each tribal has a specific mode of classify of the domestic animals, which are common for almost all, and the use is similar kind. Despite that, the differentiations are different which shows the different cultural set up. The illustration may be lengthy if each dimension is taken into consideration. The result is same i.e. tribal groups are reflecting the differential types of worldview where caste groups are uniformly patterned.

One question might be raised why the tribal groups are more closure in respect of religion while far away in case of worldview. Apparently, this is a puzzle but if we look intricacy of the religious beliefs and behaviours, pertaining to supernatural affinity it is observed that the two kinds of external force are operating in this area. The first one is the restriction imposed in this biosphere reserve, which compelled the communities to come closer irrespective of their distinctive culture. The religious functions are the platforms where people can interact and unite. The second one is the influence of Christianity where differentiations are minimized and invite uniformity. Thus, the external conditions are forced them to minimize the supernatural beliefs and practice uniformity to some extent, while worldview is traditional and external force has a limited scope to penetrate in deeper cultural set up and even they could not create a threat towards maintenance of their cultural practices. On contrary, the caste groups are sharing an almost common religious and culture greatly influences by Hindu doctrine and this commonness is reflected in their religious beliefs and worldview.

The relevant point is that how this study may help others in light of burning issues in respect of Similipal Biosphere Reserve. Firstly, it is found from the study that people are not aware of the present concept of biosphere reserve and their role within that set up. Secondly, from the above discussion it reveals that their cultural ethos and doctrine could not permit indiscriminate destruction of forest and wild animals. Thirdly, they want better future. If an approach on confidence and awareness building taken towards them the evils of the forest, (timber merchants and poachers) will be defeated by help of the people of SBR.

The objective of the above all discussion is that to overview in general the interaction of the people with the forest eco-system. In the Biosphere Reserve the people is the one of the major elements and could not be ruled out with ignorance. The interdependency in respect of people and forest and vice versa could not be interpreted in few words. From the time immemorial, man and forest are the integral part of the world. A simple conclusion that eliminates people to save forest is the oversimplification of the entire problem. The present exercise is not to deal with this controversy. Despite of that this will emerge automatically. The above results are based on the primary field data. The results are categorized into three aspects viz. the dynamics of the religious affinity of the communities inhabited in the buffer zone villages of SBR, social background story of the people and the worldview expressed through folk taxonomy and folk concept.

In nutshell, it may be concluded that the population pressure on the buffer zone is not alarming. At present the tight rules and regulation restricted unnecessary movements of the people inside the protected zone. Undoubtedly, this is a positive step towards the protection and preservation of the pristine ecology. Apart from that the forest authority needs to look on the development of the social and economic sphere of the people of the buffer zone villages to combat unscrupulous activities of the destructive outfits. The field data reveals the fact the only strict protective measures on the villagers may not help to achieve the goal. There is a need to look into entire gamete of the developmental aspects from the idea of shift of paradigm. The core and buffer area of Biosphere Reserve is devoid of many facilities such as electricity, metal road and public transportation etc. where as other villages at transition zone can get the above-mentioned facilities. This kind of dichotomy may lead to create a feeling of alienation. To get rid of that kind of antagonism a specific developmental measures for the buffer zone villages is necessary, which will help to eliminate the unwanted fear psychosis. The following examples of the sustainable developmental schemes viz. education, health, eco-friendly cash-earning plantation and self-help group etc. may enrich their economic position. No body can deny the fact that unless and until through economic progress the villages of the buffer zone could build a confidence the dependency on the forest could not be minimize. This kind of eco-friendly welfare approach will help to alienate the enemies of the forest who in many occasions may take the help of the villagers by taking the advantage of their poverty.

The above-mentioned data on religion and worldview on the buffer zone villages will focus on the entire situation. The data will pin point on the hard facts of their marginalize condition which will be detrimental towards the future of the forest. The proposition discussed above is based on the entire depicted cultural and socio-economic background story.

4

Impact of Development Plans and Programmes in the Buffer Zone Villages

Gautam Kumar Mallik

Introduction

The people all over the world are living in divergent geographical areas maintaining varied degrees of cultural identity. These typical geographical differences left a profound impact on their life and functioning pattern. Various geo-physical situations differentiated by the climatic condition, soil condition, vegetation, water resources, etc. create diverse types of economic activities, which vary from each other, tune the cultural contour of the people. So it appears that man always depends on their surrounding environment to accomplish their needs and requirement for their survival.

Man is depended on the biological resources for food, energy, construction materials, medicine and so forth. Indeed man and environment have a close and symbiotic relationship from time immemorial. Throughout the world man has dominant influence on their surrounding environment. Recent study has estimated that today man uses 40% of the terrestrial components of total primary productivity (Vitousek, *et al*, 1986). The study shows that human being uses the natural resources as much they can exploit it. Growth of human population and increasing resource consumption has a profound impact on global resources. This phenomenon arise the question of carrying capacity of the planet and whether we can continue to increase our demands on it without limit (Heywood, 1995).

The term Biosphere Reserve means conservation of resources like land, water and biodiversity of a particular area and its sustainable use through the relationship of man and environment (UNESCO, 2005). To protect the biodiversity all over the world the programme of biosphere reserve was initiated under the man and biosphere program by UNESCO in 1971. The endeavor of the programme is to minimize the use of forest resources and launch alternative developmental activities for the people living there, which are sustainable to them. Nation-wide conservation movements were launched after the enactment of the Wildlife Protection Act in 1972.

The Ministry of Environment and Forests launched the biosphere reserve programme in our country in 1986. Accordingly, the Similipal Reserve Forest was declared as a Biosphere Reserve in 22nd June 1994, with a total area of 5324 square kilometer, which is our area of study. The Similipal Biosphere Reserve (SBR) is situated in the 31°88′ to 21° 27′ north latitude and 86°17′ east longitude and located on the southern extension of Chotanagpur plateau in the district of Mayurbhanj, Orissa. The Similipal Biosphere Reserve is divided into three distinct zones (a) Core zone (845 sq km area), (b) Buffer zone (1904 sq km area) and (c) Transition zone (2575 sq km area).

Several research works done on Similipal Biosphere Reserve among those the most renowned research workers are H.R. Bustard (1975), B.K. Behura and L.A.K. Singh (1979, 1997) and S.R. Choudhary (1999) but most of the research works have been done on plants, wildlife and ecosystem. Later on several research works are also done on preservation, conservation of forest resources and people's participation in different development programmes. Sagar, S.R. and L.A.K. Singh (1993) conducted research work on socio-economic positions and people's participation in different eco-development programmes of Similipal Tiger Reserve. These works will be our benchmark information.

The aforesaid studies were mostly related to the plants and animals and the level of participation of the local people in the different development projects did not include much emphasis on the impact of the development plans on the life of the local people. The present studies will therefore, highlight the impact of development plans and programmes on the socio-economic life of the local people and how they adjust themselves with the new development plans and programmes.

Keeping the above facts in mind, the present study conducted fieldwork in the buffer area of SBR covering nine villages' eight villages in the buffer area and one in the transition area. To assess the impact of development activities in the buffer and transition area covering two Blocks i.e. Jashipur and Bangripasi. The study villages selected from all the sides of the SBR. During the selection of villages it also taken into account that the characteristics of the villages such as well exposed, uniethnic, multiethnic, villages situated inside the forest i.e. less exposed. Emphasis also given to cover all the religion practiced by the people and it also cover almost all the ethnic groups living in the Similipal Biosphere Reserve. The fieldwork was conducted in first week of November to last week of December 2006.

According to Prof. P.K. Bhowmick (1994, p. 12) "Development is a process that means movement from a condition towards another process of gradual unfolding or growth or stage of advancement".

Development is a process that improves the quality of life through the cultural life of a particular community. It depends on so many factors such as ecological setting of the area where the community lives, their economic activities and dependency over the surrounding natural resources. It also depends on the socio-religious activities and the action patterns, inter community relation and psycho-emotional level of a particular community. So "development is a multi dimensional process of upgrading of social, cultural, economic and human conditions of the people necessary for a balance growth of society" (Sachchidananda 2006, p. 333).

The colonial period the British administrator adopted several measures for the tribal problems. The first policy was the policy of isolation or leaves them alone. It is not that there were no interventions in tribal affairs by the British administrator. From time to time there was unrest even violent rebellion in different areas and they controlled that by applying force, side by side some reform measures were taken up on land, forest usurious practices as well.

The second approach were taken for the assimilation of the tribal people into the dominant Hindu society since the tribals were absorbed as backward Hindus, who had been pushed into the hills and forest by the pressure of non-tribal peasantry.

The third approach was of integration of the tribal population with the regional political, economic and social levels.

After independence various plans were taken for the development of the tribal people in our country such as:

1. Jawaharlal Nehru's Five-Point approach popularly known as "Tribal Panchsheel".
2. S.C. Dube recommended the 'Time bound area development approach' that conceived through Tribal Sub-Plan.
3. Establishment of Panchayati Raj in 1993 latter on some amendments made and Extension of Panchayat Act in Scheduled areas (PESA) introduced in 1996.
4. To get better results and allow greater freedom to the people of grass root level National Rural Employment Grantee Act (NREGA) introduced in August 2005. Now the act is renamed as Mahatma Gandhi National Rural Employment Grantee Act (MGNREGA).

In the present context the 'development' has been considered as the development of Biosphere i.e. Conservation of biodiversity in the Similipal Biosphere Reserve in general and the development of the community, who are living within the SBR, in particular. To conserve the biodiversity of the SBR forest developing agencies have taken several plans and programmes to protect the plants and animal resources of the BR. On the other hand to minimize the use of forest resources, the District authorities have taken several plans and programmes through the National Rural Employment Generating Programme (NREGP) in short form - Wage Employment Programme, and Sarnajayanty Gramin Swarojgar Yojana (SGSY) Self Employment Programme and the Forest Department have also taken several Eco-Development Programmes through the Self Help Groups (SHGS) and formation of Eco-Development Committee (EDCS) for those people who are living inside the SBR.

Objectives of the Study

- To study the basic economic activities, emergent economy and utilization of Biosphere Reserve area.
- To study the traditional method of preservation of their surrounding resource base through the customs, beliefs including ritualistic behaviour, taboos, etc.
- To know the plans, programmes and development activities taken by the various agencies and its implementation in the Buffer area of the Similipal Biosphere Reserve.
- To judge the level of people's participation in the context of plans and programmes (state agencies, others, including tourism).

Methodology

To fulfil the above objectives data collected on the development plans and activities related to Buffer region of the Similipal Biosphere Reserve. Data collected from Forest and Block development offices about the name of the villages where the development plans and programmes had been initiated. Again from the local Panchayat offices data was collected about the name of the particular villages where the development plans implemented and name and number of beneficiaries. Data collected on earlier and present economic activities, house type, settlement pattern and also the religious activities related to their economic activities. Data collected on their traditional methods of preservation of the natural resources, on the impact of various development plans and their participation and perception on it. Data collected through standard Anthropological techniques like- Household census, case history, structured questionnaire, interview and observation.

Cultural Profile of the Communities Studied

Table 1. Family and population: Community wise

Sl. No.	*Name of the Community*	*Name of the village*	*Number of Family*	*%*	*Pop-ulation*	*%*
Tribal Groups						
1.	Kol / Ho	Gudgudia Nenjaghosra/ Lanjighosra Garh-Similipal/Gad Similipal Chakundakocha Kuljhari/Koljhari	104	35.27	504	36.04
2.	Kheria	Gudgudia	36	12.20	142	10.15
3.	Bathudi	Gudgudia Kumari	39	13.24	205	14.67
4.	Bhumij	Kumari	13	4.40	72	5.16
5.	Munda	Saruda Garh-Similipal/Gad Similipal	24	8.15	141	10.07
6.	Santal/Majhi	Basilakacha/ Basilakocha Garh-Similipal/Gad Similipal	13	4.40	69	4.93
7.	Mahali	Gudgudia	3	1.01	10	0.71
8.	Makdiya/ Birhor	Durdura Tanda (Makdiya Tanda)	18	6.14	68	4.86

Sl. No.	Name of the Community	Name of the village	Number of Family	%	Pop-ulation	%
Caste Groups		Gudgudia				
9.	Goud	Kumari Garh-Similipal/Gad Similipal	20	6.79	85	6.07
10.	Kudmi Mahato	Garh-Similipal/ Gad Similipal	15	5.08	63	4.50
11.	Lohar / Bindhani	Gudgudia	5	1.69	21	1.50
12.	Teli	Gudgudia	3	1.01	12	0.85
13.	Kaibarta	Gudgudia	1	0.33	4	0.28
14.	Sodgop (Gobora)	Gudgudia	1	0.33	3	0.21
Total			295	100.00	1399	100.00

Source: Fieldwork data.

Bathudi

They believe that they are one of the original settlers of Bhanjabhum, which consist of the Mayurbhanj and Keonjhar district of Orissa and the Singhbhum district in Bihar. The Bathudi claimed that they migrated from the village Bakua of Similipal hills. The members of the tribe hold that the zamindars of their community originally reign the Similipal range and established their citadel at Jashipur.The probable homeland of the Bathudi is Panchpir and Sadar Subdivision of Mayurbhanj District (Singh, 1994, p. 96). The Bathudi are mostly agriculturalist and also work as labourers. In our studied villages they are mainly settled in Gudgudia and Kumari villages having 39 families with a population of 205 as well as male and female. Bathudi follow community endogamy and clan exogamy as their marriage rules.

The Bathudi are divided into a number of totemic clans. Several clan names that we have gathered during fieldwork such as *Andarjuria/ Kancha, Andhajodia, Biswal, Dhanakula, Dhanfula/Dhanphul, Ganasara/Senapati, Jamadia, Khandai, Khatua, Kulapadia, Kunora, Madhai, Matia, Matian, Nageswar/Madhai, Nakaro/Nakuar/Sinnokia, Parasar/Senapati, Patra, Paturia, Purti, Raut/Senapati, Saluka/Ransia, Surya.* Age of marriage for girls is 15-18 years and boys between 18-24 years. Monogamy is generally practiced but polygamy is permitted in special cases. The residence is patrilocal. Widow and divorce remarriage is allowed. The sons inherit the parental property equally and the eldest son succeeds his father. They either bury or cremate their dead bodies. They have their own traditional council controlled by the hereditary heads called "*Pradhans*". They worship *Sarna (Sal grove) or Gram than* (Singh, 1994, p. 97).

Bhumij

The Bhumij are mainly found in Jharkhand, Orissa and West Bengal. They are also found in Assam where they work in the tea garden as labourers. The Bhumij speaks their own language "Bhumij" which, belongs to Austro-Asiatic family of languages. Oriya and Hindi are also spoken for their inter group communication (Singh, 1994, p. 168).

The Bhumij are mainly agriculturalist and also hunt and trap birds and animals in the jungles. Some time they work as labourers. The Bhumij practised community endogamy and are divided into several exogamous clans or gotras. The Bhumij *gotras* or *killis* has a traditional ancestral village of its own, the name of which it bears along with the name of the *gotras or kili*". The Bhumij gotras-names indicate some kind of bird, beast, insect, plant or some inanimate object and are totemistic in nature. The members of a *gotra* are required to observed taboos in connection with the object after which they are known. Generally it is not to be touched or used as a food if edible, and should not be injured in any way (T.C. Das, 1931, pp. 22-27). In our studied villages they are divided into several clans like- *Barda, Barde, Hansda, Kariakuchi, Kauri, Kuchi, Narayanisamar, Pachri, Patri, Samara, Sandi, Sauti/Samar.* Monogamy is the general marriage rules. Marriages are commonly arranged by negotiation and bride price is customarily paid. Divorce, widower and widow remarriage are permitted. Nuclear family is common but joint extend families are also exist. Inheritance and succession followed the mail line. Bhumij perform Hinduism and simultaneously continue to propitiate some traditional deities. The village priests worship village deity called *Baram.* Goats and fowls are generally sacrificed. They also worshipped *Dharam Devta, Singh Bonga* and *Jahubura* for the welfare of the village and people.

Kol/Ho

The Ho are numerically dominant community of our studied villages in the SBR. The Ho of Orissa migrated from the Kolha region, which, is now in Singhbhum district of Jharkhand. A large number of Ho populations migrated from Singhbhum and settled in the Similipal Biosphere Reserve area, which was under the control of Mayurbhanj King and they pay all kind of taxes to the Mayurbhanj King. They are marginal and small farmer; they also work as agricultural, construction and forest labourers.

The Ho are divided into number of exogamous totemic clans *(kili)*. Adult marriage is practiced and mates are acquired through negotiation and mutual consent. Monogamy is the general rule but polygamy is also practice. They follow the patrilocal rule of residence. Widow remarriage is permissible. Both nuclear and extended families are found in their community. All the sons inherit parental property equally, while the eldest son succeeds to the position of authority. In the studied villages we found several clans among them such as *Adwasoy, Alda, Badujankho, Bajra, Bagh/Baghe, Bandra/Bhandra/Sonia Bandra (Birhor), Bangra/ Bhengra, Bangsing, Barda, Bari/Bhari, Birua/Birwa, Bodhra, Boipai, Chakki, Chatar, Dangil, Darai/Dorai/Durai, Deogam, Dubraj, Gagrai, Ghoshra, Gundua, Guia, Hambira, Hansda, Hanṣda, Hembram, Hara, Jamuda, Jarikapat, Jiwanbalmuchu, Kalondia/Kalundia, Kara, Karwa, Kerai/Khirai, Khedai, Khuntia, Kora/Koda, Korwa, Laguri, Mouchel, Patralamai, Pingua/Gama Pingua, Pirua, Purti, Saluket, Samar/Samad, Sainya, Saoni, Simku/Sinku, Singkuta, Sirka, Soy, Suri, Taki, Tariba, Tintwa, Tiriya/Tiria, Tiu, Tutti.*

They worship clan deities like *Thakurani* and *Marang Bonga* and village deities like *Gram Sin, Baram and Basuki* are their sacred specialists are the *Deheri*, who belong to their own or neighbouring communities (K.S. Singh 1994, pp. 404-409).

Munda

The Munda believe that they are the descendent of *Sing Bonga*, The supreme god of the Munda. According to S.C. Roy, (1912) the Munda were originally inhabitants of northwest India, who latter moved to Chotanagpur and settled there. Cultivation is there traditional as well as primary occupation, while hunting and collection of minor forest products are subsidiary occupation. They also work as labourers. In the present context the Munda community lived in a uniethnic village-Saruda in the Astakumar G.P. of SBR and all the villagers are practising Christianity.

Though they practiced Christianity but in their social aspect they are divided into exogamous totemic clans but endogamy also practiced. In our studied villages we found such clans among them such as *Baghe, Balmuch, Birua, Chaki, Champia, Hansda, Hembram, Hinduar, Horo, Kalundia, Kandir, Marandi, Mundey, Mundri, Mundu, Odeya, Purti, Samad, Sanga, Sirka, Soy, Tiu, Tuti.* Monogamy is the general marriage rule but polygamy is also permitted. During our field survey we do not found any case of polygamy practiced by the Munda, in the Saruda village. Remarriage is permitted for widows, widowers and divorces. The Munda's are patrilocal in nature. They are mainly nature worshippers. They also worship their ancestors. *Sing Bonga* is the main deity, who is the source of all power and energy. *Marang Buru* is another deity. They perform festival like *Sarhul, Karma* etc. (K.S. Singh, 1994, pp. 842-848).

Santal/Majhi

The Santal's are mainly found in Jharkhand, Bihar, Orissa and west Bengal. They are divided into twelve patrilineal totemic clans (*Pari*). In our studied villages most of Santal's are practising Christianity. Settled agriculture is the main occupation. They also work as industrial labourers, daily labourers, and white colour job. Clan exogamy and community endogamy is practiced. The following clans or *Killis* are found in the studied villages such as *Baske, Hansda, Hembram, Marandi, Murmu, Soren/Dandi Soren, and Tudu.* Monogamy is the general rule. The dead are cremated and a period of 5 days pollution period is observed. They have good functioning village council headed by headman called *Manjhi.* Community solidarity is very strong.

Marang Buru is the supreme deity in addition to that they have clan and family deity or Spirits (*Bonga*). Each Santal village has a grove of Sal trees, known as *Sarna*, which is worshipped during various rituals (K.S. Singh, 1994, pp. 1041-1042).

Mahali

They are migrated from the Singhbhum district of Bihar and Midnapur district of West Bengal and settled in the district of Balasore, Mayurbhanj and keonjhar district of Orissa (Singh, 1994, p. 707). The Mahali is a basket making community. Basket making is their traditional and primary occupation, while agriculture and daily-wage labour are subsidiary occupation. They are divided into a numbers of exogamous clans in the studied villages of SBR such as *Tudu, Baske, Hembram, Kulara, and Marandi.* They followed the clan exogamy and community endogamy as marriage rules. Marriages take place through negotiation or mutual consent. Monogamy is the general rule, polygamy is allowed, incase the first wife is barren. Divorce and remarriage are allowed.

Makdiya/Birhor

The Birhor is a primitive tribal group in Orissa, mainly concentrated in the state of Jharkhand and West Bengal. They are an Austric speaking people. The name of Birhor is supposed to have been derived from Bir-

Jungle and Hor-man i.e. the man who lives in jungle. We get an elaborate description of these people from the book of S.C. Roy (1925), who describes their lifestyle and other details of their socio economic structure. They generally live in their traditional small settlements of conical type of huts (*kumbo*) prepared out of grass and creepers and the settlement is known as *tanda*. They are basically a gathering-hunting group and they wonder around the sources of foodstuffs. They have a well-organized hunting group and their catches are distributed equally. Ray told about them (1925) in the following.

They have a good number of totemic clans, which, we have found in the studied villages these are *Bhagrati, Chaule, Dadujal, Golowar, Goud, Hembram, Jaksaria, Kherwar, Kumbha, Mahali, Murmu, Nagkuriya/Naikuriya, and Tikriya/Galowar*. Their marriages take place outside the clan. There are formalities rites and rituals associated with such marriages. They believe in the presence of various spirits and supernatural elements, which control the destiny of the people. *Sing Bonga* regarded as the supreme deity. He is supposed to be a benevolent God. Once in a year he is worshipped. *Chanda Bonga* may be regarded as another benevolent deity. The Birhor categories two types of ancestral spirits - *Hapram* and *Churgin* and animal are sacrificed at the time of their worships.

Development activities were taken long back and a rehabilitation colony was established at Durdura Tanda (Makdiya Tanda). Durdura Tanda (Makdiya Tanda) is a multi ethnic big village and the Birhor settlement existed at the boundary line of the village and the land around them are arid and merciless. Various development programmes were taken up for the Birhor through Hill Kheria and Makidiya development agency.

Kheria

Originally the Kherias lived in the hilly and forest clad area of Orissa, West Bengal and Jharkhand. In the Similipal Biosphere Reserve out of our studied villages the Kherias are mainly concentrated in the village Gudgudia with 36 families having a population of 142 individuals. The Kheria life is inextricately related to the forest ecosystem. Their adaptation to the environment is direct and simple. Their level of technology is just enough to cope with their gathering economy. In Similipal Biosphere reserve they mainly depend on minor forest produce, daily wage labour and a few of them are practising agriculture. In different season they collect different forest produce like, Honey, Resin, *Paro*,

Sal seed, Sal leaf, Firewood etc. and sold in the market or Sambandh (a NGO). Ray and Ray (1937) wrote an elaborate ethnographic account of the Kheria. In the SBR they are divided into number of totemic clans these *are Benedia, Bhalumia, Bhaluoriya, Bindia, Duarbandha, Duarsaini, Duorobandhi, Kusamghatia, Kusmaria, Kusumghatia, Mirindia, Nag, Nageswara, Naiko, Panibonia, Patrabichha, Paturia, Phiriria, Pichiria, Pradhan, Sanghatia, Sankaro, Sankiyaro, Seoro, Siyar, Suar.*

They believe in the existence of many mysterious spirits, some are benevolent and some are malevolent. They consider the Sun, the Moon and natural calamities as the attributes of some mysterious power to them. The hills of the locality are considered to be the abode of gods and goddesses. They have an elaborate ritual and they worshipped god and goddesses with the help of the village priest in their sacred grove.

Lohar/Bindhani

In Orissa they are mainly distributed in the district of Mayurbhanj. They mainly use Oriya language and its script. The traditional occupation of the Lohar is ironwork. Some of them own land individually and some cultivate land belonging to the Santal, Kol and Bhumij. There are a number of *gotras* among the Lohar based on certain totemic object. In Similipal in our studied villages it is observed that they are divided into number of gotras such as *Braman Baikar, Mehendiar, Kachhima, Kaithar, Sonari.* They are free to marry their mother's brother's daughter or father's sister's daughter. Junior levirate and junior sororate are allowed in the community. Sons inherit the property equally and the eldest son succeeds as the head of the family. In addition to the household work the women supplement the family income by earning wages. After childbirth the ritual *Sathia* and *Ekosia* are observed on the sixth and twenty- first days respectively. The marriage rituals are profound both at the brides and bridegrooms residence. The dead bodies are buried and death pollution continuous for ten days. On the twelfth day *Saradh* ceremony is performed with the help of a Brahmin priest.

The Lohar worshipped *Vishwakarma* as their main deity. Traditionally they accept water and cooked food from the Mahato, Goud, Teli and some other community (Singh, 1998, pp. 2021-2022).

Kaibarta

Etymologically, the word Kaibarta is derived from Ka (water) and *Varta* -livelihood i.e. those who maintain their livelihood from water i.e. fishing. Their traditional occupation is fishing and agriculture but a considerable number of them are also engaged as agriculture and other labourers. The community is segmented into several sub groups and the subgroups are further divided into several *gotras* (Singh, 1998, pp. 1450-1451). In our studied villages of SBR we found only one *gotra* i.e. *Nagasa.* Monogamy is the general marriage rule. They practice patrilocal in residence and patrilineal in descent. The remarriage of widows and divorced women is permissible. Junior sororate and junior levirate has been reported among them. Both nuclear and extended families are found among the Kaibarta. The dead bodies are cremated and the pollution period is observed for a period of ten days.

Kudmi Mahato

Kudmi Mahato also designated as Kurmi Mahato. They have been living with the tribals of Chotanagpur plateau for centuries and have been instrumented in disseminating agricultural technology and craftsmanship among them. According to Singh and Mahato (1983, pp-109) the Mahato homeland consists of a part of Chotanagpur and the adjoining regions of West Bengal and Orissa. The Kudmi Mahato live in the north and south divisions of Chotanagpur, Santal Parganas of Jharkhand, Mayurbhanj, Keonjhar, Sundargarh, Balasore and Cuttack of Orissa, The speak Bhanjabhumia or Kurmali at home and Oriya and Bengali with outsiders. Agriculture is the main occupation of the Mahato. Beside these they also engaged in service, petty business and other works. In the studied villages we found following clans among the Mahato, these are *Bansria, Banwar, Bir Banwar, Bomshiar, Chilbindha, Dongria, Hindua, Jalbandha, Kankatiar, Katial, Katiar, Makdia, Nengtuwar, Sankhuad.* In the regional caste hierarchy they place themselves below the Brahman, Karan, Khandayat, Barik, and Bathudi but place themselves above the communities such as Dom, Kamar, Ghasi and Thethri (Singh, 1998, pp. 1929-1933).

Goud

The Goud of Similipal biospheres Reserve are migrated from Singhbhum district of Jharkhand. They came here as palanquin bearer of the king of Mayurbhanj. Later on they settled in the Buffer region of SBR. The primary occupation of the Goud is agriculture. Some of them work as daily-wage labourers. Four endogamous groups like-Krishna Gour, Nand Gour, Mogda Goud and Jharua Goud have been identified among them. They have a number of lineages locally called *kili.* In our studied villages we found five killis among them, these are - *Bada/ Chadiya (Goud), Chainsa, Mantri, Sindura, Suya.* They practice lineage exogamy. Marriages are settled by the modes of negotiation, mutual consent and intrusion. Junior levirate, Junior Sororate, Divorce and remarriage are permitted among them. All sons inherit property equally, while the eldest son succeeds the authority. They practice both burial and cremation during disposal of dead bodies.

They worshipped *Hathikheda, garam, Kudra, khanarnurrho, Barpahari, Panchbahini, Satbahini and Dharam.* They also participate in the *Shiva Gajan danch, Monasa Puja, Tusu, Karam, Lakshmi, Durga* etc. festivals. Social control is extended by the *Gour Jati Sava.* The *Paramanik* or *Manjhi* of a village and the *Mahakur* are the height authority (Singh, 1998, p. 1092-1093).

Teli

The Teli derive its name from the Sanskrit word '*Talika*' or *taila* used for the oil extraction from Season or Mustard i.e. their traditional occupation. They are distributed in one hundred and twenty one districts of our country (singh, 1998, p. 3462). In Orissa the Teli have two sub groups namely *Ekadasi* and *Haldia.* They speak in Oriya language and use the Oriya script. The community is divided into several clans (gotra) that regulate marriage alliances. The *Nagashya* clan or *Killi* is found in our studied villages. Monogamy is the general marriage rule. Divorce is sanctioned with the judicial approval. Remarriage for the Widow, Widower and divorces is allowed in their society. Most of the families are nuclear followed by the vertical extended ones. All children inherit family property equally. The eldest son holds succession. Land is the economic source for the Teli of Orissa. Both land holding and land less family are found in the community. They are Hindu and worshipped their family, village and regional deities' like-*Ganesh, Monasa, Kali, Thakurani and Gomata* (Singh, 1998, p. 3468).

Sadgop

The name is derived from two Sanskrit word "Sad" or maens good and "Gope" means Milkman. They are mainly found in West Bengal, Jharkhand, Bihar and Orissa, The Sadgop is divided into several gotras (Singh, 1998, p. 3063). The *Kashyap* clan or *Killi* is found in the studied villages of SBR. Marriages are negotiated and monogamy is the common rule. Earlier divorce was not recognized but it is accepted at present. Extended families are common among them. The eldest son succeeds to the father's responsibility. The women participate in winnowing, husking and storing the crops but they have no direct role in agricultural activities.

They are land-owning community, it is their primary occupation, where as business, Service, and agricultural labours are the secondary occupations. They follow Hinduism and perform puja of *Lakshmi* and *Hari* as their family deity. Dharmaraj, Manasa, Shiva are their village deities.

Economic Development

In studied villages various communities are living side by side in the buffer region of SBR. They exploit their surrounding resources with their existing technology for their survival, which is not sufficient enough to strengthen a sound economy. So it requires various plans and programmes to develop their economy. But before going to commence a development plan it requires critical analysis of the cultural background of the community and finds out the various forces, which operates them during various activities. The development plans and programmes that would be initiated in the SBR different from other areas. In buffer region of SBR 61 villages are situated. In those villages various communities like Kol, Munda, Santal, Bhumij, Bathudi, Mahato, Teli, Kaibarta, Gond etc. are living side by side. These communities are utilizing the forest resources for their survival from historical past. The main objective of the Biosphere Reserve is to conserve the biodiversity of a particular area. To conserve the biodiversity of SBR we cannot ignore the people who are living inside it. But it is our primary duty to conserve the biodiversity of SBR at any cost without ignoring the people living within it. To minimize the use of forest resources the development plans and programmes should be initiated for the people of SBR to be prepared in such a manner that are sustainable to the people of SBR.

Table: 2 Occupational categories: Community wise

Sl. No.	*Name of the Community*	*Occupation*	
		Primary	*Secondary*
Tribal Groups			
1.	Kol / Ho	Agriculture	1. Labour, 2. Business, 3. Cattle hearding 4. Collection of minor forest product
2.	Kheria	Collection of minor forest product	Labour

Sl. No.	*Name of the Community*	*Occupation*	
		Primary	*Secondary*
3.	Bathudi	Agriculture	1. Labour, 2. Service, 3. Business
4.	Bhumij	Agriculture	1. Labour, 2. Service 3. Collection of minor forest product
5.	Munda	Agriculture	Labour
6.	Santal	Agriculture	1. Labour 2. Collection of minor forest product
7.	Mahali	Basketry	Labour
8.	Makdiya/Birhor	3. Collection of minor forest product	1. Labour 2. Agriculture (very few)
Caste Groups			
9.	Goud	Agriculture	1. Labour, 2. Service, 3. Business 4. Fishing
10.	Kudmi Mahato	Agriculture	1. Service, 2. Labour, 3. Business
11.	Lohar/ Bindhani	Blacksmith	Labour
12.	Teli	1. Agriculture 2. Business	Service
13.	Kaibarta	Service	Agriculture
14.	Sodgop (Gobora Caste)	Service	Nil

Source: Fieldwork data

Economy plays a great role in the development of a particular community. The other social factors of a community are generated through its economy. In a community of a particular place economy has a profound impact on the surrounding environment. People with their technological devices act on it for their survival and this action depend on various seasons and guided by the customs and super natural beliefs around which the whole community revolve.

In our studied villages most of the tribal communities like Bathudi, Bhumij, Munda, Santal, Kol/Ho, mainly depend on agriculture except Makdiya/Birhor and Kheria, these two gathering hunting communities mainly depend on the collection of minor forest products. Where as Mahali is mostly depending on their traditional occupation such as basketry.

In the caste community Goud, Kudmi Mahato and Teli are mainly depending on agriculture, where as Lohar/Bindhani depends on their traditional occupation as blacksmith but Kaibarta and Sadgop mainly depend on service.

All the communities irrespective of tribal and caste work as labourers as a secondary mode of occupation except Teli and Kaibarta who depend on service and agriculture, respectively as their secondary occupation. All the communities to some extent collect forest products like-wood for house construction, firewood, grasses for thatching of roof, *Sal* leaf, *Sal* seed, Medicinal plants etc. from the surrounding forest for day to day life. The Goud is the only community practise fishing as a secondary occupation.

Table 3. Dependencies on Forest Products

Sl No.	*Forest products*	*Purposes*	*Domestic use*	*For sale*
1.	Wood	Fuel	Own consumption	Nil
2	Wood	House construction	Own use	Nil
3.	Honey	Consumption /Sail	Own consumption	Sale
4	Resin	Consumption /Sail	Own use	Sale
5	Pado	Medicinal use	Own consumption	Sale
6	Sal seed	Collection of Sal seed	Nil	Sale
7.	Sal leaf	Prepare plates, bowl etc.	Own use	Sale
8.	Bamboo	House const. fencing	Own use	Nil
9.	Sabai grass	Preparation of rope	Own use	Sale
10.	Grass	Thatching	Own use	Nil
11.	Grass	Broom stick	Own use	Sale
12.	Bark	Preparation rope	Own use	Sale
13.	Palm leaf	Preparation of mat	Own use	Nil
14.	Mashroom	Consumption /Sail	Own consumption	Sale
15.	Medicinal plants	Medicinal use	Own use	Nil

Source : Field work data

Forest Dependency

In the SBR we studied 14 numbers of communities in the nine villages, eight from buffer zone and one from the Transition zone. Out of the studied communities 9 are tribal communities and 5 are caste communities. Both tribal and non-tribal communities are residing side by side in the SBR. Cultivation is the main economic activities of most of the communities. The two communities, Makdiya and Kheria are basically gatherer-hunter. At present hunting is strictly prohibited in SBR and government is trying to shift them from their original occupation to cultivation and plantation. These two communities are mostly depending on forest products for their survival. Though the other communities are mostly depending on cultivation but they are also to some extent depending on forest products to strengthen their economy. The Makdiya and Kheria collect different types of forest products in different seasons from various places throughout the year. They consume their collected items and also sold it to the market. The Kheria sold the collected forest products to the District Supply and Marketing Society, Mayurbhanj through Sambandh - a N.G.O. The other communities consume or use the collected items and also sold it to the market. They are permitted to collect the wood, which are top broken, wind fallen and standing dry only.

Kheria

Supply of bullock for the Kheria families in connection with the development of agricultural activities is observed. The ITDA supplied six pairs of bullock to them but it is wonders that most of the bullock is not found now, except in a few particular families. It requires an attitude to maintain the bullocks to get work out of them. During the lean months it is difficult for the Kheria to maintain the bullocks, and it requires an attentive care and sustaining efforts to do so.

Agricultural implements were distributed to the Kheria and Makdiya. These implements consist of spade, digging sticks, sickle etc. But it was told by the informant that all those are not in existence. Most of the Kheria families sold the item and destroy the money to take liquor and fulfill family requirements.

Goatery was subsequently included as a family oriented scheme. It was allotted to Kheria and Makdiya from the very beginning. Due to introduction of family oriented schemes it was almost compulsory that time for all the families to keep goats. Actually as the Kheria told, they were given 10 goats for two families some of which died or sold.

Table 4. Development Plans and Programmes: Present Status (P.T.G.wise-Kheria).

Sl. No.	*Name of the GP*	*Name of the Villages*	*Year of Implementation*	*Name of the project*	*No. of Beneficiaries*	*Present condition*
1.	Gudgudia	Gudgudia	1990-1991	Indra Abash Yojana	33 Families	Need Repairing
2.			1996-1997	Supply of Agricultural and Horticultural implements	33 Families	Most of the Kheria families sold the item and destroy the money to take liquor and fulfill family requirements.
3.			1997-1998	Supply of Household utensils	33 families	
4.			2001-2002	Supply of Plough Bullock 6 pair	6 Families	Four bullock died, two families sold the bullock and two families
5.			2002-2003	Nil	Nil	
6.			2003-2004	Nil	Nil	

(contd...)

Sl. No.	*Name of the GP*	*Name of the Villages*	*Year of Implementation*	*Name of the project*	*No. of Beneficiaries*	*Present condition*
7.			2004-2005	Formation of SHG group and Sanction of revolving fund Rs. 5000. (Purchase of MFP and sell it)	For 15 members group	The SHG purchase MFP from the villagers and sold it to the District supply and marketing society through a NGO-Sambandh
8.			2005-2006	Supply of Goatery 5 numbers of goat each	2 Families	1. Sacrificed 2 goats to perform puja and sold 3 goats to fulfill family requirements. 4. Two goats died, three alive with good health.

Source : Hill Kheria and Makdiya development agency and fieldwork

Table 5. Development Plans and Programmes: present status (P.T.G.wise- Makdiya/Birhor)

Sl No.	*Name of the GP*	*Name of the Village*	*Year of Implementation*	*Name of the Schemes*	*No. of Beneficiaries*	*Present Situation*
1.	Durdur a Tanda (Makd-iya Tanda)	Durdura Tanda Makdiya Tanda Makdiya settlement		Indra Abash Yojana	All the 18 Families	Not in good condition
2.				Electrified the settlement	All the families	No electricity at present due to stolen of wire.
3.			2000-2001	Irrigation cum pisciculture tank	All the families	Pisiculture is on progress and one family irrigated his land
4.				Rope making machine	All the families	At present all the machine is out of order
5.				2 Sal Leaf plate making machine	For all the families	One machine is out of order and one is running condition.
6.				2Well	For all the families	One well is abundant and another dry during summer
7.			2001-2002	1. Supply of goat unit (5 Nos. per family) 2. Supply of raw jute fiber for rope making.	1. 5 Families 2. For all the families	2. Prepare rope and sold it

(contd...)

Sl No.	*Name of the GP*	*Name of the Village*	*Year of Implem-entation*	*Name of the Schemes*	*No. of Benefi-ciaries*	*Present Situation*
8.			2002-2003	Formation of 1 SHG group with a revolving fund of Rs.16000. (Rope making and stitch of Sal leaf)	For 16 members.	The SHG running well at present Rs. 31000 in their account
9.			2003-2004	Supply of 5 HP pump set	For all the families	One family using it
10.			2004-2005	Supply of raw jute	For all the families	They prepare rope and sold it
11.			2005-2006	1. Supply of cashew seedlings. 2. Development of kitchen garden. 3. Supply of delivery pipes (9 Nos.) 4. Supply of goat (5 Nos. per family)	1. All the 18 families 2. Do 3. Do 4. Four families	1. All the cashew seedlings dried up due to lack of cashew rearing knowledge 2. Only one family continues it. 3. One family using it. 4. Due to lack of knowledge they do not take their cattle to the Vetenary Hospital at Durdura Tanda (Makdiya Tanda) and some cattle died of diseases.
12.			2006-2007	Supply of goat (5 Nos. Per family)	2 Families	

Source: Hill Kheria and Makdiya development agency and fieldwork data

Indira Abash Yojana also provided to 33 Kheria families of Gudgudia village. They are utilizing the houses for their residential purposes but at present all the houses are in dilapidated condition and need some repair work.

Makdiya/Birhor

Rope making project was initiated among the Makdiya of Durdura. These Makdiya are traditionally adopted in preparing ropes out of the bark of trees and these have a good market in the locality. To stop the bark collection they have been provided raw jute fibers for preparing the rope and for twisting, a machine has been provided to them. But the Makdiyas are not capable enough to use the machine, as they have no training and no such facilities have been extended to them.

Another interesting way of encouragement to the Makdiyas that training has been extended for the preparation of Sal leaf plate. Two machines have been purchased, it was reported that the machines were kept idle because the Makdiyas are stitching the Sal leaf as a raw material for the preparation of bowl at present and sale it to the businessmen @ Rs. 11 per thousand. The businessmen come to their settlement and collect the material.

In case of Makdiya/Birhor in the year 2001-2002 the ITDA supplied 5 goats per family for 15 families but most of the goats either died or sold. Some goats existed in the few families.

For the development of the Makdiya a rehabilitation colony was established at Durdura Tanda (Makdiya Tanda) as usual all the rehabilitation facilities were extended to them like *Pucca* huts with corrugated roofing, Electricity, kitchen garden, well and tube well etc. Basically they are gathering –hunting tribe and with their nomadic nature they move here and there and exploit the natural resources available around them. Due to their nomadic nature of movement they do not stay continuously at their houses for two weeks at a stretch and ultimately the houses become run down condition. It is noted that traditionally they use to live in a hut made up of leafs called *Cumbha* which they prepare now a days at the time of performing rituals. At present electricity is absent in the settlement due to the steeling of wire.

Table 6. Irrigation scheme village wise

Sl. No.	*Name of the GP*	*Name of the Villages*	*Name of the Schemes*	*Year of Implem-entation*	*Number of Beneficiaries*	*Present Situation*
1	Gudgudia	Gudgudia	Renovation of GP tank (Irrigation and Pisciculture)	2005-2006	All the villagers	Completed
2	Gudgudia	Kumari	1.Cheakdam 2. Const.of canal (part) 3. Const.of Canal	1. 2001-2002 2. 2002-2003 3. 2005-2006	All the Villagers	Completed
3.	Gudgudia	Nanjighesra	Const. of check Dam	2002-2003	All the Villagers	Completed
4	Barheipani	Chakundakocha	Const. of irrigation canal	1988-1989	5 Families	Completed
5	Barheipani	Kuljhari/Koljhari	Repair of Check Dam and canal	2004-2005	All the villagers	Completed
6	Astakumar	Garh-Similipal/Gad Similipal	Const. of Gad Similipal irrigation canal system	2006-2007	All the villagers	Completed
7	Astakumar	Saruda	Saruda Irrigation canal system	2006-2007	All the Villagers	Not yet completed
8	Sarispal	Basilakacha/Basilakocha	Const. of canal The villagers themselves completed it.	2003-2004	All the villagers	Completed

Source : Block Development Office, Panchayat Office and Fieldwork data.

Infra Structure Development: Present Status of the Schemes (Village Wise)

Irrigation facility have been provided to all the studied villages except the village Basilakacha/Basilakocha. The villagers themselves constructed a check dam on the stream and channelised the stream water through the canal into their fields. After the construction of canal at Kumari village most of the cultivable land of the village was brought under irrigation facility, this facility also extended to a portion of lands of the village Gudgudia, because the cultivable land of both the villages situated in one area. During monsoon (June – August) and in the late part of monsoon (Sept. - Oct.) much water flows through the stream and the villagers easily get water from the canal. During dry season flow of water reduces in the stream and the level of water could not reach up to the level in the check dam and can not creates much pressure that lead the water released through the canal in the last plot of their cultivable land. A portion of the newly constructed God Similipal irrigation canal had broken in the last monsoon, for that the canal water could not reach up to the last point of the canal. It would require some repair work. In the village Saruda the construction of irrigation canal had been initiated in the last year and at present the work is not yet complete. The people of Saruda by applying their traditional knowledge channelised the stream water in the every corner of their cultivable land. Prior to the introduction of the irrigation facility most of the cultivable lands of the studied villages were monocrop in nature and depended on rainwater for their cultivation except a few villages where the villagers by applying their traditional knowledge prepare check dam on the stream and channelised the water into their fields. This irrigation facility brought some changes in the studied villages and at present they practice bicrop cultivation in their land. They cultivate paddy, wheat, pulses, oil seed and vegetable in their land.

Roads and bridges are an important part of development. It connects people from one place to another. It also develops communication network between villages and urban centres. Roads and bridges are the prime factors for communication of all other development activities. Progress of any development plans can not be expected without at least a workable network of roads is present in a particular area. Through the roads and bridges people can easily contact to the other villages, weekly market places and sale their produces and purchased required goods for their daily use and exchange ideas whatever they gathered. These communication network have a profound impact on their social aspect

Table 7. Constructions of Roads and Bridges

Sl. No.	*Name of the G.P.*	*Name of the village*	*Name of the scheme*	*Year of implementation*	*No. of beneficiaries*	*Present condition*
1	Gudgudia	Gudgudia	1. Const.of cc road 2. Improvement of road from Gudgudia to Kondabil 3. Const.of bridge over river Khairi	1. 2001-2002 2. 2006-2007 3. 2005-2006	All the villagers of Gudgudia G.P.	All schemes completed
2	Gudgudia	Kumari	1. Improvement of road from Kumari to Kondabil 2. One k.m.c.c. road	1. 2006-2007 2. 2006-2007	1. All the villagers 2. All the villagers	1. Com-pleted 2. Inco-mplete
3	Gudgudia	Nanjighesra	Repair of road from Nanjighesra to Gudgudia	2006-2007	Both the villagers	Completed
4	Barheipani	Chakundakocha	Nil	Nil	Nil	Nil
5	Barheipani	Kuljhari/Koljhari	Improvement of a metal road inside the village	2001-2002	All the villagers	Completed

(contd...)

Sl. No.	*Name of the G.P.*	*Name of the village*	*Name of the scheme*	*Year of implementation*	*No. of beneficiaries*	*Present condition*
6	Astakumar	Garh-Similipal/Gad Similipal	Garh-Similipal/Gad Similipal c.c. road	2003-3004	All the villagers	Completed
7	Astakumar	Saruda	1. Un metal road inside the village 2. Const. of c.c. road with drainage system	1. 2000-2001 2. 2006-2007	All the villagers	1. Completed 2. Work started
8	Sarishpal	Basilakacha/Basilakocha	Const. of 300' c.c. road	2006-2007	All the villagers	Completed
9	Durdura Tanda (Mankdiya Tanda)	Durdura Tanda (Mankdiya Tanda) Mankedia Settlement	Construction of c.c. road with drainage system	2006-2007	All the villagers	Completed

Source : Block Development Office, Panchayat Office and Fieldwork data especially on marriage ceremonies, it also felt some impression on the educational aspect and health sector. In the studied area most of the villages are situated at a long distance with each other.

The main road starts from Jashipur and it crosses two forest check gates and approaches Gudgudia. From Gudgudia the road bifurcates and one goes towards Core region and other touches Barheipani, Godsimilipal and Saruda and again goes towards the Core region. On the way to Godsimilipal near Nawana the road again bifurcates and passes through the village Kukurbhuka and extended towards Baripada. So the road starts from Jashipur and passes through the buffer region of Similipal Biosphere Reserve and reached to Baripada.

The department of forest maintains all the roads of the SBR especially in the Buffer and Core region. All the roads in the SBR are non-metal, but exceptions are there. In our studied villages we found six villages where concrete roads are there inside the villages. These may be constructed to cheque the soil erosion inside the villages.

The entire SBR are constituted with hills and valleys. In this hilly area with a non-metal road it is very difficult to communicate with the other villages during monsoon. A good amount of rainfall recorded in the SBR during monsoon for that from 15th June to 15th November the SBR remain closed to the outsiders. Due to the heavy rainfall most of the non-metal roads got damaged and it requires maintenance work in every year during winter. The villagers use to communicate with other villages within the SBR and outside SBR by Motorcycle, bicycle and by walk.

The villagers cover a long distance when they want to approach Jashipur (nearest market place). They start early in the morning with their produces, sale it in the hat at Jashipur and purchase whatever they want and come back to their villages in the evening. The people of Saruda face much difficulty in this regard; the villagers set out for Jashipur one day before, stay one night at Jashipur, meet their necessities at the hat, stay again at Jashipur, and come back to their village in the next evening.

In Similipal most of the river and streams are perennial in nature, for that to cross the river, the Similipal authority construct permanent bridges on them. They do not construct bridges on the streams (Jeep and other four wheel vehicle cross the stream through the knee deep water). During monsoon when the water level increases in those streams it is very difficult to cross them.

Table 8. Indira Abash Yojana village wise

Sl. No.	*Name of the G.P.*	*Name of the village*	*Year of Imple-mentation*	*No. of benefi-ciaries*	*Present situation*
1.	Gudgudia	Gudgudia	1990-1991	33 (Kheria families)	Most of the houses are dilapiated condition need repairing
2.	Gudgudia	Nanjighesra	2001-2002	2 Families	Completed
3.	Gudgudia	Kumari	2001-2002	10 Families	Completed
4.	Barheipani	Chakund akocha	2000-2001	2 Families	Completed
5.	Barheipani	Kuljhari/ Koljhari	2000-2001	4 Families	Completed
6.	Astakumar	Garh-Similipal/ Gad Similipal	2002-2003	12 Families	Partly completed
7.	Astakumar	Saruda	2002-2003	5 Families	Completed
8.	Sarishpal	Basilakacha/ Basilakocha	2002-2003	1 Family	Completed
9.	Durdura Tanda (Mankdiya Tanda)	Durdura Tanda (Mankdiya Tanda) Mankheria Settlement	1990-1991	18 Families	Completed

Source: Block Development Office, Panchayat Office and Fieldwork data.

In the studied villages people are living in their traditional houses or huts that are constructed by the available resources collected from the nearby forest. Wood is available in the forest for that wood is used as a major item to construct the houses and thatched with the grasses collected from jungle (Tiles and corrugated sheets also used as roofing materials). They use to collect top broken and wind fallen trees for this purpose, which have no restrictions from the forest department to collect and use them. In Similipal heavy rainfall recorded in the monsoon. Due to the heavy rainfall thatched roof are to be changed every year. Change of thatching in every year is a giant task- such as collection of grasses from jungle, drying up properly, making bundles of it and then thatching the roof. It requires lot of endeavour and human labourers. So every year it hampered their economic activities and they face a monetary loss.

For that the government provided shelter to the poor villagers through the Indra Abash Yojana schemes in the SBR. Out of the 9 studied villages the Kheria of Gudgudia and the Makdiya of Durdura Tanda (Makdiya Tanda) have been provided small houses (Brick built wall, earthen or cemented floor and corrugated sheet roofing) through ITDA, Karanjia long back. Complete small shelters (Brick built wall, earthen floor and tile roof) have been provided to the villagers of Chakundakacha, Koljhari, Basilakacha/Basilakocha and Godsimilipal where as money supplied to the villagers of Kumari and Saruda only for the tile roofing purposes on their existing houses. This situation happens due to the number of beneficiaries increase in the villages and at the same time these single room small houses could not accommodate their family members. The authority provided tiles and labourers, which, they utilized for roofing purposes on their existing houses with some modification and changes. This has helped a lot to the people of Saruda and Kumari, but villagers of other villages are not ready to accept this opportunity. They want ready built small shelters constructed by the contractor.

Most of the constructed small shelters provided to the 33 Kheria families of Gudgudia long back are not in a good condition and it needs repairing. Some houses are badly affected and the people of those houses are presently residing in a temporary shelters construct near the early one. So the repair work is very urgently required.

Small brick built and corrugated roofing shelters were also provided to the 18 Makdiya families at Durdura Tanda (Makdiya Tanda) at the time of rehabilitation. The Makdiya are a hunter- gather community and they move from one place to another in the forest and exploit the natural resources for their survival. Due to the movement they are not continuously staying in their provided shelters and ultimately these shelters became very bad condition. So repair work of those shelters also required in the Durdura Tanda (Makdiya Tanda) in Makdiya settlements.

Table 9. People Oriented Development Programme village wise.

Sl. No.	*Name of the G.P.*	*Name of the village*	*No. of family*	*No. of person worked*	*No. of mandays created*	*Average man days created*	*Name of the work*
1.	Gudgudia	Gudgudia	46	52	2707	52.05	Improvement of road from Kondabil to Gudgudia
2.	Gudgudia	Kumari	16	24	1560	65.00	Improvement of road from Kumari to Kondabil
3.	Gudgudia	Nanjighesra	21	12	180	15.00	Repair of road from Nanjighesra to Gudgudia
4.	Barheipani	Chakundakocha	11	6	84	14.00	Renovation of tank at Barheipani
5.	Barheipani	Kuljhari/Koljhari	18	8	216	27.00	Const. of c.c. road at Nuniagoda
6.	Astakumar	Garh-Similipal/Gad Similipal	37	62	2012	32.45	Completion of Garh-Similipal/Gad Similipal irrigation canal system

(contd...)

(contd...)

Sl. No.	*Name of the G.P.*	*Name of the village*	*No. of family*	*No. of person worked*	*No. of mandays created*	*Average man days created*	*Name of the work*
7.	Astakumar	Saruda	23	55	1350	23.72	Const. of c.c. road and drainage system at Saruda
8.	Sarishpal	Basilakacha/Basilakocha	12	Nil	Nil	Nil	Nil
9.	Durdura Tanda (Mankdiya Tanda)	Mankheria Settlement	18	28	392	14.00	Const. of c.c. road with drainage system at Durdura Tanda (Mankdiya Tanda) s.t. basti

Source: Block Development Office, Panchayat Odffice and Fieldwork data

Table 10. SHGS and EDCS in the Studied Villages.

Sl.No.	Name of the G.P.	Name of the village	Name of the SHGS and EDCS	No. of person involved	Purpose	Fund allocation
1	Gudgudia	Gudgudia	Athradeoli Mahila Sanchay Samity	15	Purchase of minor forest product and sell it to the District Marketing Federation	Rs.=5000
2	Gudgudia	Nanjighesra	Ma Durga SHG	11	Paddy processing	Rs= 5000
3.	Gudgudia	Kumari	Sahajogi SHG	12	Paddy processing	Rs= 5000
4	Barheipani	Chakundakocha	One EDC	7	To protect the local frost from poachers	Rs =5000
5	Barheipani	Kuljhari/Koljhari	One EDC	7	To protect the local frost from poachers	Rs =5000
6	Astakumar	Garh-Similipal/Gad Similipal	Bapuji SS Dal Sanchay Sanity	10	Not registered	

(contd..)

(contd...)

Sl.No.	*Name of the G.P.*	*Name of the village*	*Name of the SHGS and EDCS*	*No. of person involved*	*Purpose*	*Fund allocation*
7	Astakumar	Saruda	Santi Morium EDC	12	To protect the local frost from poachers	Rs =10000
			Martha Morium EDC (through NGO)	12	Cash crop cultivation and regeneration of mango seedlings	Supply of seed and seedlings
8	Sarishpal	Basilakacha/Basilakocha	Sagun Supari	11	Plantation of medicinal plants	Rs= 2500
9	Durdura Tanda (Mankdiya Tanda)	Durdura Tanda (Mankdiya Tanda) Mankheria Settlement	SHG	16	Rope making and stitch of Sal leaf	Rs= 16000

Source: Block Development Office, Panchayat Office and Fieldwork data

So government provided brick built and tile roof houses to the villagers, which strengthen the economy of the villagers, this phenomenon lead them to minimum exploitation of the forest resources.

After independence various development measures have been taken for the betterment of the economically backward people of our country especially for the tribal people. After implementation of these plans and programmes the government tried to find out its lacuna time-to-time and rectify it and amended new plans and programmes. Some times it shows that the people have no freedom to say any thing about the plans and its implementation. Some plan has its own shortcomings.

In August 2005, the Ministry of Rural Development, Government of India, with the help of State Government introduced National Rural Employment Grantee Act (NREGA). Now the act is rename as Mahatma Gandhi National Rural Employment Grantee Act (MGNREGA) in some selected district of India, with an aim to provide 100 days work guarantee in the rural areas, with a freedom to the rural people that according to their requirements the work would be initiated. In the studied villages 202 families out of 295 families and out of a total population 1399, 247 persons as well as males and females directly participated in the development work implemented in their own areas. In the 9 studied villages in the Buffer and Transition zone of SBR various projects have been initiated to develop infrastructure such as construction of Concrete Road within the village territory, renovation of inter connected nonmetal village road and construction of irrigation canals etc. in the financial year (2006-2007). The villagers worked in those development projects in their own areas and created 8501 man-days in the year (2006-2007) except the village Basilakacha/Basilakocha, where no such development work was initiated.

In the year 1994 the Similipal reserve forest declared as a Biosphere Reserve. To conserve the biodiversity of this area the Similipal authority impose some restrictions on the use of forest resources, hunting of wild animals and the movement of tourists etc.; side by side they have taken several plans and programmes for the development of Biosphere itself and the people also, who lived inside the SBR. A good number of people are living inside the SBR from the time of Mayurbhanj king and they are using the forest resources for their survival. To minimise the use of forest resources the development plans and programmes should be formulated in such a way that it should be sustainable according to the needs of the people of that locality.

In this regards government has taken several Eco development projects through the formation of Self Help Groups (SHGs) and Eco Development Committees (EDCs). The villagers themselves formed the SHGs and the decision taken collectively, what they want to do through the SHGs. Accordingly after formulation the villagers send the proposal to the Block Development Office through the Panchayat office. The Block Development Authority after registering provides the revolving fund, which the SHGs and EDCs used as a capital. In the village Gudgudia 15 Kheria females formed a SHG with a revolving fund of Rs. 5000. They purchase minor forest products from the villagers and sale it to the District Supply and Marketing society, Mayurbhanj through a NGO named Sambandh with a minimum profit. In this way they earn money and distributed the profit to the members of the SHG that give the economic support to them.

In the village Nanjighesra and Kumari 11 Kol and 12 Bhumij women respectively formed two SHGs with a revolving fund of Rs. 5000 each for the purpose of paddy processing. The women of both the villages purchased paddy from the other villagers and process it into rice with the help of SHG members and sale it to other villagers and local hat with a minimum profit. During our field investigation while verifying there documents it shows that very recently they started it in spite of that they are earning profit from it.

In the Chakundakacha and Koljhari villages the villagers formed two 7 members Eco development committees with a revolving fund of Rs. 5000 each. The two villages are situated inside the forest. The aims of the EDCs are to protect the surrounding forest from the poachers those who cut trees and hunt animals from the forest. The seven members committees with the help of village elders try to convince the villagers of their own and other villages to get rid of such things. Both the EDCs want to utilize the revolving fund for the betterment of the villages not for the EDCs alone.

The people of Gadsimilipal village formed 10 members SHG and send a proposal to the Block Office through the Panchayat office but the proposal could not be registered at Block office.

Munda inhabiting Saruda village formed a 12 members EDC with a revolving fund of Rs. 10000.They kept the amount in the Post Office as a fixed deposit. They kept the amount for the construction of their new Church building.

Another Eco development programme undertaken by the Department of Forest (Similipal Tiger Reserve) in the same village very recently as a test case and the programme implemented and monitored by a NGO, named Sambandh. The NGO formed a 12 members group and provided 100 kg. ginger, 8 kg. Arhar Pulses and 40 Mango seedlings for the group. The members cultivated ginger in their homestead land and sowed Arhar Pulses in the four sides of the ginger cultivated lands and planted the mango seedlings in the hill slopes. During our field survey it was observed that the ginger grow well in their homestead land but it would require another one or two months to harvest. It was told by one of the members of EDC that ginger would give a good production. The Arhar, pulses started to give its fruits that yield a bumper crop. They maintained the Mango seedlings, which they planted and they are very happy to think about the future when the plants will give its fruits. So the Eco development project, cultivation of cash crop ultimately comes to a success with the implementation and regular monitoring of Sambandh. The NGO succeeded to convince the people to switch over their traditional cultivation to cash crop. The people accepted it and make it into practice. It is possible because the Saruda is a uniethnic Munda village and totally converted into Christianity, most of the people are educated with modern education system. With the Eco development project they initially get the economic benefit and simultaneously they would also get benefit by regenerating the Mango seedlings that would give the Eco balance in future.

The Basilakacha/Basilakocha village is situated in the Sarispal G.P. of Bangriposi Block i.e. eastern side of the SBR. The village is totally inhabited by the Santals. The Santal women formed an 11 member EDC with a revolving fund of Rs. 2500.The villagers planted medicinal plants like-*Amla, Safed Mushli* etc. which were supplied by the Similipal authority. The revolving fund that has been provided to them utilized for the rearing of the plantation at present. When the plants would started to give its fruits then they will collect fruits and sale it to the market. That will give an economic support to them and simultaneously it will help also Eco balancing.

Another SHG formed in the Makdiya settlement at Durdura Tanda (Mankdiya Tanda) with a revolving fund of Rs. 16000 with an aim to rope making and stitch of Sal Leaf. The Mankidia are very expert in making ropes. They are preparing ropes out of the bark of trees and these ropes have a good market in locality. To stop bark collection from forest they have been provided raw jute fibers for making the ropes but the jute is not sufficient enough by which they can continuously make

Table 11. Development activities taken by the Forest Department

Sl No.	*Name of the G.P.*	*Name of the village*	*Name of the schemes*		
			Cattle immunization	*Supply of Chullah*	*Organization of health camp*
1.	Gudgudia	Gudgudia	Implemented	All the families	Mega health camp organized on 10th December 2006 cover all the villages
2.	Gudgudia	Nanjighesra	Implemented	All the families	--------------
3.	Gudgudia	Kumari	Implemented	All the families	Same
4.	Barheipani	Chakundakocha	Implemented	All the families	--------------
5.	Barheipani	Kuljhari/Koljhari	Implemented	All the families	--------------
6.	Astakumar	Godsimilipal	Implemented	All the families	Health camp organized
7.	Astakumar	Saruda	Implemented	All the families	--------------
8.	Sarispal	Basilakacha/ Basilakocha/	Implemented	All the families	--------------

Source: Block Development Office, Panchayat Office and Fieldwork data

it and supply to the market. For that they are collecting bark for making ropes. So the aim of the Eco development project is failed but economically they are making profit from it. At present the revolving fund raised from Rs. 16000 to 31000, through the rope making and stitch of Sal leaf.

The Similipal Tiger reserve authority has also taken some Eco development project, like immunisation of the village cattle and supply of 'Harsha Chullah' to all the families of studied villages in the SBR. In the studied villages all the cattle have been vaccinated by Anthrax vaccine through the CDVO, Baripada. The Similipal authority immunised the cattle not for strengthening the agrarian economy but for the protection of the wild animals from Anthrax. They might have thought that Anthrax could be extending to wild animals from the village cattle. My informant told me that after the Anthrax vaccination death of the cattle occurred in all the studied villages. The view of the villagers is that this incidence happens due to vaccinating their cattle.

The communities reside in the SBR are collecting firewood from the forest for their daily uses. To minimise the use of firewood the SBR authority supply "Harsha Chullah" to all the families of studied villages. It is a portable one and one can easily carry from one place to another. To make fire in it small twigs are required, which are found fallen plenty in the forest. So the villagers use the Chullah for preparing their food and automatically use of firewood reduces, ultimately the forest will remain intact.

Beside the above development activities the forest department had also taken other development activities like-Establishment of Green Brigades, Water harvesting, Bee-kipping, Supply of Bio-Gas plant, Distribution of Sabai rope making machine, Organization of Natures camp, Training programme on preparation of paper bags.

Development of Education in the studied villages of SBR

The tribal population consists of 22.21% of the total population of the state (1991 census). The existing education method of SBR is an imposition, which the tribal communities are neither able to accept nor reject due to restriction of medium in Oriya only and this does not fit with their cultural context. Most of the schools were established in this area prior to the declaration of Biosphere Reserve. After the announcement of SBR a few of schools are also established and upgraded in the buffer area of SBR.

Table 12. Literacy Rate in the studied villages

Name of the villages	*Illiterate*				*Literate*			
	Male	%	*Female*	%	*Male*	%	*Female*	%
Gudgudia	56	4.00	110	7.86	169	12.08	99	7.08
Kumari	27	1.93	66	4.72	120	8.58	79	5.65
Nanjigeshra	19	1.36	21	1.50	34	2.43	6	0.43
Koljhari	20	1.43	23	1.64	23	1.64	21	1.50
Chakundakacha	10	0.71	17	1.22	14	1.00	10	0.71
Gad similipal	25	1.79	38	2.72	82	5.86	43	3.07
Basilakocha	10	0.71	24	1.72	21	1.50	9	0.64
Durdura	16	1.14	21	1.50	21	1.50	10	0.71
Saruda	25	1.79	40	2.86	50	3.57	20	1.43
Total	**208**	**14.87**	**360**	**25.73**	**534**	**38.17**	**297**	**21.23**

Table 13. School in the studied villages

Sl. No.	*Name of the G.P.*	*Name of the village*	*Availability of school and Ashram Hostel (no.)*	*Remarks*
1.	Gudgudia	Gudgudia	Upper primary school (1) Ashram hostel (1) G.P. High school (1)	Boys - 31 Girls - 9 22 girls are staying in the hostel (Kol, Bhumij, Bathudi, Santal and Kheria)
2.	Gudgudia	Nanjighesra	Absent	Children are studying in neighboring school at Gudgudia
3.	Gudgudia	Kumari	Absent	Children are studying in neighboring school at Gudgudia
4.	Barheipani	Chakundakocha	Absent	Children are studying in the school of Barheipani and Barsia
5.	Barheipani	Kuljhari/Koljhari	Primary school (1)	U.P. Schoolat Barheipani or Barsia
6.	Astakumar	Garh-Similipal/Gad Similipal	Upper primary school (1) Ashram hostel at Nawana (1)	Boys-13 Girls-8 (Mahato, Kol and Santal)

(contd...)

(contd...)

Sl. No.	*Name of the G.P.*	*Name of the village*	*Availability of school & Ashram Hostel (no.)*	*Remarks*
7.	Astakumar	Saruda	E.G.S. School (class i-iii) (1)	Boys-10 Girls-15
8.	Sarishpal	Basilakacha/Basi lakocha	Absent	Boys –3 Girls-7 (Santal, Studying in the school of Amdapani)
9.	Durdura Tanda (Makdiya Tanda)	Durdura tanda (Makdiya Tanda) Makdiya settlement	One Nursery education school	Boys-8 Girls-5 Karanjia-1 boy - vii Hatibari-2 boys- vii and viii

Source: Block Development Office, Panchayat Office and Fieldwork data

At present the District authority is trying to establish Education Guarantee Schemes (EGS) School in all the villages through District Primary Education Programme where no school exists.

The literacy rate is high in comparison to other rural and tribal populations of the state. This higher literacy rate is due to school going children of nursery level are included in the literate category. Persons who are able to read and write are considered to be literate. At present, the number of school going children have increased in this area for attraction of midday meals in the form of cooked rice, pulses and vegetable. The literacy rate given in the table suggests that more females are non- literate than the males. The table shows that maximum literate persons among both genders are in the village Gudgudia, Kumari and Gad Similipal. These three villages are well-exposed multiethnic villages having primary and secondary schools. Where as rest of the villages are uniethnic villages. Saruda is a sole uniethnic village, where the highest percentage of literacy is due to the influence of Christianity among villagers.

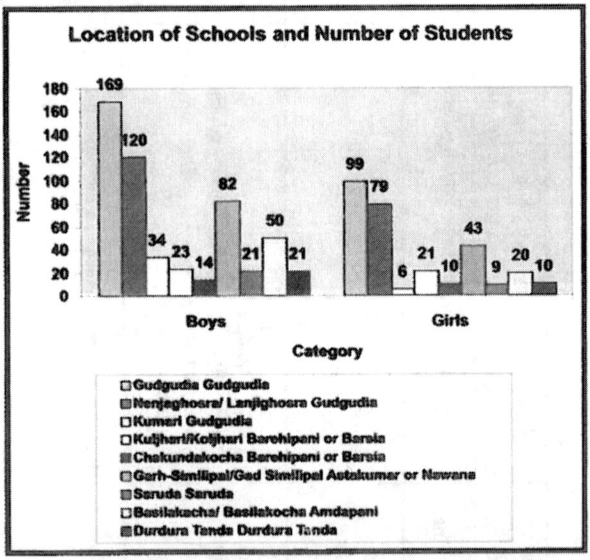

The above chart represents each village and respective nearest school as well as facts and figures. Out of the nine studied villages three villages have no schools in their village premises. The villages are Nanjighesra, Kumari and Chakundakacha. Out of the three the student of Nanjighesra and Kumari are reading in the schools of Gudgudia that is nearer to their village where as the boys and girls of Chakundakacha are used to go to the school of Barheipani and Barsia, which are far away from the village. In Gudgudia one high school is present for boys and girls who are studying in higher classes of the Gudgudia G.P. and other areas also are studying in the school. To educate the scheduled tribe and scheduled caste student with modern education system government set up *Ashram Hostel* in the rural areas especially in tribal areas in our country. In this *Ashram Hostels* students can get free lodging, food, dress and study materials. In Similipal also such Ashram Hostels have been set up long back. In our studied villages we found two Ashram Hostels one at the Gudgudia another at the Nawana, of Astakumar G.P. In Gudgudia ashram Hostel only girl's students are allowed to stay in it. At present Kheria-1, Kol-10, Bathudi-10 and Bhumij-1 total 22 girls students are staying in it. Government is providing Rs. 300 per boys and Rs. 325 per girls per month. The reading and writing materials are supplied by District Primary Education Programme (DPEP). At Nawana 13 boys and 8 girls of Mahato, Kol and Santal communities are staying in the Ashram Hostel. One Education Grantee Schemes (EGS) school is set up in the Saruda village with 25 students and one teacher. The village Saruda situated inside the forest near to the core area and the village is totally cut off from the other areas during monsoon because the River Buribalanga passes through the entrance of the village. Another nursery school has been established in the Durdura Tanda (Makdiya Tanda) Makdiya settlement at the time of colony establishment. Though they are hunter-gatherer and lead a nomadic life but they are sending their children to the school. At present the nursery school running with 13 students and 1 teacher. Three Makdiya boys are at present studying in class VII and VIII at Karanjia and Hariari and are staying in the Ashram Hostels. This phenomenon indicates that like other tribal people, the Makdiya of Durdura Tanda (Makdiya Tanda) is also getting the benefit of modern education system.

The literacy rate seems to be high as compared to the other rural and tribal populations. These higher values of the literacy rate is because of the fact that school going children reading in nursery level and above are assumed to be literate. Persons who are able to read and write are considered to be literate. Of late school going children had increased in this area because mid day meals in the form of cooked rice, pulses and vegetable are given to the school going children. The literacy rate have given in the table suggests that more females are non-literate than the males.

Table 14. Medical facilities in the studied village

Sl. No.	*Name of the G.P.*	*Name of the Village*	*Anganwari Center*	*Child Immunization*	*Health Center*
1	Gudgudia	Gudgudia	One	Satisfactory	One health center
2	Gudgudia	Nanjighesra	One	Satisfactory	Health center at (Gudgudia)
3	Gudgudia	Kumari	One	Satisfactory	Health center at (Gudgudia)
4	Barheipani	Chakundakacha	Attached with the sub center. Koljhari	Not up to the mark	Mobile medical unit once in a week at (Barheipani)
5	Barheipani	Koljhari	Anganwari sub center	Satisfactory	Mobile medical unit once in a week at (Barheipani)
6	Astakumar	Godsimilipal	One	Satisfactory	Mobile medical unit once in a week at (Godsimilipal)
7	Astakumar	Saruda	One	Satisfactory	Mobile medical unit once in a week at (Godsimilipal)
8	Sarispal	Basilakacha/Basilakocha	Attached with the Anganwari center Amdapani	Satisfactory	Mobile medical unit once in a week at (Amdapani)
9	Durdura Tanda (Makdiya Tanda)	Durdura Tanda (Makdiya Tanda) Makdiya settlement	One	Satisfactory	Health center at Durdura Tanda (Makdiya Tanda)

Source : Block Development Office, Panchayat Office and Fieldwork data

Development of health and Hygiene in the studied villages of SBR:

Good health means the human body is free from all kinds of diseases. It is quite impossible that one cannot suffer with any disease at any time of his/her life. Whenever body is present diseases are also there. To prevent and cure disease people are trying from the very beginning with the help of several methods of treatments. Lack of education, proper awareness of different diseases and their modern methods of treatment, some time lead the people to believe on the supernatural world. On their beliefs the malevolent spirits creates diseases or illness and other bad things and that could be cured by satisfying the hunger of the spirits with the help of the local medicine man. Government has provided modern medical facilities in the form of Health Centers, digging of well and tube well to supply fresh drinking water.

In the present context the Government of Orissa extended Health facilities to the people who lived in Similipal Biosphere Reserve. In Similipal the villages are situated far away from one another and some villages are situated inside the forest also. To extend the medical facilities Government set up Public Health Center (PHC) in different Gram Panchayats of SBR. With the introduction of "Mobile Medical Unit" government extended medical facilities in the remote villages side-by-side organized Medical Health Camps in different areas of SBR. Along with this the Government also set up Anganwari centers and sub centers almost in all villages of SBR. The Anganwari perform child and mother care from the time of pregnancy of mother and up to the age of 6 years of the children through its various programmes. Anganwari centers also arrange the child immunization programme in collaboration with the PHC or Mobile Medical Units. Out of the 9 studied villages the people of Gudgudia, Nanjighesera, Kumari and Durdura Tanda (Makdiya Tanda) are getting the medical facilities directly from PHC established there, whereas the people of Chakundakacha, Koljhari, Gadsimilipal, Saruda and Basilakacha/Basilakocha are getting the medical facilities from mobile medical units once in a week. Whenever any emergency arises then they referred the patient to the PHC of Jashipur or Baripada District Hospital.

To supply safe drinking water to the people Government of Orissa provided wells and tube wells in different villages of SBR. All our studied villages have either wells or tube wells by which the villagers are getting safe drinking water. Some villages have both the facilities.

Table 15. Drinking water facilities in the studied villages

Sl. No.	*Name of the G.P.*	*Name of the village*	*Well*	*Tube well*	*Present condition*
1.	Gudgudia	Gudgudia	3	6	Both in running condition
2.	Gudgudia	Nanjighesra	4	1	Both in running condition
3.	Gudgudia	Kumari	-	3	Running condition
4.	Barheipani	Chakundakocha	1	1	Well-abandone Tube well-running condition
5.	Barheipani	Kuljhari/ Koljhari	1	1	Both in abandon condition
6.	Astakumar	Garh-Similipal/Gad Similipal	2	2	Both in running condition
7.	Astakumar	Saruda	1	1	Both in running condition
8.	Sarishpal	Basilakacha/ Basilakocha	-	1	Running condition
9.	Durdura Tanda (Makdiya Tanda)	Durdura Tanda (Makdiya Tanda)	2	1	Well-one, abandone, another become dry during summer Tube well-the water is not suitable for drinking

Source: Block Development Office, Panchayat Office and Fieldwork data

Prior to digging up of wells and tube wells it is very important to select the place from where good drinking water would available. It is also very difficult to guess well in advance that the dug out wells and tube wells would provide fresh drinking water. It is not a easy task to dug out a well and tube well in this hilly region and huge money would require for it, so the people are totally depending on Government for such development work. It is also very important that the dug out tube wells and wells would supply fresh drinking water throughout the year and for a long time. The wells and tube wells dug out in different villages are at present in running condition except the villages Chakundakacha and Durdura Tanda (Makdiya Tanda) where the wells are at present in abandoned condition, another existing well at Durdura Tanda (Makdiya Tanda) becomes dry during summer season and the tube well water is not suitable for drinking. So the people of Durdura Tanda (Makdiya Tanda) are totally depending on the tube wells and wells of other villages for their drinking water.

Observation

Prior to the introduction of Biosphere Reserve the whole area were under Reserve Forest. Later on some portion of the reserve forest area designated as Tiger Reserve. After the declaration of the area as Biosphere Reserve in the year 1994, the whole situation changed. The entire areas are divided again into three administrative zones i.e. Core, Buffer and Transition. With the establishment of BR, some restrictions are imposed on the area and the people what they enjoyed earlier.

- No metal road allowed inside the BR.
- No Electricity.
- No Bus services.
- No Telephone services
- Setting up of new Tourist rout
- No six wheelers vehicle are allowed inside the BR and simultaneously to conserve the biodiversity of the area, some restrictions are imposed to use of the forest resources.

It is shown from the table that in the mid part of nineties, when the BR established out of the studied villages, except Gudgudia, in other villages, very few development activities come in picture. The mono crop cultivation, low labour charges and lack of employment generation in the local areas forced the people for seasonal migration for work to out side area.

To conserve the biodiversity on the one hand and launching of the development activities for the people without disturbing the forest resources on the other, careful attempts should be taken into account.

But it is feasible only when the people living inside the SBR will utilize the forest resources. To overcome this situation and minimize the use of forest resources, development activities are required in a manner that is sustainable to them. To conserve the biodiversity for our future generation, we cannot ignore the people who are living inside the SBR and for that development of the people are also required. If they are not getting proper alternatives for their subsistence like economic development, education and health facilities, they would utilize the easily available forest resources to fulfill their requirements.

Keeping in mind the Forest Department and other forest development agencies have taken several measures that ensured the conservation of animals and plants of this area. Side by side various development activities are also taken by the government through the District authorities and forest department for the people, who lived inside the Biosphere Reserve.

In the last part of nineties the development activities started to come in picture in a better way with the identification of people who live below poverty line (BPL) and provided free ration to them, old age pension and increased labour charges which brought some effective changes in the village life, ceased the direction of the people towards the seasonal migration and integrated themselves to the new development activities launched in their village area. So the increase of sustainable development activities reduces the destruction of the forest resources by the people.

With the introduction of wage employment programme NREGP (100 days work guarantee in the rural areas) conceived by the district authorities and self-employment programme (SGSY) through the formation of Self Help Groups, Eco-Development programme through the SHGS and formation of EDCS, the Forest Department, creates new job opportunities in the village area.

The programmes have encouraged the people to respond whole heartily with the self-sustaining development activities in which they themselves can select, take active participation to formulate, organize and implement the programmes sustainable to them.

Due to this long term sustaining programmes, the condition is not mature enough to count any concrete feedback of the people.

5

Ethno-Medical and Health Culture of Buffer Area Villages

Anupam Dutta

Introduction

All knowledge and wisdom ultimately flow from practices, but their organization differs among the different streams of knowledge. Folk knowledge is maintained, transmitted, and augmented almost entirely in the course of applying it in practice; it lacks a formal, institutionalized process for handling. Folk ecological knowledge and wisdom are therefore highly sensitive to changing relationships between people and their ecological resource base. Today, both are eroding at a fast pace for two reasons: firstly, people now have access to newer resources such as modern medicines and are no longer as dependent on local medicinal plants and animals as before; and secondly, people are increasingly losing control over the local resource base, with takeovers by state and corporate interests (Gadgil and Berkes 1991). However, folk knowledge and wisdom, with their detailed locality- and time-specific content, are of value in many contexts. They must therefore be supported in two ways: by creating more formal institutions for their maintenance and most importantly, by creating new contexts for their continued practice (Gadgil et al. 1993). The programme of "People's Biodiversity Registers" (PBR) is such an attempt. It is a program of documenting how people, primarily rural and forest-dwelling communities, understand living organisms and their ecological setting. The information recorded relates to present status as well as changes over recent years in distribution and abundance; factors affect distribution and abundance, including habitat transformations and harvests; known uses; and economic transactions involving these organisms.

The document also records the perceptions of local people about ongoing ecological changes, their own development aspirations and their preferences as to how they would like the living resources and habitats to be managed. We summarize here our experience of developing the concept and organizing the preparation of 52 such PBRs in different parts of India, the resultant understanding, and the interest that this program has generated (Gadgil *et al.*, 1993). We believe that the PBR process, involving a collaboration between people working in the organized sector (e.g., educational institutions, government agencies, and NGOs) and the practical ecologists, peasants, herders, fishers, and traditional healers (all in the unorganized sector), is as significant as the product: the recorded information. A subset of the information collected, especially that pertaining to medicinal and other economic uses, has been recorded by ethno-biologists working in academic institutions and for the pharmaceutical industry and other commercial interests (Reid et al. 1993, Martin 1995). In this process, however, the local people are treated as anonymous informants; they receive no particular credit for their knowledge and the information is accumulated with little reference to particular localities and times (Posey and Dutfield 1996). The PBR process, on the other hand, aims to record the information with full acknowledgment of the source; it thereby serves as a possible means of sharing of benefits that may flow from further economic utilization of such information. Another subset of the information recorded in PBRs is collected during "Participatory Rural Appraisal" (PRA) exercises (Chambers 1992, 1993) that feed into decentralized development planning. Generating good information for such participatory development is also an objective of PBRs; the PBRs differ from PRAs in their greater emphasis on recording all pertinent knowledge including changes over the recent past, and in giving specific credit for the information collected. Although we have so far completed only one round of PBRs in any one locality, we expect it eventually to become an ongoing process of monitoring ecological change and generating the necessary information for locally adaptive management of living resources.

People's dependence on living resources has declined along with the decline in ready availability of such resources to them. At the same time, people have access to new resources that can substitute, e.g., allopathic drugs in place of herbal remedies, tiles in place of thatching for roof, or synthetic dyes in place of vegetable dyes. This has led to a decline in interest, among the younger generation, in the knowledge of living resources, a decline reinforced by the modern, largely bookish system of education. Our studies reflect such decline in knowledge.

Poverty, ill health and illiteracy are the major bottlenecks resulting in under development and over population in a nation. This not only results high morbidity due to poor pathological conditions, but also generates an underutilized surplus population, which do not contribute much to national productivity. In this scenario it is difficult to think of a healthy society of good physic and sound mind. The justifications for defining a population as fit and sound or having ill health is encapsulated both within the biological and socio-cultural framework of a society. Economists view health as social capital. They measure human life in terms of economic values and performance (Schultz, 1961). This is one of the reasons the Welfare State invest to improve the health status of its people. However, it has been observed that in spite of huge investment the health-seeking behavior of a society is largely influenced by customary caring practices and adherence to various traditional coping strategies, which have been evolved long ago. This is embedded in definite socio-cultural values, beliefs and systems of meanings attached to death and disease, pregnancy and child care (Van Bolen and Van Dormael, 1999). However, over the time the situation has changed and access into the reproductive health services is also influenced by the level of knowledge or awareness level of the population on the medical facilities, and on their affordable capacity. More particularly in tribal societies subjective perceptions about perceived health needs underlay people's own knowledge (Behura, 2000).

Health Reform Policies: An Overview

Understanding the magnitude of different health problems, different national and international bodies have been implementing various health related programmes. To mention a few of them are Primary Health Care, Comprehensive Health Care, Basic Health Services, and Community Health Care etc. Besides this, in Indian context various commissions and committees like the Bhore Committee in 1966, the Srivastav Committee in 1975, and the New Draft Health Plan of Government of India in 1977, the glorious Alma Ata Declaration in 1978 and "Health for All" by 2000 A.D. has substantially contributed in bringing a revolution while providing health services to all. In order to provide the basic health needs of developing countries the "alternative health strategies" incorporated some of the components of 'Primary Health Care' approach, have emphasized the basic health services as the most peripheral stratum. In almost all the strategies emphasis have been made from 'curative to preventive' approach, from 'urban to rural', from 'privileged to under - privileged', and from 'vertical mass campaigns' to a system of 'integrated health services', forming a component of overall social and economic development (Newell: 1975). Still, the health policies so far adopted are yet to bring satisfactory results. However, most of these programmes have been lacking in considering certain human factors like values, attitudes, motivation, commitment, etc.

In subsequent period inadequacies were observed and everybody felt the need of involving the grass-root community health workers and N.G.Os, in all health packages supported by W.H.O., U.N.I.C.E.F. multi and bilateral agencies. So the International Conference on 'Primary Health Care' held in September 1978 at 'Alma-Ata' has to emphasize on the unhygienic health condition of millions. Irrespective of the gap between the 'haves' and 'have-nots', the Alma Ata declaration called for a new approach to provide health care services. The objective behind has been to achieve more equitable distribution of health services and to attain a level of health status for all the citizens of the world, which would permit them to lead a socially and economically productive life (W.H.O. and U.N.I.C.E.F.: 1978). Again in 1979, the thirty-second World Health Assembly declared the International Conference on "Primary Health Care" and launched the global strategy which is popularly known as "Health for all by the year 2000 A.D." It invited the member States of W.H.O. to formulate their national policies, strategies and plans of action to attain this goal and to act collectively in formulating regional and global strategies (W.H.O.: 1979).

UNICEF in its child survival programme has been giving a lot of efforts to improve the conditions for child survival. Among various strategies implemented by UNICEF has significant impact which stands for Growth Monitoring; Oral re-hydration treatment of acute diarrhea; Vector borne diseases, Breast feeding for good infant nutrition, protection against infection and immunization against the common communicable diseases during childhood. To achieve this, another strategy of 'FFF' i.e., Food, Family spacing and Female education has been adopted. Crucial to the above strategies is to make sure of social control of the health infrastructure through a high degree of community involvement.

Following the agenda presented at the International conference on Population and Development at Cairo in September 1994, Government of India has started decentralizing health and family welfare programmes, and developing good quality services in a reproductive health approach. Govt. of India has launched Reproductive and Child Health Programme in October 15, 1997 which has incorporated services for the prevention and treatment of Reproductive Tract Infection (RTIs) and Sexually Transmitted Infection (STDs) as an integral part of Reproductive and Child Health (RCH) approach. As an approach it focuses on improving the quality of services provided by health system, and includes various packages for the management of reproductive health system. This approach place individuals and couples at the center of efforts to improve family health. In operational terms the reproductive health approach entails the service delivery mechanisms based on bottom-up planning and the client's needs; established standards for provision of quality reproductive health services through clients perspectives and accountability in the system; strengthens capacities for the management of decentralized programmes which ultimately aims at Right to information and transparency in public systems. However, there are certain critical constraint still operates in rural context seems to be more important for consideration. It has been commonly observed that the clients consider the quality of health services as poor for use, encountered by the negative attitude of the service providers, and very often it is also presumed that the rural people consider the quality of services as unaffordable.

More than five decades of planning in our country to improve people's health is yet to yield satisfactory result to the tribal communities in particular, who have been explaining their ways of life in imaginary, spiritual or supernatural terms, not unlike the complex modern life, attributing to biological factors. Though the supernatural beliefs in the form of magic, religion implied by witch-doctors or sorcerers, quacks (medicine men), and the chanting of charms for *talisman* lies outside the range of scientific observation, still it prevails among the tribals. It is largely because expansions of modern health infrastructure during more than five decades have resulted in the medicalization of the entire treatment process which could not bring satisfactory result to villagers. Secondly, there are communities in both tribal and rural pockets, who have been strictly adhering to their ethno-method of disease treatments, largely based on the use of herbs, roots, barks, etc. is a part of their reproductive health system. Thirdly, there is ample potentiality within the community for their self-treatment of different type of diseases, which can be enriched keeping the reproductive health requirements of the population. By and large it is observed that the use of herbs, leaves, roots, barks, juice and other herbal ingredients, known differently as folk or domestic medicine in different communities are more scientific by nature and have also been a part of tribal and non tribal life.

Health Scenario in Orissa

The ST composition in Orissa population share almost 22.13 per cent (2001 Census of India). Out of total geographical areas of the State 44.21 per cent of land area has been declared as Scheduled Area, which includes 6 districts declared as fully scheduled and 7 districts declared as partially scheduled. Of the total 62 Scheduled Tribe communities, 13 communities have been declared as Primitive Tribal Groups (PTG). The Scheduled Areas of the State covers 118 (37.3%) Community Development Blocks (Statistical Abstract of Orissa 1991-92). The sex ratio in the State accounts to 971 females per 1000 males, where as among STs it is 1002. The Infant Mortality Rate (IMR) in the State of Orissa is 81.0 (India 67.6). Under five Mortality Rate (UFMR) in the State is 104.4 (India 94.9). Mothers' receiving at least one anti-natal checkup in Orissa is 79.5 (India 65.4), delivery in medical institutions in Orissa is 22.6 (India 33.6) and the deliveries assisted by a health professional in Orissa is 33.4 (India 42.3). The percentage of children aged 6 months to 36 months suffering from anemia in Orissa is 72.3 (India 74.3) and the percentage of children under 3 years of age suffering from under weight is 54.4 (India 47.0). The children with stunted growth in Orissa are 44.0 (India 45.7), and wasted children are 24.3 per cent (India 15.5). Percentage of women within the reproductive age group (15-49 years) falling with anemia in Orissa is 63.0 (India 51.8).

The health care facilities available in Orissa are much behind many States of India. This proves when one see that there are only 40 beds in Orissa per one lakh population is the lowest in India after undivided Madhya Pradesh. Again, more than 95 percent of these beds are mainly located in urban and semi-urban areas. Private medical practitioners account for 31.39 per cent as regards out patient's treatment in the State Public Hospitals, where as Primary Health Centers and public dispensaries account for 52 percent as regards out patient treatment in Orissa (Directorate of Health Services, Govt. of Orissa 2002). The delivery of health care services in Orissa is basically consists of three tiers. The first tier at primary level consists of Primary Health Center (PHC) or new PHC at block level. This hospital having six beds is looked by single doctor, which cater a population of 30,000 in plain areas, and 20,000 in tribal hilly areas. The Sub-Center looks upon 5,000 population in plane areas and 3,000 in hilly tribal dominated areas. Paramedics like a female Multi Purpose Health Worker or Auxiliary Nurse Mid-wife (ANM) and a male Multi-Purpose Health Worker, who primarily provide reproductive health care services and family planning services, serve these centers. The upgraded PHC with 16 beds and two specialists, and CHC with 30 beds and six specialists respectively operate at block level to deliver health services are known as secondary tier. The PHC cover a population of 80,000 to 1,20,000. Besides, area hospitals located in semi-urban or urban centers of subdivision and district head quarters are also covered under this category. The tertiary health units consist of 3 medical college hospitals and 2 specialized hospitals, which provides supportive services to primary and secondary level health institutions.

Table 1. Sensitive Health Indicators, 1999

Sl. No.	*Indicators*	*Orissa*	*India*
1.	Maternal mortality rate per lakh life births, 1997 SRS	361	408
2.	Crude birth rate per 1000 population – 1999 SRS	24.1	26.1
3.	Crude death rate per 1000 population – 1999 SRS	10.6	8.7
4.	Life expectancy at birth (in years)	61.64	63.50
5.	Child mortality rate per 1000 children	25.5	29.3
6.	Perinatal mortality rate per lakh life birth – 1998 SRS	61.4	42.5
7.	Infant mortality rate –1994-98, (NFHS-2, 1998-99)	97.0 (81.0)	70.0 (67.6)
8.	Total fertility rate – NFHS-2	2.5	3.2

Source: Government of Orissa, Health and Family Welfare Department, Bhubaneswar

According to the latest sensitive health indicators of 1999, the average health status of people in Orissa is not so encouraging as compared to the average health status of people at the all-India level. It is found that although Orissa has made relatively better progress in reducing crude birth rate, child death rate, maternal mortality rate and total fertility rate than the all-India level, its progress in reducing crude birth rate, infant mortality rate, prenatal mortality rate, and increasing life expectancy rate is worse than the country average (Table 1). This is possibly due to higher level of rural poverty and malnutrition. The body mass index (BMI), which is worked out to assess the nutritional status of population by relating weight with height is found to be low for women in Orissa. According to National Family Health Survey 1998-99 (NFHS-2), the average or mean BMI of women in Orissa is 19.2. Chronic energy deficiency is usually indicated by a BMI below 18.5. On this basis, when the nutritional status of women in Orissa is assessed according to the NFHS-2 data, almost half (48%) of women in the state have a BMI below 18.5, indicating a high prevalence of nutritional deficiency. The survey reveals that nutritional problems are particularly serious for younger women, rural women, illiterate women, and women from scheduled castes and scheduled tribes (NFHS-2 Orissa: 153). Similarly, according to NFHS-2 findings more than half of children (54%) below three years of age in Orissa are underweight (weight-for-age) and 44 per cent are stunted (height-for-age). Among them, 21 per cent are severely undernourished according to weight-for-age and 18 per cent according to height-for-age. The proportion of wasting (weight-for-height) among the children of this age is also quite evident affecting about one-fourth of the total. Interestingly, the proportions of children who are underweight, stunted, or wasted have more or less remained constant between NFHS-1 (1993) and NFHS-2, although severely underweight and severely stunted categories have decreased between NFHS-1 and NFHS-2, from 23 to 21 per cent and from 25 to 18 per cent (NFHS-2 Orissa: 166). Added to this it is found that prevalence of anaemia among the women and children in Orissa is found to be very high. According to NFHS-2 findings nearly three-fourth (72%) of the children below three years of age and 63 per cent of the women in 15-49 years age group has some degree of anaemia. While the percentages of women with moderate and severe anaemia respectively are16.2 and 1.5, in the case of children below three years of age those are found to be 43.2 and 2.9 respectively (NFHS-2:156 and 169).

Needless to say, anaemia and malnutrition of such magnitude among the small children and the women in the reproductive age group are a matter of serious concern as they cause increasing morbidity and reduce life expectancy rate of population. It may be observed from the data shown in Table 9.1 that Orissa has been successful in reducing crude birth rate, total fertility rate, maternal mortality rate and child mortality rate as compared to the all-India average figures. However, due to under-nutrition and anaemic problems of women and children, infant mortality rate and perinatal mortality rate in the state is fairly higher than the all-India level. Hence, this calls for intensification of awareness generation and sensitization of the rural mass through IEC (Information education and communication) activities and for this formulation of an appropriate and effective health policy to cater to the needs of disadvantageous category of population should get top priority for the social sector development planning in Orissa.

In this scenario, it is found that according to the latest available data of the Health and Family Welfare Department, Government of Orissa, the state has at present one medical institution/health centre for every 21580 persons and one medical bed for every 2662 persons. The doctor population ratio is 1:7560; nurse population ratio is 1:16500; and nurse doctor ratio is 1:2, whereas the suggested national norms worked out by the Mudaliar Commission in 1960 was one doctor for every 3500 persons and one nurse for every 500 persons (Mohanty 2001:6). The CSO (Central Statistical Organization) data, however, show that in the year 1992 at the all-India level there was one doctor for every 2083 population and in 1995 there was one nurse for every 1639 population. Similarly, in the year 1995 at the all-India level there was one public health centre for every 42812 persons and according to CMIE data there was one medical bed for every 1059 persons. This implies that existing public health infrastructure and amenities both in the country and the state is far from satisfactory and particularly for a backward and poverty stricken state like Orissa having higher proportions of disadvantageous categories of population strengthening of public health care measures is more crucial.

On the contrary, it is however, noticed that government spending on the health care measures in real terms are progressively being curtailed, particularly after economic liberalization measures and privatization of the state undertakings in the wake of India's resort to Structural Adjustment Programme (SAP) for faster development. Under this new economic measure the poor and the backward states of India have become the worst sufferers, although economic growth of the advance states under the capitalist production system has been much faster (Meher 2001). In the backward and poorer states the state governments are finding it difficult to strengthen social security measures for the poor and vulnerable population. Their revenue generation capability has been weakened due to accentuation of economic disparities under free market economy. It is found that in the year 1990-91, the per capita health expenditure of Orissa (Rs. 56.7) was almost equivalent to the average expenditure incurred at the all-India level (Rs. 57.1). However, in the year 1999-2000, while the per capita health expenditure of Orissa was reported to be Rs. 136.3, it was Rs. 158.2 at the all-India level (Government of Orissa 1992:242; 2001:ANX71). It is possibly due to falling health expenditure the overall prevalence rate of diseases like malaria and tuberculosis in Orissa has taken an upward trend in the recent years. Although the data relating to prevalence rate of malaria and tuberculosis collected during NFHS-1 and NFHS-2 in Orissa may not be comparable very closely due to methodological variation in the collection procedure of the data, it is found that while the prevalence rates of malaria and tuberculosis per lakh population were respectively 5200 persons and 600 persons during NFHS-1 (1993), those increased to respectively 7414 persons and 833 persons per lakh population during NFHS-2 (1998-99).

Following chart displays how different state and central organization and local organizations/service providers work together in rural areas of Orissa

Study Area and Methods

Similipal Biosphere Reserve is located in Mayurbhanj District on the northern part of Orissa. The district is located between 21°16′ and 22°34′ North latitude and between 85° 40′ and 87°11′ East longitudes. The district is bounded by the Singhbhum district of Jharkhand and Midanapore district of West Bengal in the north; Balasore and Keonjhar in the south; Midanapore and Balasore district in the east; Keonjhar and Singhbhum in the west. Mayurbhanj was an important kingdom of Orissa even after independence. This kingdom got merged with Orissa and got the recognition of a district on 1st January 1949. This is the largest district of Orissa in terms of area and area under forest. Mayurbhanj is a heavenly place that includes the river Budhabalanga, the waterfall Barehipani, Meghasini Mountain and the biosphere reserve Similipal. It is also recognized for its irrigation and power project, namely Sunei and Kaladam. Mayurbhanj district is famous for horn works, stone and clay works, Tussar silk, jute mills and spinning mills. The total area of the district is 10,418.00 sq km; out of which 4392.13 sq. km is under forest, or about 50% of the area. The predominant species in the forest is Sal (*Shorea robusta* Gaertn.f), where as the Similipal range shows diversified vegetation pattern i.e. from tropical deciduous to a virgin semi-ever green forest which express the climatic climax type of vegetation. (Braham et al, 1990). Agriculture is the main occupation of tribal people but forest and its products plays important role to meet their multifarious requirement such as-food, medicine, fibres etc. (Bal, 1942; Saxena et al, 1989). For food they are mainly dependent on agriculture but they also collect roots, tubers, leaves, flowers, fruits as supplementary food. The total population of the district is 2,221,782, out of which scheduled caste is 1,70,853 and scheduled tribe is 12,58,495. (According to 2001 census). The physiography of the district gives a perfect platform for the tribals in sustaining their ethno-cultural identity. Tribal people mostly inhabit the deep forest area, and depend on the forest resources for their livelihood. The district has been the homeland of various tribal communities with their sub-tribes, who are found in different level of development depending upon their assimilation with the so called mainstream or modern communities. There are 46-tribes inhabited in this district which is 64.86% of the total population of the district (Tiwari *et al*, 1999,) the dominated tribes of the district are Santal, Munda, Kolha, Bhumija, Bathudi, Lodha, Gond, Kondha, Saunti, etc. Although the floristic and ethno medicinal investigation have been done by various workers like Ambasta, 1986; Jain, 19791; Hajra, 1981; Majumdar, 1971, the information on plants used for health disorders is rich in these publications. The present work concentrates on the ethno medical culture and alternative medicines used by the tribals and local folks for various ailments. The soil is mostly red throughout the district and in the east there is a small patch of black cotton soil. The important minerals available in huge quantity in the district are Iron-ore, Manganese and Chromites. The temperature ranges from 10°C during winter and about 32.5°C - 40°C in summer; where as the mean annual rainfall is 1648.2 mms. The bulk of the rain is in the month of August to October, while March to May is the driest months.

The report has followed both primary and secondary methods of data collection. In order to assess the impact of the health status on tribal people living in inaccessible tribal dominated regions of Similipal Biosphere, the study has adopted both qualitative and quantitative methods. To explore the qualitative dimensions of health issues it has adopted Focus Group Discussions (FGD), key informant interviews, personal observations, etc. The annual reports of the district health department from 2005-2006, hospital records and patient cards have also been reviewed. The study was conducted during November – December 2006 in nine villages of Similipal Buffer area villages and one hamlet of *Mankedia* (Birhor) of transitional area in Similipal, covering the entire gram panchayats within the Biosphere. The villages covered under the study are Gurgudia, Kumari, Lanjighoshra, Koljhari, Chakundakocha, Garh-Similipal, Saruda, Basalikocha and Durdura (Birhor Settlement). These also covered the entire ethnic groups of the area (View map of locale).

OBJECTIVES

- Firstly, the report endeavors to illuminate various dimensions of health care practices in Mayurbhanj district vis a vis the state of Orissa.
- Secondly, it provides a brief background of the several projects Project and throws some light on the socio-demographic profile of population in the study area.
- Thirdly, the report assesses the impact of various health services provided by district health service providers.
- Fourthly, the report compares the mechanisms and the quality of health services delivered by the Government Departments.
- Lastly, the report critically examines the importance of community based approaches for the promotion of indigenous health practices in the biosphere area of Similipal.

MAP OF SECTOR AND LOCATION OF SUB-CENTRES

During the survey it was observed that Gudgudia, Nawana and Bareihpani Gram Panchayats had all the characteristics of remoteness, difficult to access hilly tribal pockets. All three Panchayats were suffering

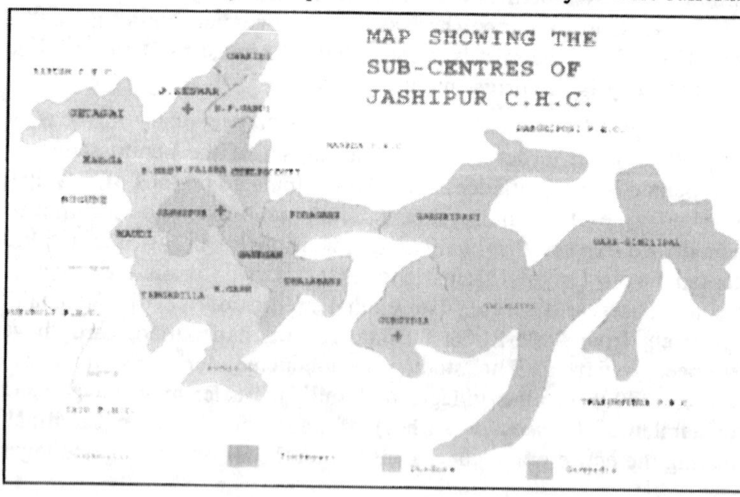

from object distress conditions induced by shortage of food during summer and rainy season, lack of minimum health services, absence of functional development and extension services, very poor literacy and awareness regarding personal hygiene, sanitation, child care and rights, crippling with influence of exploitative forces like quacks, moneylenders and so on. Though they have advance technology of agriculture now a days, most of the families of the area had to experience shortage of food for 4 to 8 months in a year. Lack of cash economy siphons a substantial proportion of food grains used to be exchanged by the people to get cash or items for day to day living. Availability of various infrastructures in the area was almost nonexistent because of the rules of forest protection act and subsequent declaration of Biosphere in 1994. The health facilities (including PHC at Gudgudia, Bareihpani and Nawana) were functional but the doctors mainly due to high staff turn over and absence due to their reluctant attitude to stay in the interior pockets. As a result, one finds high frequency of malaria, dysentery, diarrhea, maternal and infant mortality, general morbidity etc, which result loss of work/wage, indebtedness and anemic conditions. Poor awareness on personal hygiene and sanitation, irregular functioning of health facilities, along with the high influence of quacks seems to be major bottlenecks in the process of change.

Diseases

Disease is purely a bio-physical phenomenon. It cannot be isolated from the socio-cultural milieu. The perception of diseases and its cure assumes different dimensions in different societies. Cultural patterns and typical ways of living give substances to the manner in which disease is perceived, expressed and reacted. To some extent, the cultural context defined what conditions are recognized to assess and define the illness conditions. (Mechanic: 1978)

From time immemorial, human beings have found remedies within their habitat and have adopted different therapeutic strategies depending both upon climatic, pedo-logical, phyto-geographic and faunal characteristics and upon peculiar cultural and socio-structural typologies. At a time when the world is grappling with the dearth of molecules for addressing a whole range of medical indications, Ethno-medicine holds promise in addressing some of these wants. Illness is attributed to the actions of malevolent spirits, who may be ancestors who have not been sufficiently appeased or to the temporary withdrawal of soul substance from the body, etc. Shamans are frequently called in to divine the cause, often with the aid of their tutelary spirits and to effect a cure through the sacrifice of a fowl, goat, or other animal.

People of Similipal Bio-Sphere Reserve suffer from different kinds of diseases because of induction of external agents into their internal organs (effects of food on digestive tracts as well as smoke and dust on respiratory tracts) and on body surface. Cuts-minor or deep, sprain, inflammation of joints, ringworm, fungal infection, infection at the corner of the mouth and finger nails etc. that are easily identified because of their observation. These ailments are most common among the tribals of the Similipal Bio-Sphere Reserve. Other common diseases are malaria, jaundice, respiratory tract infection, diarrhea (both acute and bloody) hookworm and round worm infections. Skin disease, dropsy, tuberculosis, leprosy and fileria are not so common in the forests of Similipal Bio-Sphere Reserve. Though of late some stray cases of tuberculosis are found in this area, this might be of excessive labour not in commensuration with the food intake at the time of summer and rainy season. It is observed that Kharia and Mankedia are mostly prone to such diseases because they are mostly hunter-gatherer and work as agricultural labourer and wage labour in nearby brick kilns of the area. Cases of Sexually transmitted diseases are not so common in Similipal belt as the tribes and communities do not indulge in sexual excesses (extramarital or un-natural) secretly in the lonely places in the midst of jungle for the fear of god (bonga/ jaher) and thereby inviting their anger in the shape of misery and disaster. Occasional bout of diseases like measles, malaria and diahhorea etc. in the form of epidemic are felt in the forests of buffer area villages of Similipal Bio-Sphere Reserve because of the filthy environment of some of the villages. Recently in the month of June 2006 some cases of measles outbreak occurred and about 22 children were died in village Kumaribil at that time. The tropical climate and poverty help them to produce a great numbers of diseases. The warm moist climate of the area helps to breed the insects and parasites and other factors like food habits of the people also responsible for outbreak of various diseases. Apart from that in most cases foods are taken un-hygienically and most cases they take uncooked/ raw foods. This indirectly promotes occurrence of diseases in epidemic or endemic form. It is also observed that some that some social and economic factors like beliefs, practices, education, occupational pattern; food habits etc. cause the spread of diseases.

Table 2. Epidemiological data on Malaria of Mayurbhanj District, Orissa for the years 1998-2002

Year	*BSC/BSE*	*Positive*	*Pf*	*Death*	*ABER*	*API*	*SPR*	*STR*	*Pf percentage*
1998	305719	40709	36621	31	14.9	19.8	13.3	11.9	89.9
1999	252064	29794	26544	48	11.8	13.9	11.8	10.5	89
2000	296613	31161	27956	24	13.8	14.5	10.5	9.4	89
2001	311676	23536	21201	21	14	10.6	7.5	6.8	90
2002	407019	24163	21692	21	18.3	10.8	5.9	5.3	89.7

Source: District Malaria Officer, Mayurbhanj, Orissa

1. BSE - blood slides examined;
2. Pf - Plasmodium falciparum;
3. ABER - annual blood examination rate;
4. API - annual parasite index;
5. SPR - slide positivity rate;
6. SFR - slide falciparum rate.

*The following tables show the occurrence and prevalence of various diseases during our field study. This is the record of the PHCs and CHC as reported by various Gram Sewaks and village level Health workers.

Table 3

August- September 06											
	Gudgudia		*Badkasira*		*Bareihpani*		*Garh Similipal*		*Death*		*Total*
	<5	>=5	<5	>=5	<5	>=5	<5	>=5	<5	>= 5	
Simple Diarrhea	5	6	5	7	12	12	10	8	0	0	65
Severe Diarrhea	0	0	0	0		0	0	0	0		0
Bloody Diarrhea	1	1	0	0	0	0	0		0	0	2
Suspected malaria	14	45	12	29	6	21	11	17	9	3	155
Acute respiratory Infection	25	18	15	12	19	19	11	13	3	1	132
Measles	0		0	0	0	0	0	0	0	0	0
Neonatal tetanus	0	0	0	0	0	0	0	0	0	0	0
Acute jaundice Syndrome	0	0	0	0	0	0	0	0	0	0	0
Suspected Meningitis	0	0	0	0	0	0	0	0	0	0	0
Unusual syndrome	0	0	0	0	0	0	0	0	0	0	0
Others	13	17	12	27	17	14	16	20	4	1	136
Heat disorder (during Summer)	0	0	0	0	0	0	0	0	0	0	0
Skin Infection	0	0	0	0	0	0	0	0	0	0	0
Snake bite (Rainy season)	0	0	0	0	0	0	0	0	0	0	0

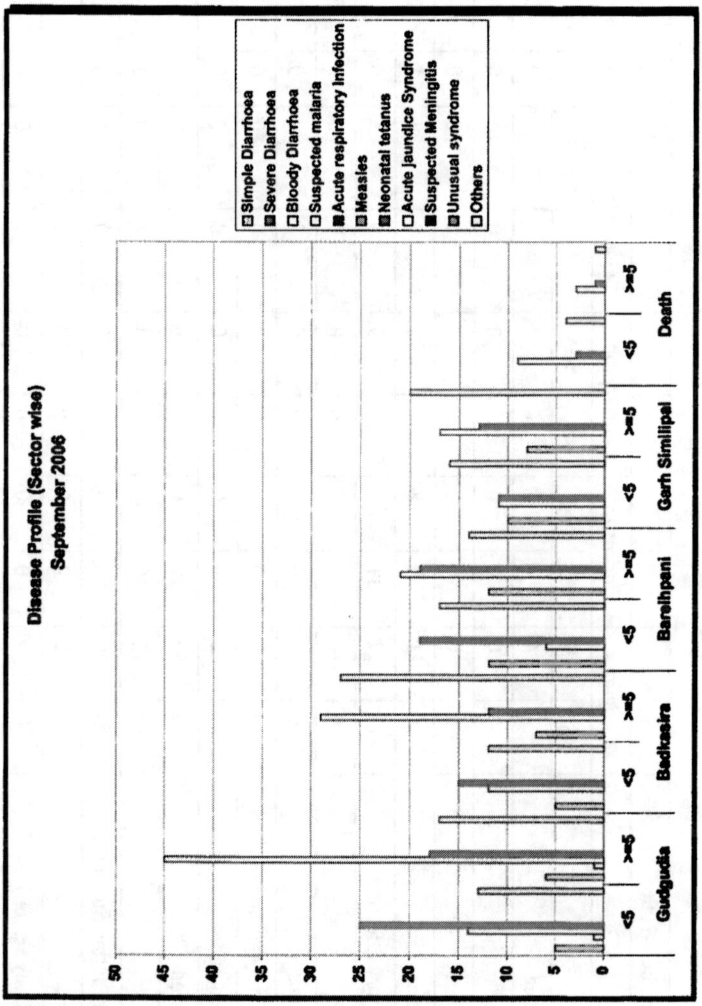

Chart 1 (Table 3)

Table 4

October--2006											
	Gudgudia		*Badkasira*		*Bareihpani*		*Garh Similipal*		*Death*		*Total*
	<5	>=5	<5	>=5	<5	>=5	<5	>=5	<5	>=5	
Simple Diarrhea	4	4	4	7	8	7	7	7	0	0	48
Severe Diarrhea	0	0	0	0	0	0	0	0	0	0	0
Bloody Diarrhea	0	0	0	0	0	1	0	0	0	0	1
Suspected malaria	16	11	9	8	14	22	7	7	1	2	97
Acute respiratory Infection	11	7	11	8	15	9	5	5	0	0	71
Measles	0	0	0	0	0	0	0	0	0	0	0
Neonatal tetanus	0	0	0	0	0	0	0	0	0	0	0
Acute jaundice Syndrome	0	0	0	0	0	0	0	0	0	0	0
Suspected Meningitis	0	0	0	0	0	0	0	0	0	0	0
Unusual syndrome	0	0	0	0	0	0	0	0	0	0	0
Others	10	12	10	30	11	13	8	9	2	3	108
Heat disorder (during Summer)	0	0	0	0	0	0	0	0	0	0	0
Skin Infection	0	0	0	0	0	0	0	0	0	0	0
Snake bite (Rainy season)	0	0	0	0	0	0	0	0	0	0	0

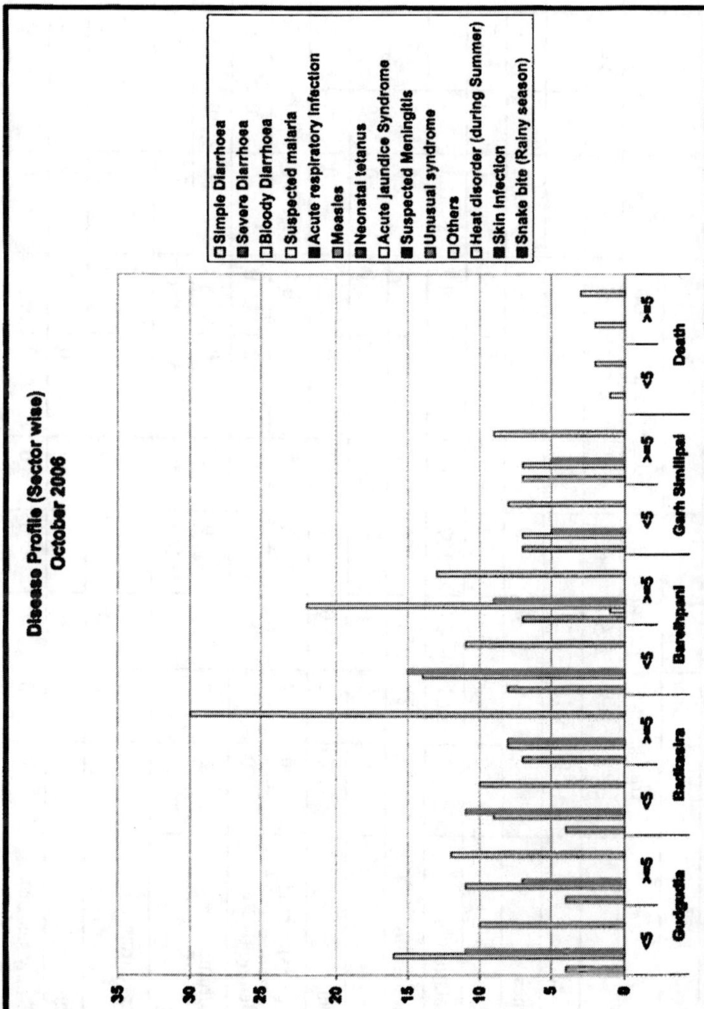

Chart 2. (Table 4)

Table 5

November-2006	*Gudgudia*		*Badkasira*		*Bareihpani*		*Garh Similipal*		*Death*		*Total*
	<5	>=5	<5	>=5	<5	>=5	<5	>=5	<5	>=5	
Simple Diarrhea	19	18	22	28	22	32	30	36	1	3	211
Severe Diarrhea	0	0	0	0	0	1	0	0	0	0	1
Bloody Diarrhea	0	0	3	2	1	1	2	1	0	0	10
Suspected malaria	37	61	31	107	15	57	19	103	5	5	440
Acute respiratory Infection	29	27	40	51	30	38	33	42	0	2	292
Measles	0	0	0	0	0	0	0	0	0	0	0
Neonatal tetanus	0	0	0	0	0	0	0	0	0	0	0
Acute jaundice Syndrome	0	0	0	0	0	0	0	0	0	0	0
Suspected Meningitis	0	0	0	0	0	0	0	0	0	0	0
Unusual syndrome	0	0	0	0	0	0	0	0	0	0	0
Others	32	43	56	106	32	57	37	104	2	19	488
Heat disorder (during Summer)	0	0	0	0	0	0	0	0	0	0	0
Skin Infection	0	0	0	0	0	0	0	0	0	0	0
Snake bite (Rainy season)	0	0	0	0	0	0	0	0	0	0	0

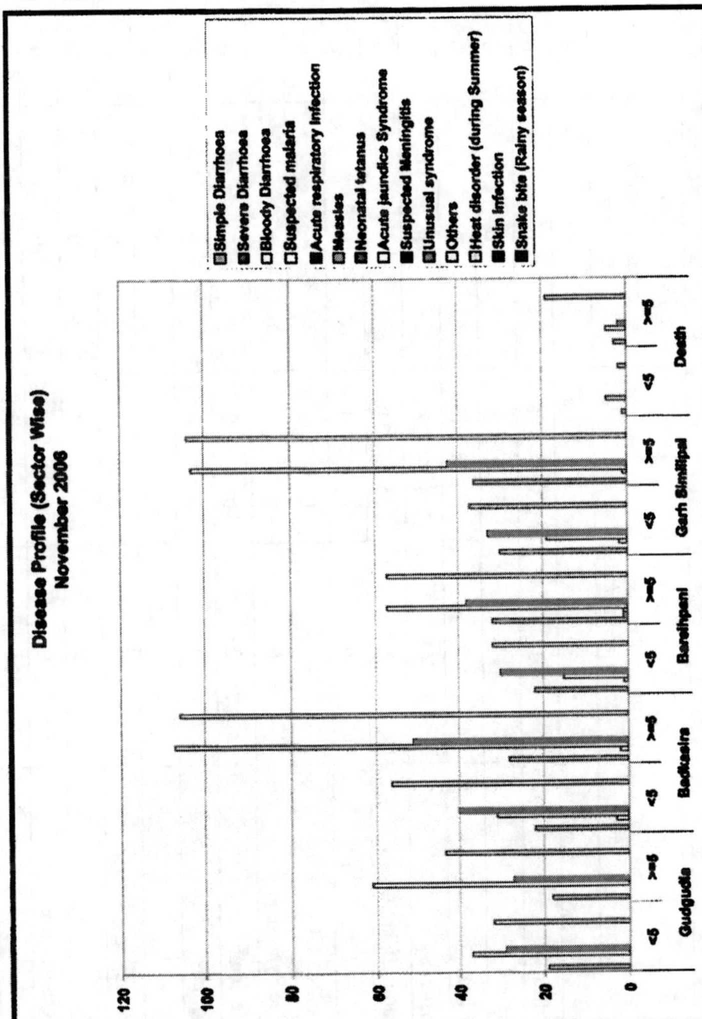

Chart 3. (Table 5)

Table 6. Health Status in Jashipur Block

November 2006	*Total number of cases in Block*			*Total number of cases of deaths in Block*		
	<5	*>=5*	*Total*	*<5*	*>=5*	*Total*
Simple Diarrhea	33	65	98	0	0	0
Severe Diarrhea	1	3	4	0	0	0
Bloody Diarrhea	2	11	13	0	0	0
Suspected malaria	44	241	285	0	0	0
Acute respiratory Infection	75	165	240	0	0	0
Measles	0	0	0	0	0	0
Neonatal tetanus	0	0	0	0	0	0
Acute jaundice Syndrome	0	0	0	0	0	0
Suspected Meningitis	0	0	0	0	0	0
Unusual syndrome	0	0	0	0	0	0
Others	88	52	610	3	8	11
Heat disorder (during Summer)	0	0	0	0	0	0
Skin Infection	0	0	0	0	0	0
Snake bite (Rainy season)	0	0	0	0	0	0
TOTAL	**243**	**1007**	**1250**	**3**	**8**	**11**

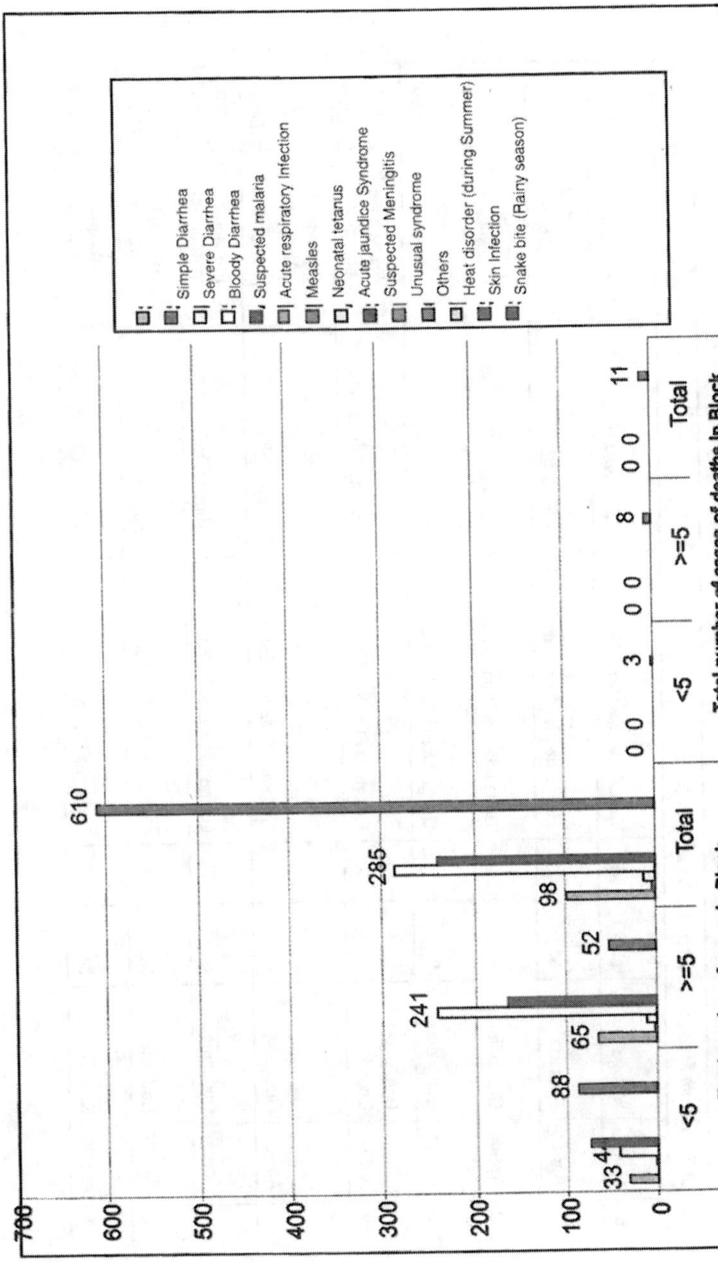

Chart 4 (Table 6): Health Status in Jashipur Block CHC, November 2006

Table 7. This table depicts the cases of only Malaria in Gudgudia G.P.

Cases of Falciperum Malaria in Buffer area Villages in 2006								
Sl. No	*Month*	*Village*	*Sex (M/F)*	*Age*	*Initial Treatment at*	*Later treatment at*	*Medicines advised*	*Result*
1	Apr-06	Kumari	M	25	Gudgudia PHC Sub Centre	Jashipur CHC	Primaquine phosphate	Cured
2	Apr-06	Gudgudia	F	8	Gudgudia PHC Sub Centre	Jashipur CHC	Primaquine phosphate	Cured
3	Apr-06	Gudgudia	F	15	Gudgudia PHC Sub Centre	Jashipur CHC	Primaquine phosphate	Cured
4	Apr-06	Bombay Chowk	M	30	Gudgudia PHC Sub Centre	Baripada dist. Hospital	Primaquine phosphate	Not Known
5	Apr-06	Gudgudia	F	2	Gudgudia PHC Sub Centre	Jashipur CHC	Primaquine phosphate (in syrup base)	Dead
6	Apr-06	Gudgudia	M	8	Gudgudia PHC Sub Centre	Jashipur CHC	Primaquine phosphate	Cured
7	Apr-06	Gudgudia	M	40	Gudgudia PHC Sub Centre	Jashipur CHC	Primaquine phosphate	Cured
8	Apr-06	Kumari	M	25	Gudgudia PHC Sub Centre	Jashipur CHC	Primaquine phosphate	Cured

(contd...)

Sl. No	*Month*	*Village*	*Sex (M/F)*	*Age*	*Initial Treatment at*	*Later treatment at*	*Medicines advised*	*Result*
1	May-06	Gudgudia	F	15	Gudgudia PHC Sub Centre	Jashipur CHC	Primaquine phosphate	Cured
2	May-06	Gudgudia	M	5	Gudgudia PHC Sub Centre	Jashipur CHC	Primaquine phosphate	Cured
3	May-06	Gudgudia	F	8	Gudgudia PHC Sub Centre	Jashipur CHC	Primaquine phosphate	Cured
4	May-06	Gudgudia	M	14	Gudgudia PHC Sub Centre	Baripada dist. Hospital	Primaquine phosphate	Cured
5	May-06	Kumari	M	30	Gudgudia PHC Sub Centre	Baripada dist. Hospital	Primaquine phosphate	Cured
6	May-06	Kandibil	M	9 month	Gudgudia PHC Sub Centre	Jashipur CHC	Primaquine phosphate	Dead
7	May-06	Kumari	M	45	Gudgudia PHC Sub Centre	Jashipur CHC	Primaquine phosphate	Cured
8	May-06	Gudgudia	M	13	Gudgudia PHC Sub Centre	Jashipur CHC	Primaquine phosphate	Cured
9	May-06	Gudgudia	M	15	Gudgudia PHC Sub Centre	Jashipur CHC	Primaquine phosphate	Cured
1	Jun-06	Jenabil	F	25	Gudgudia PHC Sub Centre	Jashipur CHC	Primaquine phosphate	Cured
2	Jun-06	Gudgudia	M	15	Gudgudia PHC Sub Centre	Jashipur CHC	Primaquine phosphate	Cured
3	Jun-06	Kumari	F	7	Gudgudia PHC Sub Centre	Jashipur CHC	Primaquine phosphate	Dead
4	Jun-06	Gudgudia	M	17	Gudgudia PHC Sub Centre	Jashipur CHC	Primaquine phosphate	Cured

(contd...)

Sl. No	*Month*	*Village*	*Sex (M/F)*	*Age*	*Initial Treatment at*	*Later treatment at*	*Medicines advised*	*Result*
5	Jun-06	Kharia Dongri	F	1yaer 6 month	Gudgudia PHC Sub Centre	Baripada dist. Hospital	Primaquine phosphate	Dead
6	Jun-06	Lanjighosra	M	2	Gudgudia PHC Sub Centre	Jashipur CHC	Primaquine phosphate	Cured
7	Jun-06	Sarpat	F	3 year 6 month	Gudgudia PHC Sub Centre	Jashipur CHC	Primaquine phosphate	Cured
1	Jul-06	Kumari	M	55	Gudgudia PHC Sub Centre	Jashipur CHC	Primaquine phosphate	Cured
2	Jul-06	Kumari	M	8	Gudgudia PHC Sub Centre	Jashipur CHC	Primaquine phosphate	Cured
3	Jul-06	Gudgudia	M	7	Gudgudia PHC Sub Centre	Jashipur CHC	Primaquine phosphate	Cured
4	Jul-06	Gudgudia	M	12	Gudgudia PHC Sub Centre	Jashipur CHC	Primaquine phosphate	Cured
1	Aug-06	Gudgudia	F	55	Gudgudia PHC Sub Centre	Jashipur CHC	Primaquine phosphate	Cured

*Medicines in all cases were given Primaquine phosphate and Chloroquine phosphate combination in blister pack supplied by Nestor Pharma Ltd.

Table 8. Results of cross-sectional opinion survey, conducted during field visit, regarding rapid diagnostic kits for Malaria and blister Combi pack in Similipal Buffer area villages

Parameters	*Similipal Buffer villages*	
Total no. 29 cases		*Percentage*
No. of respondents suffered from malaria	29	100
Average time lag between collections of blood smear and availability of report	20 Days	
Knowledge about malaria diagnosis by RDK	3	10.6
No. of respondent examined through RDK	3	10.6
No. of respondent in favour of RDK over microscopic examination	29	100.0
Knowledge about blister combi packs	4	12.0
No. of respondent who have taken blister packs for treatment of malaria	3	10.6
No. of respondents in favour of blister packs than taking loose tablets	28	97.0

*RDK: Rapid diagnostic kit

Introduction of rapid diagnostic test kits and blister packs in the operational malaria control programme was a beneficial step in reducing malaria morbidity and mortality especially in the hard-core remote and inaccessible areas. However, so far the blister packs are available only for adults, while in such malaria endemic areas children suffer more from malaria. Therefore, blister packs should also be made available for younger age groups. The pathologists at the health centers reported patient's satisfaction with rapid diagnostic kits and blister packs. They also felt that continuous supply of rapid diagnostic kits and blister packs was necessary for compliance. Although there was no perceptible improvement in the patients' health seeking behaviour because rapid diagnostic kits and blister packs have been introduced very recently, but it will be worthwhile to study this at a later date. Further, the health workers opined that the instructions for the use of rapid diagnostic kit and blister packs should be in local languages also.

The routine identification of infections with *Plasmodium* species currently depends upon microscopic examination of blood smears. However, microscopy is time consuming, labour intensive and largely depends on the expertise and diligence of the pathologist. Moreover, microscopic examination is rather impractical in remote inaccessible tribal areas of Similipal, which constitute a distinct malaria paradigm and are badly affected by perennial malaria transmission. In this context, introduction of rapid diagnostic kits and blister packs in the operational malaria control programme in remote and inaccessible highly malaria infectected areas have special significance in reducing morbidity and mortality due to malaria. Presently, the rapid test kits are being used for early and easy diagnosis of *P. falciparum* malaria but it may assume special significance in highly endemic tribal areas for mass screening of asymptomatic carriers, which are commonly prevalent in such areas due to development of natural immunity because of repeated exposures. The only disadvantage of rapid diagnostic test based on PfHRP antigen is the false positivity up to three weeks even after clearance of parasitaemia. This problem may be solved if rapid test kits detecting only viable parasites are used. In conclusion, keeping in view all the advantages, both these new tools will help in reducing parasite load in high transmission areas for achieving a sustainable malaria control.

Medical facilities in studied buffer villages

Gudgudia village of the buffer area has only health clinic i.e. Sub center cum PHC covering above 3000 populations of biosphere reserve villages. This center has only one mobile health unit and a pathological laboratory only for the test of Malaria slide test with two patients' emergency admission facility. One MBBS doctor with a lab technician and four ANMs (Anganwadi Nurse cum Midwife) are posted. Other three Sub centers are in the villages of Garh Similipal, Bareihpani and Barkasira, mostly cover up to 3000 population but no MHU (mobile health unit). Other than sub centers most of the villages are covered by additional ANMs who used to visit the remote villages and keep registers for new born, emergency cases, sudden out break of infectious diseases, record and cause of deaths etc. and send report to PHC at Gudgudia and CHC at Jashipur weekly. The medicines are mostly supplied free of cost to the patients through local NGOs by the district health department. Some medicines are also supplied by WHO immunization programme through NGOs or by Block office directly or by UNICEF's representatives. Such medicines includes immunization vaccines for DPT, BCG, polio, measles, TT, DT and Vitamin 'A' supplements. During our field visit we came across a 'Mega Health Mela' organized jointly by District Health Department and a NGO Chitta Gyani Charitable Trust, Cuttack at Gudgudia Forest rest house where 25 specialized doctors from different medical colleges of Orissa attended and the patients were treated and medicines were given free of cost. In this 'Mega Health Mela', Gynecological, Surgical, Orthopedic, Pediatric, Skin, Dental, ENT, Neuro- surgery, Pathological – ECG, ultra-sonography, consultations were done by the doctors of S.C.B Medical college, Cuttack, V.S.S. Medical College, Sambalpur and Bhubaneswar Medical College, Bhubaneswar.

Discussion

The system of social relationship, i.e. society persists as a stable structure beyond the life span of individuals. Every society has its own particular form of customary behavior and social system in which people behave in a particular way. Behavior is here considered all the activities of individual may be it physical or mental, explicit or secret. The different cultural groups are having their own sets of life style and customs. The prevalence of diseases may vary from culture to culture – community to community as well as from eco-zones to eco-zones. Many times customs of people affect their physical health, such as here in Similpal irrespective of community the village people apply cow dung mixed with some herbs (neem leaves) after cutting of umbilical cord of new born baby, this later develops tetanus for new born. The distinctive culture thus includes the distinctive pattern of health and diseases and these can produce variant clinical expression even in single geographical area. The habits of people also vary from culture to culture, such as random smoking habit of people of SBR (though it is restricted inside the forest area) in most cases leads to bronchitis, infection of respiratory tract, hypertension etc. It is also observed that cancer of cervix of uterus among the women is more frequent among the married women particularly those who borne large families with multiple issues but is less those who have less or no children. It is also observed that the habits of taking country made liquor (rice beer/ *hadia*) profusely during festival and marriage days almost in all cases leads to stomach upset and gastritis or peptic ulcer, this later leads to jaundice which is very common in this area.

Health Seeking Practices

All the villages of the study area are mainly depend on home remedies for diarrhoea and malaria in the initial stage, and then they go to Anganwadi centers or PHC centers. There is no PHC center in the remote villages of buffer area or in core area of SBR, so, the people have to depend mainly on PHC at Gudgudia or CHC at Jashipur. In severe cases they also consult village quacks or local medicine man. In Basalikocha nearest PHC center is in Bangriposi 10 km away and nearest main hospital is around 50 km away in Jashipur. Anganwadi workers are very co operative here. They have easy access to traditional healing process using materials from forest. In all the villages the physical and economic accessibility of hospitals is a distant dream. The Anganwadi workers are very regular in these villages, but sometimes they themselves go to the village quacks for some help.

Antenatal and Post-natal Care and Maternal Nutrition

There is no change in diet during pregnancy in all the villages. Their daily workload gets reduced due to the support extended by the joint family system. They work even in the last trimester of their pregnancy; however they get assistance in cooking, water carrying and other household work. The village women go to Traditional Birth Attendance during pregnancy period and for after delivery care. They do not get any special care before or after childbirth. They do not take any iron (IFA) tablets during pregnancy, as they do not have facility of any such services.

In Koljhari and Chakundakocha villages Anganwadi workers help the pregnant women by giving advice and providing before and after childbirth care. There is no change in diet during this period. But they take iron tablets because they know it increases blood and make child healthier. Here they do not get any support from the family during pregnancy.

In Gurgudia village women are aware of the importance of pregnancy care and they also get help from the Anganwadi workers as well as from the Doctor and staffs of PHC. These workers visit the village and give them medicines and injections. They also take iron tablets, as they know it is good for both mother and child. They do not get help from family members but they go to Anganwadi centers for help and advice.

Immunization

In Gudgudia, Lanjighosra and Kumari people are knowledgeable about the immunization programmes and that it prevents diseases. But they know only about the Pulse polio vaccines and not other vaccination. Some children especially children of Kharia community are not immunized because of ignorance. In Koljhari and Chakundakocha village people are aware of immunization. They know it prevents diseases. All the children of these villages are immunized by Anganwadi workers. They give vaccination of pulse polio, DPT, measles and also take vitamin 'A' supplement. In Saruda village people are ignorant about the immunization programmes and not a single child is immunized here due to ignorance and distant location. Though there is one Anganwadi to create awareness in this village.

Most of the villagers do not have any idea of taking anthropometric measurements regularly. Sometimes some of them measure their body weight. But the villagers of Basalikocha village and Birhor tanda (Mankedia) of Durdura village are totally ignorant of taking measurement or weight. Infant Mortality Rate is also at an alarming rate in these villages.

Table 9. Health Programmes of Nutrition for all groups done by NGOs and Government agencies

Pregnant Women	Prevent LBW children	• ICDS type of Programme.
Nursing Mothers	Achieve WHO goals	• ICDS type of Programme.
Infants (0-2 years)	Reach through mothers	• ICDS type of Programme.
Preschool Children (1-6 years)	Achieve nutritional security	• Special play school programme with meals served.
Youth and school going children (6 to 18 years)	Food for Eco-development	• Noon meal Programme in Schools.

Adolescent girls	Food for nutrition	• Self-employment training and noon meal programme.
Adults (18 to 60 years)	Food for Nutrition	• Food for work program Eco-friendly Self-employment training in poultry, livestock, mushroom etc. • Public distribution System and Anna Koshas to give grain on credit to the needy as well as at subsidized level.
Old and infirm persons	Food for Nutrition	• Antyodaya and Anna Poorna and free distribution of food and Old age pension scheme.

FLOW CHART 1

The flow chart shows the how information is generated and its interrelation with doctors, health service providers and patients.

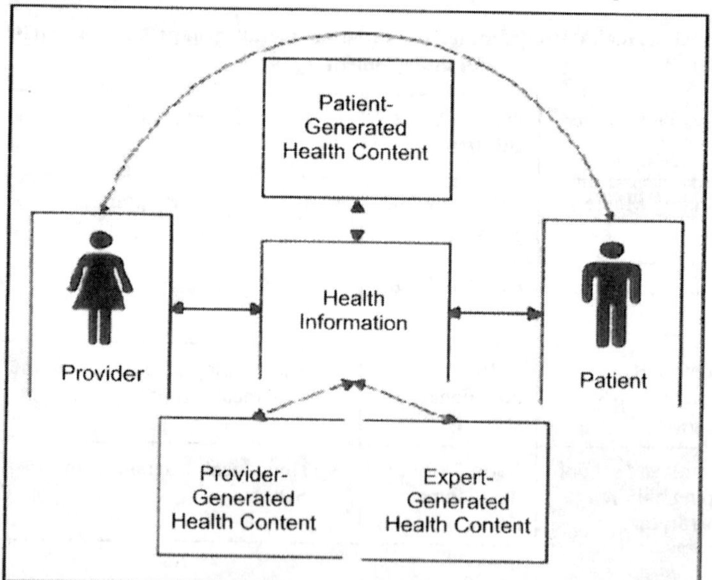

FLOW CHART 2

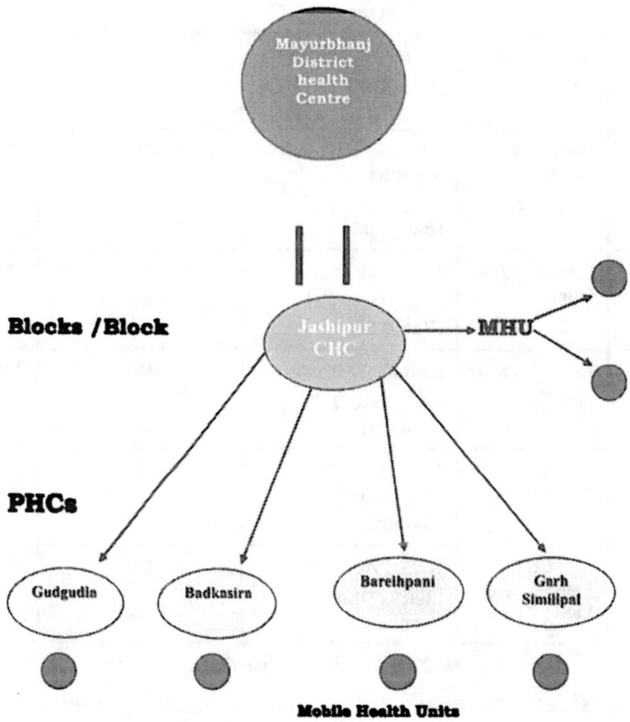

Table 10. List of Pathological laboratories in Jashipur

Sl. No.	*Name*	*Proprietor and degree*	*Test facilities*	*Established in*
1	Dayal Patho-Lab	Basanta Kr. Mahanta B.Sc., DMLT (Baripada)	Malaria (Slide and serum), Typhoid, urea creatinine, blood sugar, VDRL, montox, TB card test, Toxo test, ASO test, RA factor, sickling, serum bilirubin test, Seminal fluid, RE stool and urine	1996

Sl. No.	*Name*	*Proprietor and degree*	*Test facilities*	*Estab-lished in*
2	Sanjay Patho Clinic	Sanjay Kr. Mahanta B.Sc., DMLT (Guntur)	-Do-	1995
3	Rabi Patho Clinic	Rabindra Nath Mahanta B.Sc., DMLT (Baripada)	-Do-	2000
4	Life Patho Clinic	M. D. Parvez B.Sc., DMLT (Baripada)	-Do-	1991
5	Lalit Patho Clinic	Lalit Kr. Mahanta 10+2, DMLT (Keonjar)	-Do-	1998
6	Number One Patho Lab	Monoranjan Mahanta B.A., DMLT (Baripada)	-Do-	2005
7	Shivam Pathological Lab	K. Das Adhikary Matric, DMLT (Sambalpur)	-Do-	2005

· DMLT – Diploma in Medical Laboratory Technician.

Besides, the above private laboratories Jashipur CHC, has a pathological test laboratory where most of the OPD and admitted patient's cases referred normally but some times patients are aware of such diseases (like malaria, diahhorea etc.) and come along with the test report for consultation at CHC or to any private doctors.

Local medicines are of indigenous in nature and are classified under the sub-section of 'Folk Medicine' in the domain of 'Traditional Medicine' as distinguished from the modern medicines (Mohanti: 1996: V). It can also be termed as Oral traditional medicines. Most of the medicines are water based and are simple in terms of ingredients used, methods of preparation as well as administration to the patients. The ingredients are mostly locally available and hand picked by the Kharia people of the villages but other than Kharia, Kol, Mahali and Bhumij people also know the characters of the medicines/ herbs/ animal parts etc. and where and when they will be available in the forest. Some of the a-biotic ingredients are different types of soil, pebbles, water are

Table 11. Traditional Medicines Used For Treatment of Diseases (as narrated by the local medicine men)

Phyto-medicines used for curing diseases

Sl. No.	*Vernacular name*	*Scientific name*	*Ethno-medicinal Uses*	*Parts used*	*Remarks*
1	Kaicho, Kaincho, Karjani	*Abrus precatorius*	White seeds kept in unboiled cow milk for the period of over night and the seed is given to woman in the morning at the end of menstruation cycle for preventing conception.	White seeds	This time sexual intercourse may be avoided.
2	Heren-ba	*Ageratum conyzoides*	Boiled leaf paste is applied on skin before bed for the treatment of Khasar	Leaf	During this period daily twice bath is necessary.
3	Ashwagandha	*Withania somnifera*	Decoction of flower with honey is taken in the ratio of 5:1 frequently for one month against seminal weakness	Flower	
4	Kotadi, Karkata-daru	*Ziziphus rugosas*	Pasty mass of leaves mixed with sugar or honey (5 gm) and is taken as a preventive measure for pox	Leaf	Animal meat may be avoided.

(contd...)

Sl. No.	*Vernacular name*	*Scientific name*	*Ethno-medicinal Uses*	*Parts used*	*Remarks*
6	Begunia. Bengunia. Lunguni	*Vitex negundo*	Decoction of root bark (about 15 ml) with paste of 21 black peppers (*Piper nigrum)* is taken twice a day after food continuously seven days for the treatment of typhoid fever.	Root and bark of root	Light food may be taken
7	Arjuna, Kahua	*Terminalia arjuna*	Decoction of stem bark (20 ml.) is taken once a day in empty stomach continuously five days for the treatment of malaria.	Bark of stem	
8	Ambra. Ambada	*Spondias pinnata*	Root bark paste mixed with mustard oil used as massage for the treatment of muscular pain.	Root	----
9	Sal, Sarjam, Sakhu	*Shorea robusta*	Seed paste is applied with slight boiling before bed for the treatment of boils and carbuncle.	Seed	----
10	Dhanicha. Charpijang	*Sesbania bispinosa*	Mixture of seed paste with lime water is applied before bed as a cure for ring worm.	Seed	Non-veg food is avoided

(contd...)

Sl. No.	*Vernacular name*	*Scientific name*	*Ethno-medicinal Uses*	*Parts used*	*Remarks*
11	Gururi	*Holorrhena pubescens*	A small piece of root is grinded to a paste and applied on the spot of snake bite	Root	Local medicine man should be attended
12	Bhat-bhirabi. Puhes	*Lantana camara*	Decoction of plant is taken for the treatment of tetanus and decoction of leaf is taken twice a day for a week after food for the treatment of malaria fever.	Stem	There is strict prohibition of taking of sour food during treatment.
13	Tandibhidi. Ote-rai	*Desmodium gangeticum*	Decoction of root is taken once for seven days for the treatment of Malaria.	Root	Should be taken in empty stomach
14	Jungi-ba. Tandi kode baha	*Evolvulus nummularis*	Plant paste with Kusum seed oils (*Schleichera oleo.* is used continuously fifteen days twice a day after food for the treatment of bronchial asthma.	Whole plant	-----
15	Agora. Bakula Janum	*Argemone mexicana*	Seed paste mixed with seed oil of *Madhuca longifolia* is applied fifteen days continuously on skin for the cure of eczema.	Seed	------

(contd...)

Sl. No.	Vernacular name	Scientific name	Ethno-medicinal Uses	Parts used	Remarks
16	Limba, Neem. Nimba-daru	*Azadirachta indica*	Aqueous extract of leaves is taken once a day in empty stomach continuously one month for treatment of diabetes.	Leaf	------
17	Sweat chandi	Catharanthus roseus	Root paste is applied twice a day continuously seven days for healing of septic wounds and fresh leaf juice (few drops) mixed with a cup of water and is taken in empty stomach for the treatment of blood dysentery.	Root	Should be taken in empty stomach
18	Simal-dare, Daldara.	Bombax ceiba	Paste of fleshy roots of young plant mixed with un-boiled cow milk is taken once a day in the early morning for a week by women to regulate irregular menstruation and flowers paste is apply on boils before bed for ripening purpose.	Root	-----

(contd...)

Sl. No.	*Vernacular name*	*Scientific name*	*Ethno-medicinal Uses*	*Parts used*	*Remarks*
19	Ban-haldi, Ama-haladi	*Curcuma amada*	Plant paste with paste of peppers (*Piper longum*) is used twice for seven days after food for the treatment of piles and decoction of rhizome with common salt is taken twice for five days against cold and cough.	Whole plant	-----
20	Barun, Varuna, Pitmaiel, Banena-ba	*Crateva nurvala*	Fresh juice of stem bark mixed with seed powder of peppers, *Piper longum* is taken by women in the seventh days of menstrual cycle as a contraceptive.	Bark of Stem	Taken during bed time.
21	Hatisundha	*Heliotropium indicum*	Root paste with lime is used by rubbing the infected portion of skin as a cure for ring worms and decoction of root with honey is taken as vitamin for iron deficiency by woman against anemia during pregnancy period.	Root	-----

(contd...)

Sl. No.	*Vernacular name*	*Scientific name*	*Ethno-medicinal Uses*	*Parts used*	*Remarks*
22	Jaba-gacha	*Hibiscus rosa-sinensis*	Stem bark paste is given to woman continuously five days for causing abortion and mixture of pasty mass of flower buds with rust of iron and women take country liquor at the days of menstruation as a contraceptive.	Bark of Stem	------
23	Uli, Maska. Amba.	*Mangifera indica*	Stem bark decoction is taken in empty stomach twice a day continuously for three days as a cure for blood dysentery and pasty mass of cotyledons is applied on heads as a cure for falling of hairs.	Bark of Stem	------
24	Gangaseoli,	*Nyctanthes arbor-tristis*	Decoction of stem (3 ml) with common salt (2 gm) is taken in empty stomach for cure of dysentery and decoction of leaf (10 ml) with paste of *Piper longum* (2 gm) and honey (1 ml) is taken twice a day before food against cold, cough and fever.	Stem	This is the most common ethno medicine for early symptoms of malaria.

(contd...)

Sl. No.	*Vernacular name*	*Scientific name*	*Ethno-medicinal Uses*	*Parts used*	*Remarks*
25	Padam, Ranga padam	*Nelumbo nucifera*	Decoction of red-flowered plants (Kakananda Salukid.) about 15 ml is taken in empty stomach once a day for treatment of blood dysentery.	Flower	
26	Gandhatulsi, Dimbu	*Ocimum gratissium*	Dried leaf powder is smoked as a cigarette for getting relief from asthma and leaf juice (3 ml) with honey is given to babies once in the early morning continuously ten days for the treatment of common cold and cough.	Leaf	------
27	Muskundu-daru	*Pterospermum acerifolium*	Dried flower powder is mixed with coconut oil (*Cocos nucifera*) and is applied on head for killing of hair lice.	Flower	----
28	Susumujar, Bisalya karani	*Tridax procumbens*	Fresh leaf juice is applied as antiseptic cream for healing of cut wounds.	Leaf	----

*For putting the scientific names of the floras annotated volume of Indian plants and publications of Forest Survey of India were consulted.

mainly used for preparation of simple and complex medicines. In such medicines water is used as base. They also use honey, oil, and animal fat as base material while preparing the complex medicines. Some other medicines they use as body adoration like they wear some charm objects is also kept close to skin to facilitate cosmological effects and enhance the will power of the patients to survive. '*Mantra*' (chanting of sacred hymns) is another form of medicine prescribed often by tribal medicine man with unusual dance, scream and twist of limbs to the tune of traditional musical instruments. Illness in most cases is attributed to the actions of malevolent spirits, who may be ancestors who have not been sufficiently appeased or to the temporary withdrawal of soul substance from the body, etc. Shamans are frequently called in to divine the cause, often with the aid of their tutelary spirits, and to affect a cure through the sacrifice of a fowl, goat, or other animal.

ANNEXURE

Case studies on Child care

Indigenous knowledge is local knowledge, that is unique to a given culture and society. This knowledge is the information base for the society. Codified in the language of the society, it facilitates communication and decision making. Indigenous knowledge is dynamic; it changes through indigenous creativity innovativeness as well as through contact with other knowledge system. Indigenous knowledge is passed down from generation to generation usually by word of mouth. Today because of its oral tradition, as well as the introduction of new technologies, the preservation of indigenous knowledge is at risk unless it is documented. Being inspired by the uniqueness of the indigenous knowledge, the author tempted very much to collect some of the indigenous practices prevailing among the tribals of Santal and Munda tribe in Similipal Biosphere area of Mayurbhanj District of Orissa, relating to child care practices.

PRACTICES FOLLOWED FOR HEALTH CARE OF SANTAL AND MUNDA CHILDREN IN VILLAGE GARH SIMILIPAL AND SARUDA

1. Stem of Ocimum sanctum (Tulsi) with garlic petals and purified Kochila seeds (Strychnos nux-vomica) are worn around the neck or waist of Santal babies to prevent against cold, cough and evil spirits. Juice of tulsi leaves with honey given to children for cure against cold, cough and fever (essential oil of garlic and alkaloid

of Kochila seed probably have the medicinal value against diseases of children).

2. Against the viral infections like chicken pox, small pox and measles, seeds of Harida (Tarminalia chebula) and root of Moti-saga and Cannabis sativa tied to left or right arm of children in black thread during spring season particularly on Saturday or Thursday as a preventive.
3. The petiole of betel leaf smeared with castor oil and applied to the anus of the constipate baby to facilitate motion, i.e., passing of stool.
4. Against severe headache, acute fever and for reduction of body temperature, the paste prepared out of Dalchini (Cinamomum xylanicum) is smeared on the forehead of the baby. It gives an instant relief to the child.
5. Gall stone of a cow called "Gorochana" is used as a medicine. It is added with mother's milk or honey and given to the children to cure indigestion, hiccups, flatulence, and liver malfunction. It is a very common practice.
6. "Jastimadhu" i.e. stem part of Glycyrrhiza glabra is roughed on a clear polished stone and given with "Gorochana" for prevention against indigestion and used as a carminative, improves vigour, voice and skin colour of Santal children. It is also given with honey to cure jaundice. Decoction prepared with milk and sugar given to weak babies and anemic babies' increases iron content of the blood.
7. On an auspicious day namely, "Bata-Usha" the children of the household and neighbourhood are given gentle strokes with Bajramuli plants (Sida cordifolia). It gives as it is believed, longevity and vigor to young child and delays the aging process. The root powder, the seeds and the decoction of the entire plant also if taken continuously with lukewarm milk for a fortnight gives miraculous recovery of liver weakness. It was collected with a religious rite so that during "Bata-Usha" time it is cheaply available which we have perhaps forgotten and currently it has been reduced to gentle stroke.
8. A very useful plant to the tribals called "Puruni" (Boerhovia diffusa) is used in different forms. The entire plant especially the root are ground and given to the children with food for

rejuvenating the body cells. The leaf juice with honey, given to children for curing body weakness, cough, jaundice and liver troubles.

9. For the children suffering from the loss of memory, low intelligence, sore throat, hoarse tone, an instant improvement has been marked by chewing the rhizome of the plant called "Bacha" (Achorous calamus) and taking its powder with honey or lukewarm milk at the time of full lunar eclipse.

6

Folk Music and Dance: The Rhythm of a Cultural Milieu in Similipal

Amitabha Mitra

Anthropological Study of Music

The importance of folk music as a part of culture is universally accepted. The study of folk music in appropriate social and cultural context provides scope for understanding the culture of a people. Although in the ethnographic reports from all over world we do have some description of the musical aspects of culture, but systematic Anthropological study of music is a relatively recent development. Mention may be made here of the contributions of Kurath (1964), Lomax (1968), Merriam (1967), Nettl (1956), Slobin (1976), and Raymond Firth (1990). Kurath's "Iroquois Music and Dance" is an ethnological and choreographic description of various musical and dance forms relating to agriculture, hunting, rituals, war, etc. of Iroquois tribe. Lomax's "Folk song style and culture" is based on the song style and reinforces certain important aspects of social structure. Merriam, in his book "Ethnomusicology of Flathead Indians" has dealt with the problem of aesthetic and regionalism in the folk music of Flathead Indians. Nettl has analyzed and classified the traits of many different primitives of Africa and North America in his book "Music in primitive culture". Slobin's "Music in the culture of northern Afghanistan" presents a brilliant analysis of Pathan's music. Raymond Firth in his book "Tikopia song" has discussed the folk music of Tikopia islanders with an insider's view of their socio-cultural and socio-economic life.

In traditional folk and tribal societies of India, the importance of music as a media of communication and presentation of culture is more

Ethnic composition of the Study Villages

Ethnic Composition in Study Villages				
Block	*Gram Panchayat*	*Village*	*Community*	
			Tribe	Caste
Jashipur	Gudgudia	1. Gudgudia	Kol/Ho	Goud
			Kharia	Lohar/ Bindhani
			Mahali	Teli
			Bhatudi	Kaibarta
				Sodgop (Gobora Caste)
		2. Nenjaghosra /Lanjighosra	Kol/Ho	Nil
		3. Kumari	Bathudi	Goud
			Bhumij	
	Barheipani	4. Kuljhari/ Koljhari	Kol/Ho	Nil
		5. Chakundakocha	Kol/Ho	Nil
	Astakunar	6. Garh- Similipal/Gad Similipal	Kol/Ho	Kudmi Mahato
			Munda	Goud
			Santal/Majhi	
		7. Saruda	Munda	Nil
	Paudagarh	8. Durdura (Durdura Makdiya/ Birhor Tanda)*	Makdiya/Birhor	Nil
Bangiriposi/ Bangriposi	Sarispal	9. Basilakacha/ Basilakocha	Santal/Majhi	Nil

* This is one of the Makdiya/Birhor settlements located in transitional zone of the SBR.

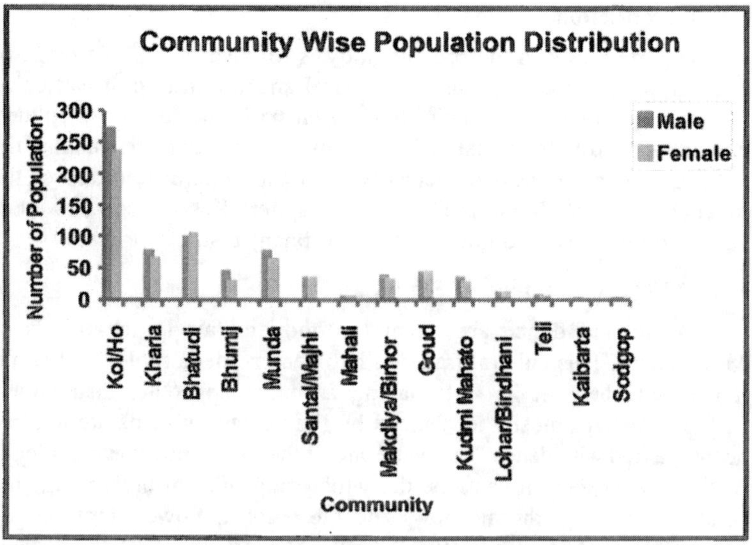

significant. But, very little attention appears to have been paid to this aspect of music. However, Sinha (1957), Bhattacharya (1968), Ray (1973), Vatsyayana (1976), and Onkar Prasad (1987) has taken the study of folk music in India. Surajit Sinha was the first Anthropologist in India, who has explored the expression of sentiments in songs associated with the Karam festival of rural Manbhum in Bihar (Man in India, vol. 37(1), pp. 34-48). Bhattacharya, in his book "Ethnomusicology and India"has tried to show a correspondence between the level of socio-cultural development of different tribes and that ot their music. Ray's "Music of Eastern India" is a study of the traditional and modern folksongs of Assam, Bengal, Manipur and Orissa. Vatsyayana, in her book "Traditions of Indian Folk Dance" has presented the vital links between the folk and classical traditions, and its socio-cultural significances of the folk dances. Onkar Prasad's "Folk Music and Folk Dances of Banaras" has laid emphasis on the theme of Folk-Classical continuum.

Methodology

The traditional Anthropological tools are applied for the study of folk music in Similipal Biosphere Reserve. Primary data are collected through interviews, case history and through participant observation with the help of recording of songs, dance drama, sketches, photographs of folk musical instruments, performing dances. Secondary data are also collected from published materials.

(1) Introduction

The main task in the present study is to cover the whole range of ethnographic context of music in general and folk music in particular. It is assumed that a systematic study of music in the social and cultural context will provide special insights. This will help in understanding the pattern and the process of persistence of the cultural tradition of the local communities living in Similipal Biosphere Reserve and as well as surrounding areas of Similipal in Mayurbhanj district, Orissa.

Folk Music and Dance of Similipal

Music and dance are equal to food and air to the tribals of Mayurbhanj. This cultural tradition had been so deep rooted that so far it had not shed off its style having around many other distractions. Tribal songs are mostly influenced by the versatility of nature and are accompanied with dance. The twittering of the birds, movement of clouds in the sky, change of seasons, the whispering of cool breeze, rippling sounds streams in the meadows and the colorful flowers representing every aspects of beautiful nature offer the background for all their songs. Many of the songs are seasonal in frame and spirit. Accordingly the *raga* and the *taal* changes with the change of the season. The *adhanti* songs of marriage, the *yadur* of the spring, the *Karam* of the summer and monsoon are famous in the Munda tribes. The Santhali's favorites are '*doda*', *lagde sohrai, dayat riza* and *dhanta* etc. The Ho's sing *wa* in the spring and *hairon* during the harvesting seasons, and *karam* dance during Summer.

Santhal : The santhal traditionally accompany many of their dances with two drums: the *tamak* and the *tumdak.* The flute was considered the most important Santhal traditional instrument and still evokes fellings of nostalgia for many Santhal. *Banam* is the only string instrument used by the Santhals of Similipal which has two strings for simple sound makeing. Santhal music and dance both retain connection to traditional celebrations. The Santhals are musicians and dancers par excellence and have dances for every imaginable occasion. The martial dances-*golwari* and *paikha* are marked with vigour, virility and a lot of jumping and leaping. They carry bow and arrows while doing martial dances and perform mock fights and attacks. Their courtship and marriage dances are typical. These dances are romantic and lively in nature, are performed on full moon nights. The loud drumming, resembling thunder, calls the bells of the community and they come dressed in their fineries, adorned with flowers, feathers and assemble under a large tree (banyan/

mango/*mahua*/*sal* etc.). The young men come forward taking strides with drums and lilting songs of their lips, and then the dance commences in two rows, their arms interlinking in pairs. The rows surge forward like rhythmic waves and then recede with supple foot work and swaying heads and bodies. The boys in the row opposite play on flutes, drums, and large cymbals and sing songs in perfect harmony. They have their hunting and sowing dances. On *dassai* festival men folk dance from one locality to another. Then there are the *jhika* dances in which men and women dance together. Men form the outer ring and the women the inner circle. The *dhang* and *langi* are exclusively confined to women. The *largen* has many forms and variations according to the occasion, be it a marriage, a festival or social gathering. All these dances reflect their collective nature, cohesion, community feeling and social awareness. They are great spontaneous collective singers and dancers. The Santhal women and girls can be seen singing and dancing while engaged in there daily chore like sowing plantation, journey to and from the forest. They work and sing simultaneously and in between pause for a round of dances. They use song and beating of drums as a convenient tool of dancing. Dance is a super ordinate and all the rest is subordinate as they are the gay, colorful and enjoy every moment of the life for music.

The names of many Santhal tunes are-derived from the traditional ritual with which they were once associated. *Sohrae* tunes, for example, were those sung at the *sohrae* festival. The Santhals love to dance and music in each and every festival. The Santhals celebrate *sohrae* as their biggest festival. It is a harvesting festival, observed from the mid December to the middle of February. They also celebrate festival like *Baha utsav, ashare parav, fagu parav. Karam* festival is celebrated by the Santhal in the month of *Aswin* (September – October) in order to have increased wealth and progeny and to get rid of the evil spirits. During this festival, two youths after being purified, fetch two branches of Karam tree from the forest and plant them just outside the house. This worship is followed with singing, dancing and playing of drum. Santhal women dance with traditional dress which is called '*pandhat*' covering from the chest to the foot. The creation of the songs of the Santhal remains anonymous. As soon as a new song appeared, it becomes common property. There is no distinction made between the performers and the composers.

Bhumij : The Bhumij perform a ritual dance called *jadur* to propitiate their god *buru bonga.* They celebrate the occasion on the nearest hill top and offer a kind of rice beer to the mother Earth followed by a gay community dance by men and women crown the hill to the village. Such simple dances to propitiate the deities are common to all the tribes. Dancing is a most convenient, satisfying and re creative medium, through which they can express their love of rhythmic movement, commemorates their hunting expeditions or invokes the blessing of the deities. *Jhumur* is another favorite song among the Bhumij. They also observed *karam* festival and the entire community celebrate it with full of joy and passion. They sing and perform dance during the celebration and it lasted the whole night.

Kharia : The Kharias of Mayurbhanj have no specific language. They speak Oriya. Most of their songs are in Oriya language. They sing some songs in their mother tongue. Although they sing different songs on different occasions *raja* and *makar sankranti* are the most important ones. During raja festival they sing *doli* song and on the occasion of *makar sankranti* they sing *changu* song. Most Kharia songs are meant to be sung at the seasonal dances. Only a few of them are not accompanied by dancing. Most of their marriage songs are called '*durang*' besides, some songs are sung when they are involved in their games.

The subject matter of the Kharia songs relates to love and joy and sorrows of lovers. A joyful appreciation of the beauty of flower and birds would be marked in a large number of Kharia songs. The songs reveal the real soul of the tribe, perhaps even more directly than their customs and institutions do. In order to enter into the spirit of the songs with their usual accompaniment of dance and music in the natural surroundings. The songs of the Kharias are :

Marriage songs – At each stage of a marriage they sing different songs.

Doli song – Doli songs are sung during the Raja festival, the boys and girls sing this song sitting on the swing made of *siali* creeper.

Changu song – The boys and girls while dancing sing this changu song. They dance with step according to the different lyrics of a song.

Jhumar – These are sung on the occasion of *karam* festival. This is a mixture of both Oriya and Bengali language. Karam festival is not the festival of Kharias. They have borrowed it from their neighbouring Kol / Ho community.

Bathudi : The Bathudi tribes speak in Oriya language and love to sing modern Oriya song and use modern musical instrument as accompaniment in song. Only exception can be found among them in traditional dance during the *Atharo deol* festival. The *Atharo deol* festival is a traditional festival of *changu* dance. *Changu* is a rural variety of tambourine. It is played by the Bathudi. In Similipal, other tribes like the Bhumij and Kharia also use to play the *changu* in there respective festivals. The *changu* dance of the Bathudi is performed by women folk alone. The men only sing the songs and play on *changu* and move with the female dancers with simple steps, while women advance, the males recede back and on there advance the females retreat. In between males perform vigorous stunt in which they leap into air and make circling movements. Peculiarly enough the women cover up the males with long local made sarees, only their bangles and hands and feet remain visible. In a group the female dancers performed dance in a half sitting posture with swaying and sometimes in jerky movements. The dance continued from evening to dawn on the moonlit night.

Kol / Ho : The Kol / Ho of the Similipal are very fond of their traditional music and dance. Karam dance is the most favorite dance to them. Karam or karma literally means 'fate'. This pastoral dance is performed during the worship of the god or goddess of fate (*karma devta and karamsani devi),* whom the people consider the cause of good and bad fortune. It begins from *Bhadra Shukla Ekadashi* (11th day of the bright moon of the month of *Bhadra*) and lasts for several days. Karam is also popular among the Kharia and Bhumij of the Similipal. In the afternoon of the auspicious day two young unmarried girls cut and bring two branches of the *Karam* tree from a nearby forest. They are accompanied by drummers and musicians. The two branches are then ceremonially planted on the alter of worship and symbolize the god. Germinated grains, grass flowers, country liquor are offered to the deity. After completing the ritual the priest tells the story or legend connected with it. This is followed by singing and dancing in accompaniment of drum (*madal dhamsa or nagara*), cymbal etc. The dance performance is full of vigour and energy combined with charm of the youth decked with colorful costumes and ornaments brings the onlookers as well as the performers to a mood of trance and ecstasy. In this dance both men and women take part and continue to engross themselves for the whole night. The skillful movement of the young boys with mirror in hand indicates the traditional pattern of love-making in course of dancing and singing. The dance is performed sometimes by boys in group, sometimes by girls in group and sometimes both the sexes together. The subject matter of songs constitutes the description of nature, invocation to *karamsani*, desires, love, and aspiration of people. The *karam* dances continues from dusk to dawn. In the early morning they carry the *karam* branches singing and dancing and them immerse them ceremonially in a river or tank and then disperse.

The technique of the *karam* dance is a circular pattern, where men and women dance together. It is always léaded by a leader and generally the men at the head of line. Only the best dancers join in right next to or near him. Very young girls and children join in at the tail end to learn the skill. The dancers hold hands in different ways in different dances. Sometimes they simple hold hands and sometimes hands are placed on neighbor's waist band or are crossed. It is the legs and the feet which play the principal part in the dance. The dance begins lightly with simple step forward and backward, left and right, then gradually the steps grow smaller and faster, growing more and more complicated, until that dances reaches its height. Then it goes gradually to the first steps as the music leads to give dancers rest. The dancers have no special dress they dance with their usual costumes which they were daily.

Munda: The Munda tribe of Similipal accepted Christanity. It reflected on their music world. They left their traditional tribal song and accepted the songs praising Jesus Christ. They assembled in Church on every Sunday and sang the prayer song. During prayer song they do not use any musical instruments. Traditional musical drum instruments like *dhol and nagara* are used only in the marriage ceremony. But the songs are mainly composed by the Church authority. Song texts are meant to be a happy married life of the couple and they get a blessing from all mighty Jesus Christ. The marriage dance is usually held in the courtyard of a bride's house which is decorated with bamboo split arch with colorful flowers and mango leaves.

Chhau dance of Mayurbhanj: Tribals of North-Eastern Orissa have a special martial dance form known as *Chhau.* At the leisure hours of their life they were performing and enjoying this dance, which was perhaps first practiced during battles at nights and was therefore named "*chhauni nritya*". But it was subsequently adopted by the relatives of maharajas in a sophisticated style to serve the purpose of elites, which was mostly performed at the time of "*chaitra parav*" a festival of the ex-states of Mayurbhanj, Kharsuan and Seraikela being patronized by the rulers. A number of ex-ruler states before their merger in Orissa state where practicing *chhau* dance and some prominent among them are – *chhau* of Baripada in Mayurbhanj, Asanpal in Keonjhar, Nilgiri in Balasore, Arnapal in Bhadrak, Bonei in Sundargarh, Talcher in Angul, Balaramprasad in Dhenkanal, Bhubaneswar in Khurda district. Traditional Mayurbhanj *chhau* has lived through *Uttarshia* and *Dakshinsahi* since time immemorial. Dancers of both these segments worshipped *bhairava,* an aspect of anger mood of lord Shiva and perform during *chhau* festival, every year which also signifies the Oriya new year i.e. 14th April *panasankranti.* The dance was at first performed with masks but Mayurbhanj adopted these dances without masks. This gave scope for transformation of the erstwhile tribal dance to a semi classical dance form and with the passage of time it has now enriched itself to be called a classical dance, the glory it richly deserves.

The *chhau* festival of *chaitra parav* also comprises two other rituals i.e. Bhaktas and Udajatra which signal the completion of chhau festival. The unique association of *Bhakla nata,* with *chhau nata* is usually presented in one stage in honour of lord *Bhairava* and *Shiva* respectively. The devotees of lord *Shiva* maintain strict discipline by lord *Shiva* maintain strict discipline by fasting and devotees of *Bhairava* offer sacrifice of goat for having extra energy for the martial chhau dance.

The basic structure and the style of *chhau* get itself modified in course of time according to distinct style of its own with its religious culture and local environment. Prof D.K. Sarangi opines, "*chhau* is undoubtedly a tribal Orissan art. This famous art form not only originated in a predominantly tribal belt but that the basic posture, the accompanying music and the aesthetic mood of the *chhau* quite often resemble different dances and songs of these tribes".

Mayurbhanj *chhau* is performed in solo, duet and group. The subject matter has been brought thematically and substantially from the Ramayana, Mahabharata, Krishna cult and the purans. In the beginning probably it had no themes but subsequently a thematic base was provided to this dance form. *Chhau* in its present form is a fine blending of various systems. Elements of martial, folk, traditional and classical art have been nicely woven into grand mosaic of *chhau.* But element of martial art is absent from *Odissi.* Vatsayan opines "Thus the dance style is folk form the point of milieu and environment but is classical form the point of view of body technique and its internal structuring".

Martial form of *chhau*

In Mayurbhanj *chhau,* there are uses of *Hindustani, Odissi, chaupadi, chhanda, champu* etc. and the old forms i.e. tribal and *jumuras.* There has been such a beautiful and fine assimilation of classical and *jumur* in Mayurbhanj *chhau,* that enhancing its beauty, grandeur and performance, it has maintained it distinctive style without any decorative masks. Rather the dance has become more attractive and enjoyable. There is no extensive and wide use of songs in Mayurbhanj *chhau,* but the limited uses of songs have come down from generation to generation without any knowledge of the composers.

A close looks reveals that the *chhau* dancer builds movements from a variation of *tribhanga.* This posture of *tribhanga* and *chauka (*square), the open grand pile of *mandalasthana,* are quite akin to those motifs of *Odissi* dance. Sanjukta Panigrahi mentions that "the *tribhanha* of this dance which is absent from other Indian classical dances, has been imitated from daily life of Oriya women folk not from any *nritya shastras.*" This is what happens in case of Mayurbhanj *chhau* in respect of movements.

The facial expression of *chhau* dancer of Mayurbhanj where masks are not used, are not as important as in other Indian classical dances. Even the hands and fingers do not play a major role. One can notice identical *mudras* and *karanas* in both *Odissi* and *chhau* dance under different titles. In both the dances one can also observe same intricacy of foot work, walks, poise, gaiety, whirr and jumps. The artistic composition, music and graceful performance in *nishith milan, maya savari, garuda bhahana* dance of *chhau* can be noticed in their basic stances, *chauka* and *tribhanga.*

Mayurbhanj *chhau* is the cultural heritage of Orissa for its own stylistic movements. It is different in style from Seraikela and Purulia *chhau* dance. It has various movements of eyes, face, body and legs. In this Mayurbhanj style of *chhau* dance, the artists use various *mudras, abhinaya* and *navaras.* It has the classical elements, but till date it has not yet been recognized as a classical dance form. The Mayurbhanj *chhau* has six (6) *tabaka,* thirtysix (36) *ufuli* and two hundred sixteen (216) *upa-ufuli.*The dance is learnt in a *guru – sishya parampara* for a minimum period of six years to complete the training in all aspects of dance form. The chhau dance of Mayurbhanj has also adopted some other dance styles and forms by imitating *kathak, bharatnatyam, kathakali, pharkanda kela, amdalia jamdalia, jhumar* and Udaya Shankar Oriental dance forms. In the 18th century the chhau dance was modified by late Krishna Chandra Bhanjdeo.

Mayurbhanj *chhau* does possess not only classical *raga* and *ragini* but also definite *taal* and *laya.* One finds extensive use of *taals* like *kaharba, ektal, dadra, rupak, jhamptal* of Hindusthani classical music. The use of techniques like *topaka, ufti,* and use of *mudras* in Mayurbhanj *chhau* take the dance from on the road of classical touch. In the words of Vatsayana,"once enter into the dance style itself we find that it has shed almost all the features of tribal, folk and village dance and has acquired a distinctive stylization which is a characteristic feature of the dance forms commonly termed as classical".

The musical instruments like *dhol, mahuri, chadachadi, dhumsa* are the main musical instruments used in *chhau.* These musical instruments are primarily used by the tribal people in their dances. The music of *chhau* at the *nataki* stage i.e. final stage is that of *jhumar,* is a folk music. The vocal songs at times used in Mayurbhanj *chhau* is also written in folk lyrics. It is clearly known that Mayurbhanj *chhau* has acquired dance styles of different dance forms, some of which are tribal dance forms. The dances have freely imbedded the techniques of prevailing folk dances such as *chadheria, patna, kathi, changu nacha,* etc. Mayurbhanj *chhau* is a unique combination of the characteristics of both folk and classical dances.

The *chhau* dance of Mayurbhanj is not at all inferior to any classical dance of India and no doubt it is similar to Odissi dance. Form the comparative study of both the dances one may draw conclusion that Odissi and *chhau* are only dances style of India, which without being too much sophisticated still return a unique prestige, simplicity combined with graceful posture and ritual sanctity. The only difference between the two is that, Odissi represents the unmutilated form of '*lasya*' dance, where as the *chhau* represents a form of martial art. The style, posture, gesture, dress arrangement, presentation, musical tone and rhythm are highly excellent and entertaining in both dances of Odissi and Mayurbhanj *chhau.*

Jhumar: The *jhumar* is a most popular song style in Mayurbhanj. It is also known as Mahato song because Kudmi Mahatos are specialist in this form of song. *Jhumar* is prevalence in Mayurbhanj, Keonjhar, Sambalpur and Sundargarh districts of Orissa, Purulia, Midnapur and Bankura districts of West Bengal and Singbhum, Ranchi, Hazaribagh, and Palamau districts of Jharkhand state. The languages of the jhumar song are a mixture of Oriya, Bengali, Hindi and Chhota Nagpuria. But those are not in pure form, but in colloquial language as practiced in these regions.

Jhumar song on marriage

Chhau dance has a special position in Mayurbhanj. As jhumar is inter-linked with *chhau, jhumar* enjoys distinctive position too. Unlike chhau, *jhumar* is performed by female dancers. *Chhau* is mainly in dance form along with its music, but *jhumar* is mainly in song form along with dances. When the *chhow* dance reaches its climax the *jhumar rangas* are played. There are two types of dances in *jhumar* dance. When a single lady dances, it is called *nachuni* and when dance is performed in a group it is called *panta* dance. *Jhumar* songs are tune based. Its main element is tune, that means *swara pradhan* and its rhythm is typical of its own. Lyrics are descriptive mainly of Radha Krishna's love, melody about separation from beloved one and from mythological stories. One of the main lyricists in *Jhumar-Kabi Bhushan* Bhabaprita Nanda Ojha named his first edition of his book *Jhumar rasmanjari* published about 100 years back contains mainly valuable famous *jhumar* songs. He was from old Manbhum district.

Dancing in a circle to the rhythm of *jhumar taal* in a group of 10 to 15 maidens holding one another side by side is called *panta naach* (Dance). Male folks sing and play *dhol*, *maadal* and flute while female folks dance to the tune and *taal* putting their feet in unison to form a chain like scenery which is beggar's description and throws the audience into trance. Maidens also sing *jhumar* songs while dancing. The dancers are so engrossed in their performance that they cannot know when evening passed by and night came and night made room for morning.

The prevalence of *jhumar* dates back to 18th century but not in organized form. Local Kudmi Mahato and other tribal folks of Orissa, Jharkhand, Bihar and West Bengal used to practice *jhumar* songs and dance to beat boredom and monotony after day's hard work which was energizing for further work. Even when they were on their way to distant places in quest of work their lips were stuck with melodious lyrical and colloquial *jhumar* songs. Gradually a *nachuni* group was introduced which consisted of eight to ten people, including one *rasika* (the leader of the group) one *nachuni* and instrument players for flute, harmonium, *dhol, chad-chadi, maadal, mahuri* etc. 18th century downwards *Raja , Maharaja, Zamindar* noticed *jhumar* and they became so charmed by the language, sweet tunes and dance that they could not but patronize it which helped remove the financial constraint of *nachuni* groups which further paved its way to earn recognition in upper level of society. A *nachuni* practically lives with *rasik*, learning all types of *jhumar* songs and a dance performs with the professional group. They are frequently called for performance in festivities and marriages. The group moves from place to place for performance in return they charge certain amount. The *nachunis* perform specially *jhumar* dances and songs. The *rasik* (leader) selects talented girls, and other interested artistes to form a group and they are imparted training for professional performances whenever they are called for. The calls are proportional to their efficiency.

Jhumar song in *chhau* dance

Each dance and songs are accompanied by some musical instruments but for which the real essence of that would not be enjoyed. Similarly *jhumar* songs and dance are performed with instruments like *dhol, maadal, chad-chadi, ghungroo,* bamboo flute, harmonium, *mohuri, nagara, judi nagara, tuila* etc. The classified types of *jhumar* songs are mythological songs about Radha-Krishna's sacred love, mythological stories in song form mainly from epic Ramayana and Mahabharata, *bhakti bhava* or devotional songs, love songs treating social themes concerning common people, social subjects, sorrow and happiness of the people and songs related to various festivals such as *makar sankranti, karam ekadashi, raas purnima* etc.

Conclusion

Kol/Ho: Though they are migrated from Jharkhand, they maintain their own traditional culture in respect of music The song related with festivals like – *makar parav, magho parav, baha parav, jamna parav* and *karam*. Musical instruments are mainly *dumeng* (a drum); *dhamsa/ nagara* (single membranous drum) Young boys and girls take part in dances. Mask dance is also found among them. Ritual songs like marriage song are also reported.

Bathudi: *Changu* a Traditional dance of Bathudi. *Changu* a single membranous drum played by men. Male members only sing the songs and young boys and girl dancers' move with simple steps. It is performed during *Atharo Deol* festival. Apart from this; Bathudi tribes are largely followers of modern Oriya music and use modern drum instruments.

Munda : The Munda tribe of Similipal accepted Christianity. It is reflected on their music world also. They left their traditional tribal song and accepted the song praising Jesus Christ. They assembled in Church on every Sunday and sang the prayer song. Musical instruments are traditional– *dhol* and *nagara.*

***Nagara*, Munda**

Santhal: The Santhals of Similipal are both Hindu and Christan. They follow their traditional songs related with rituals and festivals. Santhals those who are accepted Christianity, sang the prayer song at Church on Sunday only. *Tumda* (a drum) and *banam* (a string instrument) are the main accompanying instruments.

Chhau dance : *Chhau* means six faces – fore head, eyes, nose, cheek, lips, chin and a mask bears the six parts of the face. *Chhau* dance depict stories from myths, history, folklore, etc. Three styles of *chhau* exist, originating from Seraikela, Purulia and Mayurbhanj. In Seraikela *chhau* the mask is the center of attraction. Purulia *chhau* with towering head gears (mask) have a very dramatic influence. Mayurbhanj *chhau* does not use any masks. The movements are quite vigorous maintaining the basic spirit and style of *chhau* dance. The Mayurbhanj *chhau* is a beautiful mosaic comprising elements from folk, martial and classical art. Mayurbhanj *chhau* has become popular as a medium of choreography, with its wide range of postures and movements that adapt well to modern as well as traditional treatment.

Jhumar: The prevalence of *jhumar* dates back to 18th century as local folks of Orissa, Jharkhand, and West Bengal among the Kudmi Mahato and tribes like Bhumij. The language of *jhumar* songs are mixture Oriya Bengali and Hindi in colloquial form as practised in this region. *Jhumar* is mainly in song form along with dances. Male folks sing and play musical instruments, while female folks dance to the tunes. When dance is performed by single lady it is called *nachuni* when dance is performed in a group it is called *panta* dance. Its main element is tune and its rhythm is typical of its own. Lyrics are mainly Radha Krishna's love, melody about separation from beloved one and from mythological stories. *Jhumar* songs and dances are performed with *dhol, maadal, ghungroo*, Flute, *mohuri, nagara*, etc. Jhumar songs are also related to various festivals such as Makar Parav, Bandana Parav, Karam Parav and Raas Purnima. Jhumar song is also known as Mahato song in similipal and it is regarded as the most popular song in Mayurbhanj district of Orissa.

7

Tourism in Similipal : An Anthropological Observation

Rabiranjan Biswas

According to world Tourism organization 1995, tourism is reputed to be the world's largest economic sector. Throughout the world, the governments of different countries try to enhance tourism to earn sufficient revenue. Tourism get impetus to national income and tourism development has a great impact to community development, their economic benefits and their employment. Tourism is an excellent revenue earner and also a significant source of foreign exchange.

Sharma (2002) stated that the Geography of Tourism development comprises a complete system of nature - the universe, the space and their galaxy i.e. the men and their activities. The tourism industry affects with both positive and negative impacts. The tourism has a positive contribution to the local people for sustainable development and creating various employment opportunities. On the contrary, any development work somehow causes environmental degradation, which is negative impact of the development.

According to Chawla (2003) "the word 'tourism' is typically used as a single term to designate a variety of concepts, partly because it is an area of study in a range of disciplines (geography, economics, business and marketing, sociology, anthropology, history and psychology), and the different conceptual structure within these disciplines lead inevitably to contrasts in perspective and emphasis."

According to Pigram (1997 op.cit. Wahak), 'Tourism certainly can contribute to environmental degradation and be self- destruction; it also has potential to bring about significant enhancement of the environment'.

Tourism can spread everywhere across the geographical and ethnic boundary. Thus, a tourist becomes an integral part of tourism development. Tourism helps to foster a climate of piece and prosperity by bringing together the peoples of different cultures and countries and can encourage international understanding. The tourism as well as the local people can promote understanding to each other and make friendly atmosphere among both national and international masses.

The British were responsible for the large-scale deforestation for earning revenue, in spite of being aware with 'Forest Preservation' and implemented various 'Forest Policy' to save the forest. Afterward, different conservation strategies and safeguards have been taken to save the wild animals, avoid soil erosion, water resource management, etc. Later, different forests of India have been declared as the 'Reserve Forest' and the 'National Forest'.

Total area of the Biosphere Reserve to some extent also includes human settlement. Every Biosphere zone, consists of three separate regions namely Core area, Buffer area and Transitional area. Out of total 14 Biosphere Reserves, Similipal Biosphere Reserve (SBR) is one of the major Biospheres in India.

It is stated that since 1992, environmentalist were pressuring on tourism industry to boost its environmental performance in common with other economic sectors and to work towards ecologically sustainable forms of tourism development, Wahab (1997). As a result the concept of ecotourism has developed. Ecotourism is nature based tourism and ecologically sustainable. Different geographical and ecological diversity is a treasure for the ecosystem in India, which is well preserved and well protected. Bande (2006) stated that ecotourism is an amalgam of two separate concepts – ecology and tourism and viewed jointly as ecological conservation and tourism development. According to Binu R. and K. Sasikumar (2006) "ecotourism is a purposeful travel to natural areas to understand their natural and cultural history". Thus conceptually ecotourism means a type of tourism involving nature and men and not to endanger it amidst of wild life and plants. Keeping this view, the Government of Orissa has been started eco-tourism destination in Similipal Biosphere Reserve.

According to the naturalist and environmentalist, there should be at least 35 percent of forest in a county. But in India there is only 22.8 percent area of forest despite being a country of vast species of wild animals including faunal and floral species. The forests are the natural habitat of most of the animals. Therefore, Wildlife Protection Act is much necessary for the preservation of forest and to save natural habitat. Implementation of various Forest Acts has given protection and preservation to the wild animals and wildlife species in various Reserved Forests in India. Wildlife is an important natural resource bestowed by nature and it has positive contribution towards human culture, economy etc. for maintaining the ecological balance.

Importance of Similipal

Similipal is situated in the Mayurbhanj district in Orissa and has the area of 2750 sq. kms. It is covered with thick and green forest varying from evergreen to dry deciduous. South Similipal is also characterized by extensive grassland. Luxuriant sal forests are seen throughout Similipal. The most common trees are Kadamba, Champaka, Mesua, Arjuna, Bahada, Harida, Rosewood, Salai etc.

Similipal is a dreamland of nature and wilderness. It has a variety of wild animals like 12 species of amphibians, 29 species of reptiles, 264 species of birds and 42 species of mammals. The wild animals are the tiger, leopard, elephant, bison, sloth bear, sambar, spotted deer, mouse deer, barking deer, boar, chowsingha, mongoose, pangolin, squirrels, monkeys etc. More of birds are there like fowl, hornbill, hill myna, peacock, grey wagtail, serpent eagle etc. Retiles include python, cobra, viper, krait, lizard, chameleon, mugger crocodile, turtle etc.

Similipal is the richest watershed in Orissa with number of perennial rivers namely Budhabalang, Khadkei, Khairi, Bhadan, West Deo, East Deo, Salandi, Sanjo and Palpala. Out of three beautiful waterfalls Joranda and Barheipani are more attractive to the visitors. The third one Oski is equally magnificent but less in height. The visitors have access to go nearer to the falls at Oski but it is not possible in other two falls.

Similipal is full of beautiful hills and valleys. The pick of the hillocks starts from only 14 meter. Average mean elevation of Similipal hills is 900 meters. There are two highest picks – Khairiburu (1178 meters) and Meghasani (1158 meters). Chahala, Nawana, Dhudruchampa etc. are the important valleys of Similipal.

It is reported that about 4.5 Lakhs people and about 80,000 families live in 4 villages in core zone, 61 villages in buffer zone and 1200 villages in Transition zone and are depended on Similipal directly in many ways. Most of the tribal pollutions like the Santal, Munda, Kol/HO, Bathudi, Bhumij, Kharia, Mankharia or Birhor are very much depended on jungle products like honey, resin, arrowroot, mushroom, firewood, medicinal plants etc. Among the tribals, the Mankharias are traditional monkey catcher and till date they practise their traditional economic habit. The non-tribal populations of Similipal are Gour, Lohar, Mahakur, Patra, Mahato etc. The Kharia and Mankharias are mostly sustaining on food gathering. The traditional tribal hunt *akhandsikar* (group hunting) is organized by the tribals in the month of April. Though the usual practice of hunting and *akhandsikar* by the tribal people has been stopped legally but the practice is going on secretly. The Kharia and the Buthudi are the original inhabitants of the region. The Kol, Santal, Munda, Bihor and others have come to this place later on.

Orissa has the great potentiality for tourist attraction for its rich cultural heritage, geographical diversity of hills and rocks, sea, forest and diversified folk music and folk dance etc. Thus the Orissa tourism is much developed from its beginning. The state government encourages tourism very much and highlights tourism as a global importance to the tourist of Indian and abroad.

Different disciplines like forestry, ecology, zoology and botany have contributed on ecotourism and policies. But their results often ignore peoples' perspective. The social scientist specially the anthropologists have a role to play in this field and give a comprehensive picture for making micro level planning keeping in view the local culture, tradition and customs of the people in context of the biosphere area.

Biosphere and tourism are both interesting fields of study for many disciplines. Therefore, different aspects of these have been studied using different methodology and objectives. As a result, since last 20 years there exists huge documentations of data and production of many educational film and still photographs. These are of great help in planning the development.

Objectives

Tourism has its own objectives in the broad perspective for the management of SBR by the authority. But the objectives of the present study is centering around the tourism industry. The focuss would be given to make Similipal as a very attractive place of tourist interest.

(a) The attempts have been made to prepare the present tourism route map. The map indicates usually which route follows by the tourist or by the travel agencies to enjoy the scenic view of Similipal.

(b) The different tourist spots would be highlighted in the tourist map. The close viewpoint of wild animals and waterfalls are the main source of tourist attraction.

(c) The present work would highlight the tourism perspective of Similipal, its present condition and how far it would be possible to develop in infrastructural facilities.

(d) Explore the present positions of infrastructure facilities available Infrastructure means the basic needs of a tourist for comfortable journey and stay, safety and security, availability of canteen facility, fresh drinking water, medical facility etc.

(e) Scrutining the suggestions given by different authorities (tour operator, shop keeper, hotel owners, NGO's local leader etc.) regarding compatible and sustainable tourism in Similipal.

(f) Enquire the requirements of the tourists; their advices and suggestions taken for better tourism perspectives.

(g) Assessing the potentiality of tourism development in relation to local communities i.e. eco-tourism in Similipal.

(h) The focused has been given to assess Similipal as very attractive place of tourist interest.

Methodology

Keeping in mind the above-mentioned objectives, standard anthropological tools and techniques like observation, interview, scheduled questionnaire, group discussion etc. were used during the fieldwork. A good number of still photographs were also taken. Interview of the tourists visited Similipal was taken, who also helped in filling up scheduled questionnaire.

The Tourist routes along with different tourist spots were thoroughly visited and the specific tourist route map has demarcated. The present infrastructure positions of the spots have been highlighted.

Beside tourist spots, the study villages were also selected from east, west, north and central Similipal on the criteria like close to tourist center (Gudgudia and Kumari village) to look in direct and indirect involvement of the villagers.

Due attention has been paid for data collection and as far as possible and it has been compared as well.

Importance of Similipal as a Tourist Spot

Similipal provides beautiful natural scenario, evergreen deciduous forest with dense vegetables, perennial rivers, magnificent waterfalls, different topographical feature, rocks and hills etc. Wild life is also important natural resource of Similipal. Several Preservations and Reservations Act have been implemented time to time to save Similipal from environmental and ecological degradations.

Similipal is important for a Tiger Reserve which was announced in 1973 but implemented from 1979. In addition to that it was declared as sanctuary in 1979 and also declared as National Park in 1980 and work started from 1986. Similipal was declared as a Biosphere Reserve on 1994. It is also declared an Elephant Reserve in 2001. Therefore, the importance of Similipal is wildly accepted to the tourists who are natural lovers and as well as animal lovers.

According to the information of Orissa Tourism 2005 about 96,450 visitors entered Similipal, out of which 96,265 were domestic visitors and only 185 were foreign tourists. Average tourists visiting at Similipal is around 15000 heads in a year. Statistical information supports the rush in different FRH (Forest Rest House) inside Similipal.

Information required for the tourist:-

- The tourist has to enquire about SBR at government of Orissa Tourist Offices.
- Information on Similipal is also available from web site or Internet.
- For seeking reservation facilities staying in SBR the concerned officials are -
  - (a) Field Director, Similipal Tiger Reserve, Bhanjapur, Baripada.
  - (b) Assistant conservator of Forests, Dhabadiva, Khairinivas, Jashipur.
- Advance booking for staying at Forest Rest House (FRH) inside Similipal and Ramtirtha can be made between 30 to 90 days before at Baripada and 10 days before at Jashipur.
- Entree fee is Rs. 40.00 per day for an Indian tourist and Rs 1000.00 for a foreign tourist. An entry fee for an Indian student is Rs. 30.00 and for a foreign student is Rs. 450.00.
- The rental charge of a Jeep or car is Rs. 100.00 per day. Separate charges also have to be claimed for camera.
- Entry permit issued from the offices of-
  - (a) Assistant Conservator of Forests, Khairinivas, Jashipur.
  - (b) Range Officer, Pithabata Range, and Pithabata Check gate.
- Season of visit: The Sanctuary remains open to the visitors from 1st November to 15th June. Winter is the best season to visit.
- Entry time:

(a) Day Visitors can enter between 6.00 AM to 12.00 Noon.

(b) Visitors with reservation can enter between 6.00 AM to 2.00 PM.

- Malarial prevention measured should be taken according to the advice of Physician.
- A leaflet of details information and rates is available from Forest office.

Other important information to be noticed for the visitors for Dos or Donts do (Informations taken from tourist office)

Guidelines to be followed by the tourist

1. The public should not carry any firearms or weapons.
2. One should not take any pet animals inside SBR.
3. Wild animals are not to be disturbed, teased or chased.
4. Cooking of any type of flesh/mutton is strictly prohibited inside Similipal.
5. Swimming and fishing are not allowed in river, gorge etc.
6. Visitors are not allowed to play transistors, tape recorders etc.
7. Alcohol consumption is totally banned inside the sanctuary.
8. Damage or disturb any birds, animals, and plants are punishable offence.
9. Campfire is not to be organized.
10. Picnic is not permitted inside biosphere area.
11. Use of plastic bags and litter the ground with waste materials are totally banned.

Restricted Guidelines

1. One should be taken proper entry permit to enter SBR.
2. Speed of a car should be restricted within Similipal.
3. Vehicle should be parked in a parking place only.
4. For visiting inside SBR, eco-guide/guide/driver-cum-guide is compulsory in accompany to the visitors.
5. Photography is allowed with prior permission from the competent authority.
6. Accommodation facilities are availed from the appropriate authority.

7. Paper bags are used for disposal of waste materials.
8. Visitors are requested to carry drinking water.
9. Before entering in SBR one should ensure the vehicle with full of fuel.
10. One should always be careful inside the sanctuary.
11. Before entering into Similipal the visitors should be taken malarial preventive measure.

Tour Operators

Tour operator means the individual or the agency or the institution that has expertise in tour operation. The different travel agencies are also engaged as tour operators. There are both government recognized and private tour operators in and around Similipal area. Maximum 60 vehicles (40 from Jashipur and 20 from Pithabata) are permitted to enter Similipal. Thus transportation has its importance to tourism development, public involvement and as income generating medium. Tourism is manpower oriented and also provides varieties of employments. Tour operators, touring agent, transporter, and labours all are involved in tourism. The travel agencies are now more expert and competitive. Now they employ more educated and smart employees. The agencies have now taken much responsibility to make the journey successful; likewise they also arrange the hotel reservations. There are some travel agencies who take all the responsibility of traveling i.e. has to pay money and starts travel from home and returned back accordingly. The all India Travel Agents Association was established in 1952 by the travel agents to safeguard and protect the interest of its members (Sethi 1999).

In Similipal entering of vehicle is restricted as per rule without official guide but it is not followed strictly as the guides are not always available. Therefore, the tourists mostly depend on the drivers as their guide.

There are 10 official guides at Pithabata and 12 guides at Jashipur. The rate of guide is Rs. 80.00 per day and those who are worked as driver cum guide; their rate is Rs. 50.00 per day. These drivers are officially license holders. These guides have their official dress with identity cards. An orientation training programme for the drivers are being conducted with the basic objective, that they will guide the tourist as well as make the trip interesting. The drivers were also making aware about Similipal Biosphere and its conservation. A nature camp has been also arranged since 1993 for the people (18-48 ages) to make them aware in conservation and to gather knowledge on management approach. The office of the field Director, Similipal Tiger Reserve, Baripada, organizes both the orientation-training programme and the nature camp.

Usually tourists hire car by themselves or through travel agents. The type of cars use to travel in SBR are Marshall, Bolero, Travera, Sumo, Safari, Indica, Mahindra Max, Comander, Maruti, Ambasador etc. It is found that the travel agencies mostly preferred four types of cars like Marshall, Bolero, Sumo and Commander to travel the biosphere as per the demand of the tourists.

Rental charges for Sumo, Bolero and Marshall are of Rs. 2200.00 per day and the rate of commander is Rs. 1800.00 per day. These rates are calculated from Baripada. The rates from Jashipur are more or less two hundred rupees less. Except this, there is another rate of car, in which the car owner will charge Rs. 600 per day and the fuel charges will be bourne by the tourist. According to the rule not more than 10 persons/tourists are allowed to travel in a car in the SBR.

There are many tours or travel agencies in surrounding Similipal. Two recognized tour agents are at Baripada – (a) Mayur Tours and travels and (b) J.P. Tourist Service. There are other travels named Sourav Travels, Taj Travels etc. from Baripada, Sahoo Travels from Bombay Chowk. Some travels named Bajarangbali Transport, Krishna Travels; Ashish Agency etc. are at Jashipur and also conduct tour to Similipal. A few travels named Millennium Travels; Chalo Jai etc. also come from Kolkata. A few other Government agencies also come from Bhubaneswar and Kolkata. The outside travelling agents who conduct tours in the SBR have their own guides.

A few hotels of Baripada also contact with the outside tour operators or tour agencies and they inform the operators as and when necessary, about when the tourists want to visit Similipal. Sometimes the hotel owners directly arrange the vehicle for the tourist to travel inside the Similipal.

Tourist Map

Similipal, a tourist spot has a great importance to the government to earn revenue. A significant tourist map is necessary for the advertisement and for the tourist. The present tourist map of Similipal is now confined only within the buffer region. By observing the map one can know about the place. The tourist map always to follow the Forest guideline. The tourist route runs through the buffer region and touches the important spots of tourist interest. In earlier days, some places were included as tourist spots but as these places confined in core region, are now excluded from the tourist spots after declaring Similipal as a National Park. These three routs given below are followed at Similipal for tourist movement.

(a) Jashipur- Gudgudia – Oski – Barheipani – Nawana – Joranda – Chahala – Jashipur.

(b) Jashipur – Oski – Chahala – Barheipani – Nawana – Joranda – Pithabata – Baripada.

(c) Baripada – Pithabata – Lulung – Bhajam – Joranda – Nawana – Barheipani – Chahala – Jashipur.

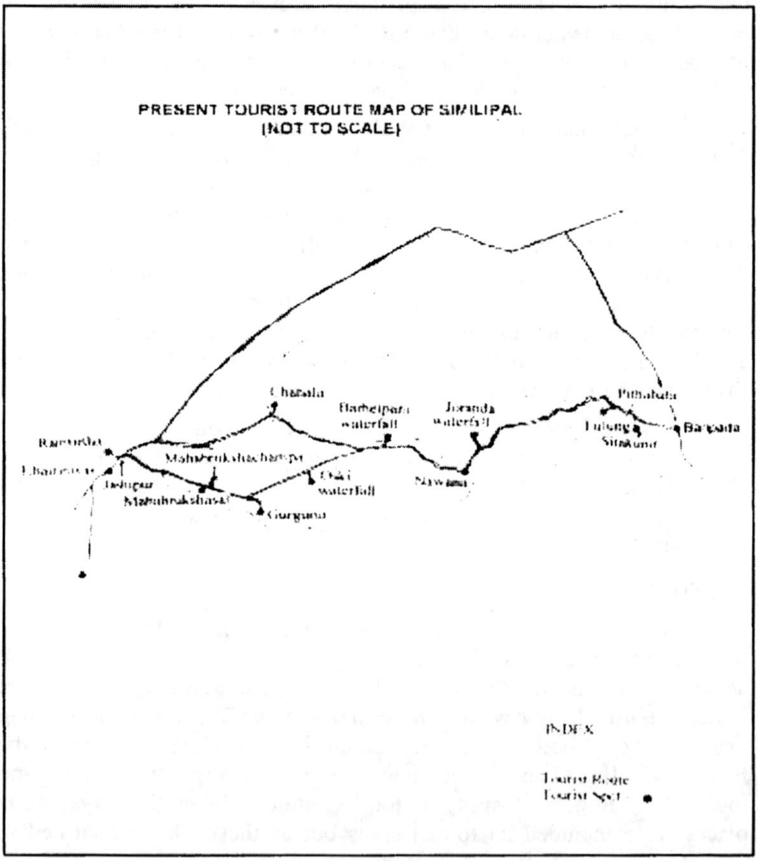

Map 1.

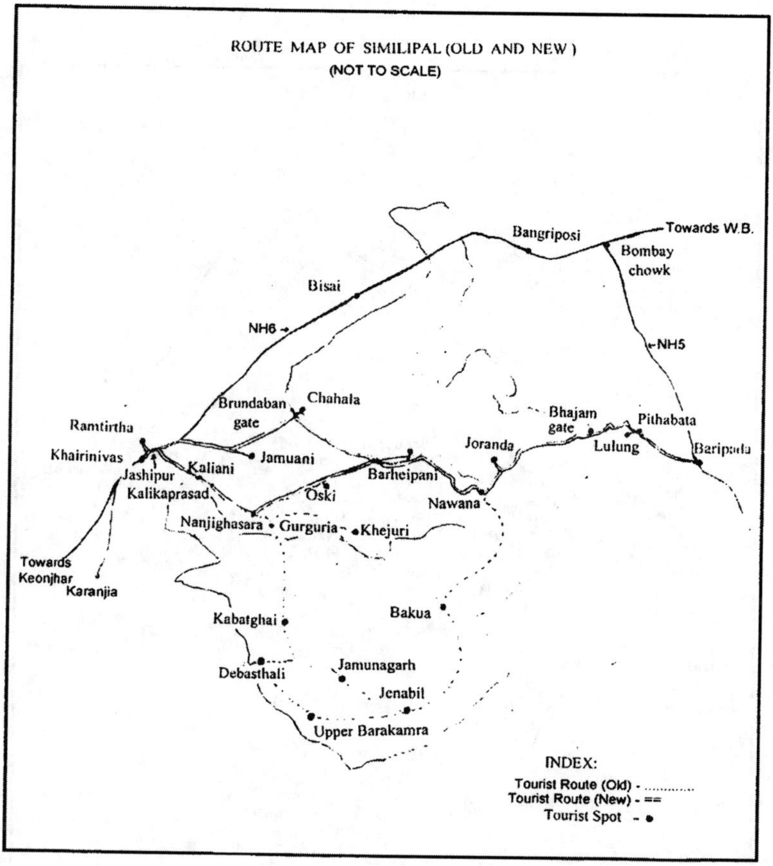
ROUTE MAP OF SIMILIPAL (OLD AND NEW)
(NOT TO SCALE)
Bangriposi
Towards W.B.
Bombay chowk
Bisai
NH6
NH5
Brundaban gate
Chahala
Ramtirtha
Khairinivas
Jamuani
Jashipur
Kaliani
Kalikaprasad
Bhajam gate
Pithabata
Joranda
Lulung
Baripada
Barheipani
Oski
Nawana
Nanjighasara
Gurguria
Khejuri
Towards Keonjhar
Karanjia
Bakua
Kabatghai
Debasthali
Jamunagarh
Jenabil
Upper Barakamra
INDEX:
Tourist Route (Old) - ....
Tourist Route (New) - ==
Tourist Spot - •

Map 2.

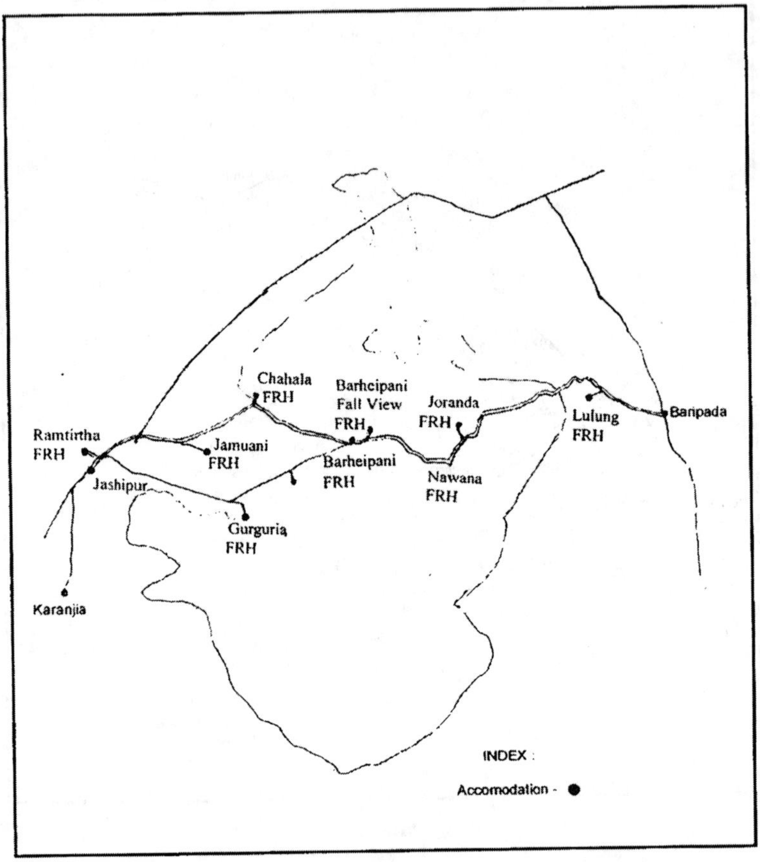

Map 3. Accomodation of FRH at Similipal.

DISTANCE BETWEEN

Khairinivas – Ramtirtha	=2 km
Jashipur – Kaliani	=11 km
Kaliani – Kalikaprasad	=3 km
Kalikaprasad – Nanjighasara	=11 km
Kalikaprasad – Mahabrukshasal	= 4.5 km
Mahabrukshasal – Mahabrukshachampa	= 1 km
Nanjighasara – Barheipani	= 19 km
Nanjighasara – Gurguria	= 3 km
Barheipani – Chahala	= 19 km
Chahala – Jashipur	=34 km
Barheipani – Barheipani Fall view	= 3km
Barheipani – Brundaban Gate	= 16 km
Brundaban Gate - Chahala	= 3 km
Nanjighasara – San Oski	= 11 km
San Oski - Oski Fall view	= 3 km
Barheipani – Nawana	= 11 km
Nawana – Joranda Fall view	= 8 km
Jashipur – Nawana	= 65 km
Nawana – Baripada	= 68 km
Baripada – Lulung	= 22km
Baripada – Pithabata	= 4 km
Pithabata – Bhajam Gate	= 20 km
Jashipur – Baripada (By NH)	=103 km
Jashipur – Baripada (Through SBR)	= 133 km

Tourist Spots

Khairinivas

Toursts start their journey to SBR by visiting Khairinivas at around 9.30 AM.The name Khairi makes nostalgia to the tourist. Khairi, a tigress cub brought here in early eighties from the deep forest in the core region of Similipal. The founder Filed Director of Similipal Tiger Reserve padmashree Mr. Saroj Roy Chowdhury brought khairi from the forest tribes and nourished her in his own quarter as a pet animal. Khairi, the tigress lived here from 1973 to 1981 till her death. The Khairinivas is an old Forest Bunglow with a beautiful garden in front.

Ramtirtha

Ramtirtha is just two kms away from Khirinivas at Jashipur. It was established as a crocodile breeding cum rearing center in 1979. There is a nursery bed of crocodile with well maintained separated small water shed for each of the infant crocodile and are kept in this nursery bed for a few years. Most of the grown crocodiles are left into the rivers of the deep forest in core region of Similipal. A few aged crocodiles are kept in big water shed inside the center.

In 2006, the Chief Minister of the government of Orissa, Mr. Naveen Pattnaik inaugurated a beautiful eco-tourism complex at Ramtirtha. A beautiful cottage named Debasthali was also inaugurated at that time. There is also a garden consisting plants relating to different zodiac. A herbal nursery or garden consisting more than 50 varieties of herbal plants are also there.

There is an interpretation center which is a small museum consisting of stuff birds, elephant's head, eggs of crocodile etc. There is also a Meghasini congregation Hall. It is a meeting or conference Hall. There is sufficient accommodation i.e. three dormitories consisting 30 beds. There is another cottage looked like Onge Hut Called Tribal Hut.

After getting permission from Assistant Conservator of forest, near Khairinivas, Jashipur the tourist party first go to Ramtirtha and after seeing the crocodile project they start their journey towards Similipal, passing over the small town Jashipur. The tourists usually take snacks, biscuits, fruits, sweet etc. from Jashipur market.

Kaliani

The first gate is 11 kms distance from Jashipur. In this gate a shop is installed in the year 2007 by Forest Division of Karanjia and supported by Sambandh, an organization of NGO works at Similipal. Varieties of items like biscuits, lodgens, mineral water, paper bags, and terracotta art objects are available here.

Kalika Prasad

The second entry gate or check post is at Kalika Prasad, 3 kms distance from Kaliani gate. The buffer region of Similipal starts from this place. All sorts of final checking regarding permission of vehicle, how many people enter into Similipal etc. are verified and checked here.

Mahabrukshasal

It is the first tourist spot inside Similipal to the tourist who enter through Jashipur. The tourist vehicle are instructed to the speed to look after the gigantic sal tree (*Shorea robusta*). The tree is aged about 300 years and situated just about 5 kms distance from Kalika Prasad gate. There was an story around this tree. Sometimes, between 1916 to 1918 the Sal tree was tried to cut for logging by the Britishers. Two labours started to cut it down but the saw blade got struck in the tree trunk. The two labours who were engaged in cutting were died, one at the spot after vomiting and the other one died in the evening at his camp. Thereafter, the tree left untouched and remains as a witness to the history of Similipal.

Mahabrukshachampa

Champa (*Michelia Champaka*) is one of the common and important trees in Similipal. The tree is erected one km. distance from 'Mahabrukshasal' and it also makes the history.

Gudgudia

This is overall excellent place of scenic beauty to nature lover. Eco-tourism project is continuing at this place. There is a very old Forest Rest House (FRH) made by the than Bhanja King named Krishna Chandra in 1855. Later the British renovated the house. There are other accommodations for the tourist like Bamboo Villa, two cottages; one dormitory and other one is Pine Villa. There are also another separate building made for a Recreation club and a restaurant, which are going to be started. There are an Orchidarium containing with varieties of orchids. Bird watching center, watchtower, Pine garden are also there. There is also a track of national trail or trekking route. The river Khairi with her excellent beauty is passing nearby. Eco-tourism project is in operating here. The village Gudgudia and the village Kumari are situated nearby and the tribals like Kol, Ho, Bathudi, Kheria and a few general scheduled castes populate the hamlets. The labour and the cook are engaged from these two villages. There is also a weekly hut or market erects attention to the tourists.

Oski Waterfall

The tourists who enter through Jashipur look after the first waterfall, Oski, near the village Son-Oski. Oski fall is about 18 kms distance from Mahabrukshachampa and 14 kms distance from the village Nanzighosara. The fall is at 3 kms away from the main road going towards Barheipani. This is one of the magnificent falls and not much in height. The approaching road is not in good shape.

Barheipani Waterfall

This waterfall is about 19 kms from Gudgudia. The approach road is going downward and is about 3 kms from the Barheipani – Nawana Road. Barheipani waterfall is the biggest one and ranging is about 400 meters. The water is falling down into the river Budhabalam. The river is originated only two and half kms away from this place. A beautiful log house and watch tower are also a place of tourist attraction. The minimum temperature of the fall is 7°C during winter and maximum 35°C during summer.

Nawana

There is a Forest Range Office at Nawana and from where the road is bifurcated in two directions, one is going to Baripada via Pithabata and the other one is going towards the core region of Similipal via Garsimilipal. There is a FRH bungalow in a small hillock and a watchtower. A good locality is there in and around. Nawana is the center place and the distance of Baripada is 60 kms from here and Jashipur is 65 kms.

Joranda waterfall

Joranda is 8 kms away from Nawana and is situated 3 kms away from the road going towards Baripada. The waterfall is of great feeling and attractions of a panoramic view. This fall is most beautiful and wonderful one and the water falls under 150 meters below in a gorge at a stretch. There is no way to reach nearly the falls; the tourists look it just opposite site of the gorge. A small water shade is there and also uses as a bathing place. A watchtower stands to watch the surrounding.

There is also a beautiful FRH Bungalow for staying the tourist inside the forest. Just outside the boundary fencing there is also a salt lick, where the wild animals use to come in the afternoon and in the evening. But at present salt is not used regularly. Two years back Joranda was also more attractive to the tourists for its fall viewing, loneliness, accommodation locality and salt lick with wild beast. Sufficient salt is not supplied at present.

Chahala

After seeing Joranda waterfall a few of the tourist party directly go back to Baripada via Pithabata. But most of the tourist party usually travels to Chahala via Barheipani. Chahala is the most and beautiful tourist spot in Similipal. The spot is surrounded by dense forest and no human habitat nearby. It is a beautiful place to night stay and to close view of wild animals. It is the only place where the wild animals could be observed in close view from a watchtower, as there is a salt lick. The animals like elephant, sambar, bison, deer, boar, peacock etc. could be viewed easily from this spot. There is a beautiful garden with blooms of seasonal flowers make the environment more attractive. There are sufficient accommodation facilities like FRH, villa, dormitory etc. are present.

Brundaban Gate

It is 3 km. distance from Chahala. This gate is to be crossed for returning back to Jashipur or other destination by the tourists. A salt lick is there and wild animals are looked after.

Orissa is endowed with cultural heritage, beautiful nature and biodiversity like wildlife, forest, beaches, monuments, tribal life, classical folk music and dances etc. Therefore, the state of Orissa has a great potentiality for tourism and the government of Orissa is aware of these. Circuit tourism is also encouraged by the Department of Tourism, Government of Orissa. The tourists are preferred to visit in a long tripped in two trails in and around Similipal.

(a) Travel from eastern side of Similipal- Chandipur, Chandaneswar, Panchalingeswar, Romuna, Deokund, Sitakund, Khiching and Similipal.

(b) Travel from western side of Similipal- Ghagra Waterfalls, Ghatgaon, Sitabinjh, Khiching and Similipal.

Chandipur: Chandipur is 16 kms distance from Baleswar or Balasore. It is a unique beach extended about 5 kms long at the time of low tide and happens twice a day. It is a marvel of nature and rarely found.

Chandaneswar: It is a famous temple of Lord Siva.

Panchalingeswar: It is 30 kms distance from Baleswar. Panchalingeswar, a place of unique panoramic view and where the perennial stream flowing over five Lingams.

Remuna: It is nearer to Baleswar and is famous for the shrine of Khirachora Gopinath. Sri Chaitanyadeva once visited the place.

Deokund: It is about 65 kms distance from Baripada. It is also an enchanting place with a series of waterfalls and a Shakti shrine is placed here against the outer periphery of Similipal.

Sitakund: It is only 2 kms away from Pithabata check gate periphery of Similipal. There is a small holy waterfall in the name of Sita of the Ramayana.

Ghagra Waterfalls: There are two beautiful waterfalls in Keonjhar District nearer to Keonjhar town. The beautiful park of the place is there another attraction.

Ghatgaon: About 50 kms distance from Keonjhar town. The place is situated inside the jungle area nearer to Baitarini River and is famous for a temple of Ma Tarini. It is also a powerful living folk deity of this area.

Sitabinjh: It is nearer to Keonjhar town. A mythological place related to the Ramayana and the stone inscription evicts the archaeological importance.

Khiching: The place is about 30 kms distance from Jashipur. There is a very old temple. Mata Kichakeshwari, the other name of Goddess Durga, is the main shrine. Khiching is a famous place for religious tourism. There is also a shrine of Lord Siva. In the month of Falgun (Feb-March), Sivaratrimela runs for 15 days. Thousands of Devotees visit this place through out the year from the locality or from far away.

Impact of Tourism at Jashipur

All sorts of favourable atmosphere are there at Jashipur to grown up and impact of tourism development in relation to Similipal. About 80% tourists enter into Similipal via Jashipur, thus tourism has a great impact in the locality of Jashipur. Thirty years back Jashipur was not a popular and developed place as it is now. After approaching tourism by Saroj Roy Chowdhury, Ex-Director, Tiger Reserve, Similipal became the place of tourist attraction. A small place is automatically grown up into a township by the impact of urbanization based on its infrastructure in different angles. Thus Jashipur also gets its present shape of a small town. A Youth Hostel has been constructed. To accommodate more tourists, a tourist center at Ramtirtha has been built. Three more private hotels were also started. The State Government has provided adequate transport services connecting Similipal via Jashipur. There are also needs of the guesthouses or bungalows of different Government offices like irrigation, PWD etc.

Various new economic activities have been grown up related to tourism like tourist operators, tourist car drivers, tourist guides, shops of dry foods, restaurant, medicine shops etc.

Jashipur is easy communicable and well connected by bus or cars. The NH-6 is passing through the place. Similipal is easy accessible from Jashipur and it is hardly 15 kms distance from here.

Jashipur is connected with Bhubaneswar, Baleswar, Keonjhar and other places of Orissa by buses and also with Kolkata in W.B. The distance of Kolkata is about 280 kms, Bhubaneswar is about 252 kms, Bariapada is 103 kms, Baleswar is 160 kms and Tatanagar is about 107 kms. Both Baleswar and Tatanagar are well connected by South – Eastern and Central Railways.

Jashipur is a block under the sub-division of Baripada and the district Baripada. All sorts of Government offices like BDO and other offices, Municipality, Hospital, Schools, College are there.

Similipal is nearer to Jashipur and more than 80% of tourists enter Similipal via Jashipur after getting the entry pass. The tourists sometimes get night halt at Jashipur. There are inadequate infrastructures in accommodation. Shopping complex is another attraction for tourists and it is not up to expectation at Jashipur, rather it is required for necessary of essential commodities. There is no good quality of restaurant or hotels for taking their foods by the tourists. For the entertainment and tourist attraction cinema hall, hotels, motels parks etc. are necessary.

Political Perception

- Most of the Panchayat of SBR is under the control of the political party of SUCI. The party has some suggestions and demands.
- The first and foremost demand is to give back the forest to the forest dwellers.

- Tourist's travel in core region has been stopped after declaration of SBR, but in some cases this rule is too relaxed and not properly maintained.
- Poaching is going on but it is not always possible to identify and punish the culprits.
- The administration should have given the dignity of living to the local people without harassment by using laws.
- Nearest sawmills that are situated in outer periphery of SBR should be closed.
- There should be wild life safari at Similipal and this would be more attractive to the tourist.
- Advertisement on SBR is much important.
- Entry fee is much for the poor to bear.

The other major political parties like the National Congress, BJD and BJP etc. are also concerned over these and they are also in favour to develop tourism and conservation of the biosphere of Similipal. They are also very much interested in sustainable development of local people.

PERCEPTION OF NGOS

The 'Gram Saraj' of Baripada, 'CREFTDA' (Center for Regional Education of Forest and Tourism Development Agencies), 'Sambandh' of Jashipur are concerned with the developmental activities of SBR. According the NGOs of Jashipur, the place has the potentiality as a tourist center adjoining Similipal. The distances from Jashipur to Kolkata are within 300 kms. on road and the nearest railway station are at Baleswar and at Tatanagar. Their suggestions are as follows-

- Similipal tourism should be given importance globally.
- If tourism is developed the people of different professions enter into SBR and the importance would be given to eco-development and thus the ecotourism would be properly developed.
- No proper guidance for the tourist is available. The guides have a little knowledge about the local pharmacopoeia as well as the local history and mythological background of the tourist spots. So a proper guideliness should be developed.
- Inter and intra departmental co-ordination is necessary.
- The forest dwellers are living hundreds of years in the forest and have been sharing their age-old experience. The competent

authority should give respect to their indigenous knowledge about how they are living with wild experience in such a remote area and without hampering the local ecology. The Government should have taken the responsibility to educate the local people and not to impose on them about the conservation of SBR.

- They believe that the SBR is much destroyed and more harmful by the outsiders like the illegal intruders - the poachers, woodcutters etc. rather than the local people.

PERCEPTION OF HOTEL OWNERS

- The tourists are less at present; only 10 to 15 families come in a year at Baripada. There are only two or three hotels where the tourists visit more specially in the season.
- More package tours are needed. There should be good relation between the tourists and the tour operators. Then the tourists revisit the place once more.
- Tourists are less coming due to various reasons like ban on picnic, can't operate music system, hiking of entry fees and vehicle charges etc.
- At present, tourists are less interested to visit SBR, as no animals could be seen throughout the day except at Chahala. Therefore, automatically tourists become less in Similipal.
- The entry fees are much for foreign tourists in comparison to local people. Therefore the foreign tourists use to come less in numbers than earlier.
- In many cases the tourists directly go to the SBR after getting permission from the Forest Department and they usually stay in FRH bungalows in SBR.

PERCEPTION OF LOCAL PEOPLE

The local people, the shopkeepers, the reporters and the local agencies of Jashipur and Baripada also prefer to develop Similipal region for tourism purpose. More facilities should be provided to the tourist in terms of accommodation, comfortable journey and easy access to Similipal. At the same time, they are in doubt how long it would be survived, as there is no tourist in lean season. There are instances of shut down of a hotel and one restaurant at Jashipur. They are not well informed by the authority that what programme they should be taken in future for the development of SBR.

CENSUS ANALYSIS

In relation to the study of Similipal total 50 individuals of tourists have been interviewed and shared their experiences after visiting Similipal. Various categories of tourist mostly national i.e. Indian and International (foreigner) are the students, service holders, businessmen etc. From the interview and from the filled up scheduled certain information come out. (Table No. 1, 2 and 3)

In age group the categorizations are 16 to 25 ages, 26 – 50 ages and 50 and above ages. A good number of the tourists belong to 16-25 age groups and mostly they are the college students coming for the excursion or education tour. 26 persons belong to middle age group i.e. in 26-50 age groups and three persons in 50and above age group.

Among the visitors interviewed, 34 persons are males and 16 persons are females. 22 persons speaks in Oriya and 23 persons in Bengali, 3 persons speak in Hindi and two others are foreigners, one is a American i.e. English speaking and another one is in Spanish.

From the information it is assumed that not a single illiterate person visit Similipal. 9 persons studied up to Class VIII to XII standard, 29 persons are graduate level, 7 persons are master degree holders, two persons are Physicians and three are engineers. 28 persons are married, 25 persons involve in service, three in business and rest are the students.

Except the students the visitors belong to income generating group. 8 persons earn Rs. 3000.00 to Rs. 10000.00 monthly, 15 persons earn more than Rs. 10000.00. Five persons did not given their income.

Most of the tourists prefer to visit in a party or group. Only two persons came alone and five persons had come with their families.

Among the visitors 17 persons belong to rural background and 33 persons have their urban Origin. Most of the tourists i.e. 27 persons are of Orissa and others come from outside Orissa. Out of these, 21 persons come from West Bengal, one from America and one from Spain.

The visitors are mostly motivated by their friends circle to visit Similipal. Out of 50, 35 persons are motivated by friends, 5 from family members, 5 from college or teacher circle and other 5 persons got motivated after reading books or getting information from TV or internet.

21 persons mostly student visited Similipal for their study tour or excursion to gather Scientific knowledges on flora, fauna, geography etc. About the frequency of visit to Similipal 39 persons visit the area for the first time, 4 person for second time and 2 persons each visit for

Table 1: Basic Information of Tourists

Age Group			*Gender*		*Mother Tongue*					*Education*				
16-25	*25-50*	*50 above*	*M*	*F*	*Oriya*	*Ben-gali*	*Hindi*	*Eng-lish*	*Spa-nish*	*Up to Class XII*	*Grad-uate*	*Master degree*	*Doctor*	*Eng-ineer*
21	26	3	34	16	22	23	3	1	1	9	29	7	2	3
42%	52%	6%	68%	32%	44%	46%	6%	2%	2%	18%	58%	14%	4%	6%

third, forth and fifth times respectively.

Mode of transport availed are only jeeps and out of 50 visitors' 31 people prefers to travel in train and 19 people in car. 31 persons also did not take the help of a guide but 19 persons availed the guide. Almost 21 persons are satisfied with local infrastructural facilities availed in Similipal and 29 persons are not satisfied.

Data have been collected by interviewing fifty individuals and have filled up census scheduled covering the relevant points like the name, gender, age, place from where they come, income, what and why motivate to them visit Similipal, suggestion to improve tourism and the problem faced.

The important facts that came out

- So far the Census data only literate persons visit Similipal (18% studied up to Class XII, 58% are graduate, 14% have master degree, 4% are doctor and 4% are engineer).
- Economically well up persons do visit (16% earn up to Rs. 3000/- per month, 30% more than Rs. 10000/- per month).
- Most of the tourists favor to travel in-group or party (86%).
- Oriya speaking are 44% and Bengali speaking are 46% (out of these 6% Bengali are settled in Orissa).
- 70% visitors are motivated by friends to visit Similipal.
- 42% student visit Similipal for their educational tour or excursion.
- 40% tourists are satisfied with local infrastructure facilities.

Roy Lodge (Ref. Table 4)

Established in 2003 at Raruan Chowk at Jashipur: going on for two years only after the death of the proprietor. His sons restarted again but close down due to some family problem in 2006.

Another restaurant started at Matiagarh Chowk on NH-6 in 2003 but it was also close down after running for only one month.

Suggestion to Promote Tourism in Tourist Perspective

The tourists were interviewed mostly for knowing how they observed or faced trouble in Similipal and to get suggestions from them to promote tourism.

Youth Hostel at Jashipur

The rent is higher in youth hostel in comparison to other places.

Table 2. Basic Information of Tourists

Marital status		*Occupation*			*Monthly Income (Rs)*				*Place*		*Coming from*				*Who Motivate*			
M	*UM*	*Ser-vice*	*Bus-iness*	*Stu-dent*	*No income*	*Rs 3000-10000*	*Rs 10000 Above*	*Not given*	*Rural*	*Urban*	*Ori-ssa*	*West Bengal*	*Ame-rica*	*Spain*	*Fri end*	*Family*	*College teacher*	*Book /TV*
28	22	25	3	22	22	8	15	5	17	33	27	21	1	1	35	5	5	5
56%	44%	50%	6%	44%	44%	16%	30%	10%	34%	66%	54%	42%	2%	2%	70%	10%	10%	10%

Table 3. Basic Information of Tourists

What motivate				*Other places visited/to be visited*		*Frequency of visit*					*Mode of transport*			*Interest local people*		*Visit Nearly village/interact With local People*		*Satisfied with local infrastructure*	
											Prefer	*Availed*							
Excursion	*Relaxation*	*Travel*	*Wild life*	*Only Similipal*	*Other place*	*1st*	*2nd*	*3rd*	*4th*	*5th*	*Train*	*Car*	*Jeep*	*Self*	*Guide*	*Interest*	*No Interest*	*Satisfied*	*Not Satisfied*
21	26	1	1	33	17	39	4	2	2	2	31	19	50	31	19	8	42	21	29

Table 4. Accommodation at Jashipur

Place	*Accommodation*	*No. of Suits*	*No. of Beds*	*Charges Per Suit (in Rs.)*	*Facilities Availed*
Ramtirtha	Tribal hat	3	6	200	Kitchen present with utensils, Electric light available
	Debasthali	2	4	200	
	Chincara (Dormitory)	1	10	300	Filtered drinking water supplied
	Chital (Dormitory)	1	10	300	
	Chowsingha (Dormitory)	1	10	300	
Youth Hostel (Jashipur)	Double Room	2	4	200	Kitchen with utensils, drinking water from hand pump, Electric light available, TV in office room
	Dormitory	5	40	400	
Sairam Holiday Home	Single Room	5	5	70	Food available, electricity available, TV available
	Double Room	3	6	200	
	Double Room (Deluxe)	1	2	300	
	A/C (Deluxe)	1	2	600	
	A/C Ordinary	1	2	500	
Sanguin Lodge	Single Room	4	4	60	Food available, electricity available, TV available
	Double Room	6	12	200	
	Dormitory	1	10	500	

Another restaurant started at Matiagarh Chowk on NH-6 in 2003but it was also close down after running for only one month.

Table 5: Accommodation at Baripada

Sl. No.	Name of Lodge	Estab-blished	Total Accommodation with Tariff									Total Staff	Facilities availed
	Hotel		Single Room		Double Room		Double Room A C		D. Room A C Dlx		Dormi-tory	Category	
			No.	Rent	No.	Rent (Rs.)	No.	Rent	No.	Rent	Rent (Rs.)		
1	Hotel Ambika	N.A.			9	S.O-175. D.O.-225.	7	700	1	750		Manager-2. Waiter-3,	Ph.Generator, Bar
						Non A/C Dlx-275						Sweeper-2. 8 Canteen Worker, Night guard-2	
2	Hotel Bishram	1994	6	100(TV)	5	150 (G.F.). 170					3 Beded 220	Manager-3, Waiter-5 Sweeper-2. Watchman-1	
3	Baripada Guest House				6	150	2	600				Manager-1, Waiter-2. Sweeper-1	
4	Hotel Durga	1984			6	160	8	600	2	750			
					8	200							
5	Ganesh Lodge	1972	10 C.Bath	40	5	100					3Bd-2Room 120	Manager-2. Waiter-8, Sweeper-2	TV. Generator
			14 A.Bath	60									

(contd...)

(contd...)

Sl. No.	Name of Lodge	Estab-lished	Total Accommodation with Tariff									Total Staff	Facilities availed
	Hotel		Single Room		Double Room		D. Room A.C		D. Room A.C Dlx		Dorm-itory	Category	
			No.	Rent	No.	Rent(Rs.)	No.	Rent	No.	Rent	Rent (Rs.)		
6	Kalika Lodge	1956	20 C. Bath	60	10	200							T.V. Arrange-ment of cooking
			10 A. Bath	100									
7	Hotel Mahapatra	1998			8	Single. 150	2	550				Manager-1, Waiter-2	TV. Phone, Canteen, Generator
						Double 200							
						Double Deluxe 250							
8	Krishna Lodge	Not avai-lable	8	50	2	100						Manager-1, Waiter-2	Phone, Generator
9	Santi Lodge		9	50	2	100						Manager-1, Waiter-2 Sweeper-1	TV, Phone, Generator
10	Swagata	1996	6	TV 100	4	175							
			2	75	2	140	6	100			4 Bd-3 Room 200	Manager-3, Waiter-6, Sweeper-2	TV, Phone, Generator

(contd...)

Sl. No.	Name of Lodge	Estab-lished	Total Accommodation With Tariff									Total Staff	Facilities availed
	Hotel		Single Room		Double. Room		Double Room A.C		D. Room A.C Dlx		Dormitory	Category	
			No.	Rent	No.	Rent(Rs.)	No.	Rent	No.	Rent	Rent (Rs.)		
11	Hotel Sivapriya	1998	8	120	4	220							
					2	Double Deluxe 350	2	500	2	600	3Bd-1Room 250	Manager-2, Receptionist-2, Cook-4, Helper-2,	TV,Phone, Canteen, Conference Room
												Sweeper-2	
12	Hotel Siddharth	1990			10	200							
					10	C Bath 100					3Bd-4Room 150	Manager-1, Waiter-5, Sweeper-2	TV,Phone, Generator
											5Bd-2Room 175		

N.A. Available C= Common, A= Attached

Table 6. Accommodation in SBR

Place	*Accommodation*	*No. of suits*	*No. of bed*	*Charge Per suit (in Rs.)*	*Facilities availed*
Gudgudia	FRH	2	4	400	Dhaba, Recreation Room yet to be started, separate kitchen
	Bamboovilla	1	2	300	
	Cottage	2	4	200	
	Dormitory	2	6	300	
Barheipani	FRH	2	4	400	Kitchen with utensils
	Barheipani Fall View	2	4	400	Kitchen with utensils
Nawana	FRH	2	4	300	Kitchen with utensils, Lanthan used
	Dormitory		6	200	
Joranda	FRH	2	4	400	Kitchen with utensils, Lanthan used, Filtered well's water used for drinking
Chahala	FRH	2	4	600	Kitchen with utensils, Lanthan used
	Sal villa	1	2	1200	Solar light available
	Chital villa			400	
	Eucalyptus villa		5	400	
Jamuani	FRH	2	4	200	Kitchen with utensils, Lanthan used
Lulung	Aranyanivas cottage	8	16	650	Kitchen with food availability, Aquaguard used for drinking water, Solar light for 24 hours, TV in office room
	Dormitory	2	24	125 per bed	

(a) The rent should be exempted or reduced for the students specially those who come for educational tour or excursion to youth hostel at Jashipur or at Ramtirtha.

(b) Rooms are not as good as comparison to home rent or changes claimed in Youth Hostel.

(c) Baths are not clean, shower is blocked, sink also are unclean, no proper water supply, insufficient light i.e. low voltage.

(d) No supply of drinking water. Therefore fresh supply of drinking water and aqua-guard should be used.

(e) Light facility and water supply should be properly provided.

(f) The room service and the hospitality not too much accommodated. Even the rooms are not cleaned properly. Suitable accommodation and good hospitality is the great pleasure for the tourist.

(g) There is much open place inside the youth hostel at Jashipur. A beautiful garden or park should be there. A children park also should be there, as most of the school children visit regularly at SBR for their educational tour.

Accommodation

(a) Accommodation seems to be less in peak period, the facilities should be more and of good quality.

(b) It creates a big problem when a large party wants to travel at a time. There is no much accommodation inside SBR except Chahala. Inside SBR not more than eight persons can stay at one place. For a large party, they would have to adjust by staying separately in different FRH.

(c) Hospitality management is not much available in the guesthouses in SBR.

Food and drink

(a) Arrangement of food should be there in FRH, otherwise it is horrible to carry on foods from outside by the tourist party itself and it would more troublesome if they would like to stay more than a day.

(b) All sorts of cooking arrangement including cook, vegetables, and grocer commodities should be there at FRH.

(c) Canteen facility should be provided and well maintained.

Reservation

(a) Easy method is to be procured for getting reservation.

(b) At present only one can allowed to stay three days inside SBR. The period should be extended for at least seven days to a nature lover.

(c) There is exorbitant cost of entry fees and it must be reduced.

Communication

- The inhabitant should be rehabilitated outside Similipal and inside roads should be widened and improved.
- The inner road of SBR should be well maintained otherwise the tourists who travels more than 150 kms in a day, would have terrible experience of journey.
- A direct bus-route from Baleswar to Jashipur is required.
- All sorts of facilities should be provided in FRH like canteen, drinking water, more accommodation, and malarial precaution measures etc.
- Wildlife exposure is one of the great attractions to the tourist other than natural beauty. But it is hardly be seen except the place of salt lick at Chahala.
- The tourist spots should be more attractive and decorative with garden, park, fencing, watchtower etc.
- Strong and durable fencing should be procured and well maintained in the Barheipani and Joranda waterfall viewpoints. Otherwise it would be a cause of dangerous incident.
- Conservation of Forest is necessary for keeping intact it's natural habitat and beauty and to make attractive to the tourist.
- Check deforestation by educating the local people's consciousness.
- One who likes to silent visit, Similipal is a nice place to them but it is now a terrible journey with dust moving at the time of traveling.
- Trekking facilities should be provided with proper guidance.
- The permission should be given to visit the core area of SBR with proper guides and guards and that should given the chances to view the wilderness in close and in that case the interested persons should paid more.

- One of the tourist suggested that Ram Mandir should be constructed in Ramtirtha at Jashipur.
- The first hand information in SBR and the rules to be followed i.e. what to do, what not should be handed over to the tourist early in leaflet form.
- Brochures, leaflet on SBR should be provided at the point of getting permission.
- Involve local people by which they can earn something.
- Formations of group comprising the local peoples representation of different villages of Similipal and engage them for road repairing, forest management etc.
- Educate the local people by which they can communicate with the tourist.
- Development of local handicrafts will be encouraged so that the tourists can buy it.
- The tourist should have given free roaming from place to place inside Similipal.
- Conservative is necessary and by which the natural beauty of the area will be more flourished.
- Mobile facility should be provided.
- Shopping complex is necessary in tourist places from where the tourists can collect the local products like – amla, honey handicrafts etc.
- It has been stated that a few local people who had their early experiences, about twenty years back, the wild animals were easily found by the roadside with bare eyes. At that time the forest was also quite dense to look at. But now for moving the cars throughout the day and the creating noise, the animals are not interested to come outside.
- The information regarding Hotels, Rest Houses for accommodation of the tourist is given in the Tables (4, 5 and 6).

The present situation inside Similipal

The basic needs, which are necessary for human survival, do not run thorough properly inside the village of buffer region of Similipal. Though, eco-friendly atmosphere more or less present between the local people and the administration in some respect but most of the basic needs are not fulfilled.

Accommodation: For the construction purposes the local people use the forest woods for making the house by following the Forest Rules that they use the felling woods by natural ways. But they are not always using it according to their needs and excess woods are always cut down.

Communication: Constructed roads are not in good condition. The local people only use bicycle or a few well off persons also use two wheelers. But these are not always easy communicable for bad condition of roads. Sometimes, one has to travel day long to go in other places. The actual problems they faced when they are suffering with a patient.

Health Facilities: Medical facilities are insufficient inside the Similipal. There is only one Public Health Centre (PHC) at Gudgudia. A few viral diseases are diagnosed here like malaria, phyleria, diarrhea etc. But for other serious diseases one has to go to Jashipur or Bangriposi or Baripada to check up. Sometimes, health department of Orissa Government in different places inside Similipal also organizes a few medical camps. But these are insufficient according to need.

Education: There is only one High School at Gudgudia. After getting primary education, the children have to go to Ashram Schools in other place for continuing their studies. In this regard, the tribal students of eastern Similipal are getting help from Christian Missionaries.

Local market: Local market is always necessary for the transaction of commodities according to public requirements or demands. There are many weekly *huts* (market) found in various places but these are not fulfilling their demand and they have to go to Jashipur for marketing. The people use to go to Jashipur from distant places and has to spend at least three days to return back to his village after marketing.

Firewood: Use of firewoods is their traditional habit for cooking purposes not only by the local people of Similipal but also by the outside people of that region. It is often observed that in the transition area, the people are coming from forest side towards town area by carrying fire woods bundles in their bicycle carrier.

Woodcutting: Wood cutting or wood felling always creates problems to the Forest Departments as no one actually follows the Forest Rules. Both the licensed and non –licensed woodcutters use illegal way for more collection. They engage local people for their benefits.

Poaching: Poaching is always a problem in the forest areas. The Forest Department is trying their best but it is not possible for them to protect in all levels with their less infrastructure like lacking of vehicle and insufficient manpower for covering miles after miles distance. The Forest Department also has recruited the local and outside young people in 'Sabuj Bahini' (Green Brigade) for strengthening their communicating network and to keep watch against the offences like woodcutting, smuggling, poaching etc.

There is a conflict between the Forest Department and the Tourism Department regarding the development of tourism at Similipal. The Forest Department wants to vacate Similipal from human habitation for the shake of wild animals and undisturbed natural habitat. The Forest Department always wants to implement the Forest Acts. The Tourism Department is not allowed to construct any kind of OTDC (Orissa Tourism Development Corporation) accommodation like hotels, bungalows inside Similipal except at Lulung.

Observation and Suggestion:

Indira Awas Jojona should be implemented for the poors and for the middle class people and they are bound to make the pucca (brick) houses and that should at least for the major houses. In that case repairing is not required regularly and at the same time woodcutting is no more necessary. In earlier days the plank of wood was used for fencing but at present no such new fencing are made or observed.

To check the woodcutting and wood felling or to check poaching, the local people should be involved and to be educated about the knowledge and importance of conservation. The saw mills nearer to the jungle area should be banned. LPG may be supplied to the local people on subsidized rate, so that the fire wood collection would be nominal.

Road should be properly constructed, maintained and well communicated from one village to other. An adequate planning is necessary for the improvement and extension of tourism at Similipal. Health-care is a major problem to the local people; therefore, the government should have taken proper care in this regard. At least, another two Primary Health Centers should have been there at the villages Nawana and at Barheipani. For the proper development in education two High Schools should be established one at Nawana and the other at Barheipani. Local market should have been developed and the government should look after the matter and the co-operative may be established to supply the useful commodities to the local people.

Encouragement of alternate economy should be provided for the local people. Migrant labour is practised in eastern side of Similipal. The tribal people of this area use to go in different places adjoining to the state of Jharkhand and of West Bengal. At present the people of Similipal are not much dependent on forest production or collection and they should be supplemented their needs by practising agriculture. Throughout Similipal there are huge sources of perennial water and this should be properly utilized in irrigation. Agricultural education is much necessary to the villagers so that they can utilize the materials (seedlings, fertilizer etc. supplied by the government) properly and can produce more crops.

The government and the NGOs are trying to engage the local people in different economic activities for their sustainable development in different levels like collection of jungle products, making of poultry, seri culture, collection of sabai grass and rope making, horticultural plantation and plantation of medicinal plant etc. The NGOs has involved the local people specially the women in Self Help Group (SHG).

Involvement of local people

The local people of Similipal are indirectly involved in tourism. The government has engaged the local people for various developmental works of the region. To maintain FRH campuses in different places inside the SBR, the local people work as a casual labour for cleaning, gardening, fencing, cooking etc. Before coming the tourist season (started from November) a major work is to be done on road construction and repairing. If any incident of road blockage is happened, the Forest Department clears the road by the help of the local people. In 'Sabuj Bahini' the local boys are recruited by the Forest Department on daily wage basis.

The tourist halts at Chahala for more than hours. A small museum may be opened here covering the information on Similipal, life of the tigress 'Khairi', collection of flora and fauna, knowledge of habitation, the people, their life and culture etc.

Complain and suggestion book should be kept at Chahala that records experiences of the tourists, the problems they faced, share their views and suggestions for the development of tourism at Similipal.

Usually there is no direct interaction between the local people and the tourist, only the children sell grass-garland and *amla* (*emblic myrobalan*) fruits to the tourist by standing at the roadside. In ecotourism perspective the sustainable developmental works should be operated but this is a long process and the local people should be educated first.

Conclusion

Conservation of wild life is necessary for maintaining ecological balance of living organism. Each and every species plays an important role for maintaining ecological balance. It has a specific value to study all animals and plants and research on them provides valuable knowledge and scientific understanding of wild life population.

Tourism should never be the destroyer; rather, it should be the protector. If the tourism industry and tourist share responsibility for conservation of the environment and cultural heritage, it becomes possible to preserve their value and at the same time use them as tourism resources (Chandra, 2005). Ecotourism stands for the management of tourism in such a manner that man derives maximum benefit from nature without disturbing its innate balance. It seeks to restore man's communication with nature and to ensure the fulfillment of the need of the local communities.

Active involvement is necessary of the local people to promote conservation and tourism simultaneously. According to Yadav (2006), ecotourism can develop three basic goal of conservation of biological diversity, (a) Conservation of biological and cultural diversity by strengthening protected area management. (b) Promoting the sustainable use of bio-diversity, by generating income, jobs and business opportunities. (c) Share the benefit of ecotourism developments equally with local communities.

In case of ecotourism the local people benefits more, as the products, services and the infrastructure are mostly localized. It also provides alternative means of livelihood to the local people in various ways like working as guides, catering, selling of local handicrafts, selling of souvenirs, broachers, guide map etc.

There is no doubt that the residentials in and around Similipal fully exploit the forest. The inside condition is not so favorable for living smoothly. The basic issues like roads and communication, education and literacy, health and malnutrition, lack of employment, drinking water, exploitation, not proper implementation of government policy, always terrified about displacement are their every day companion.

The government has the national policy to look after the welfare and development of the tribal people and it should be looked after properly in Similipal also. The basic amenities like housing, communication, drinking water etc. should be provided. There has been implementation of various plans and programmes among the local people by the forest department after recognition of Similipal forest as Biosphere Reserve. In earlier days as they were living in Similipal forest area and they used forest resources for fuel, fodder, herbal plants, hunting games, timber (for construction of house, agricultural implements) for their day to day life.. However, many tribal groups like the Kharia, Mankharia or Birhor, Bhumij, Bathudi, Munda, Kol and the Santal were also celebrating their traditional hunting festivals and religious festivals in the forest. After introducing the forest policy through special plans and programmes, these tribal and caste people like Gour, Kurmi Mahato and Lohar have to left their natal forest land and adjust themselves to the new situation for survive. It is also observed that the impact of tourism does not bring any economic changes in their life.

8

Maps of Buffer Villages

P.K. Guha

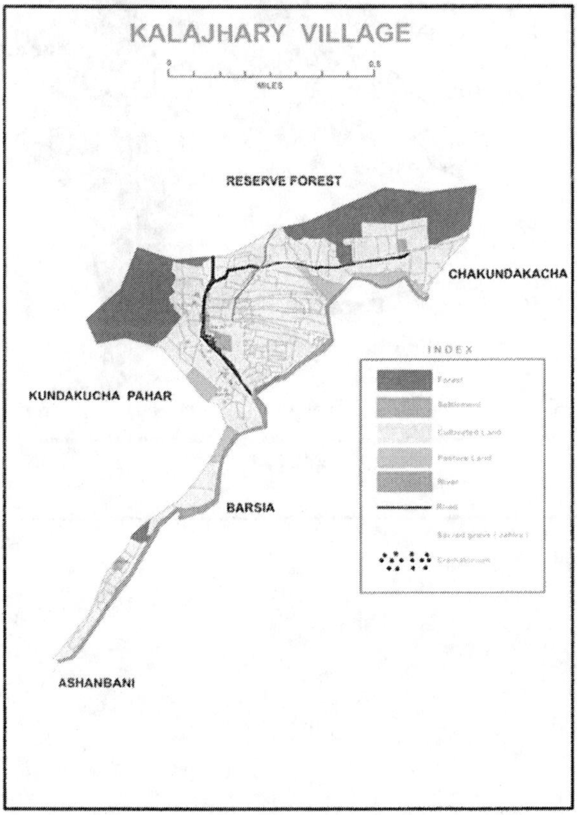

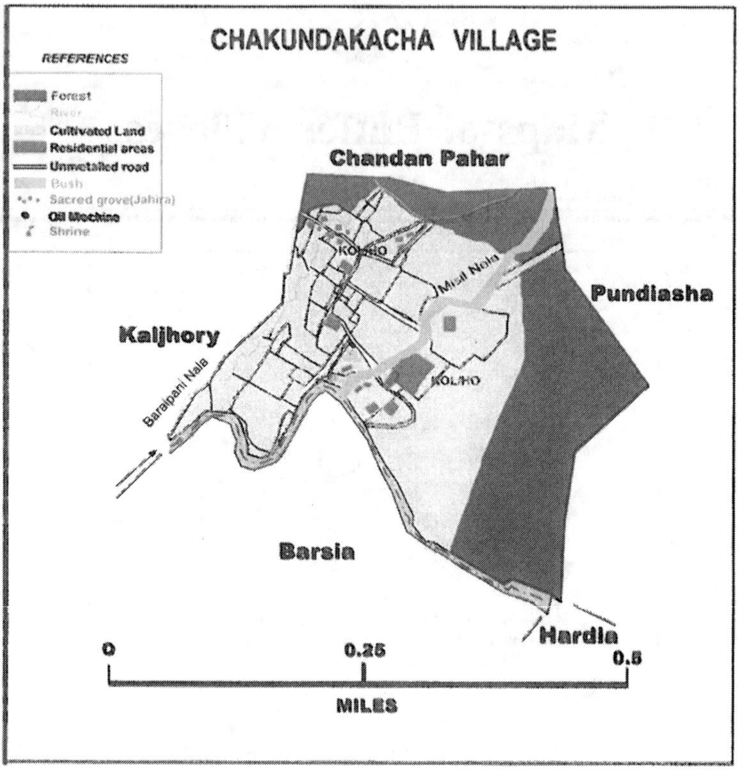
CHAKUNDAKACHA VILLAGE
REFERENCES
Forest
Cultivated Land
Residential areas
Unmetalled road
Bush
Sacred grove(Jahira)
Shrine
Chandan Pahar
Misli Nala
Pundiasha
Kaljhory
Baraipani Nala
KOL/HO
Barsia
Hardia
0
0.25
0.5
MILES

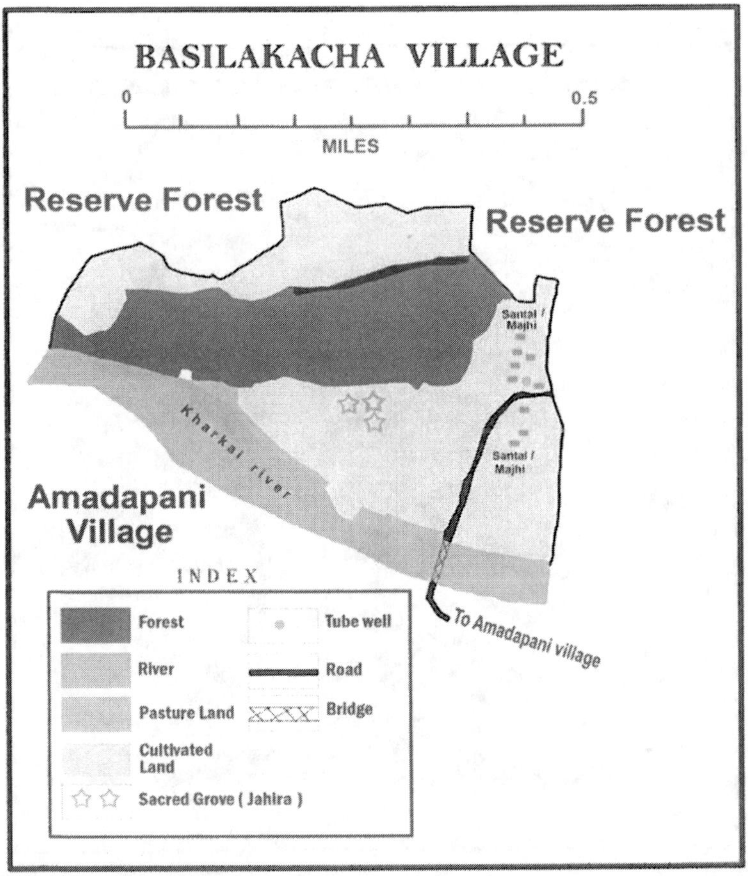
BASILAKACHA VILLAGE
0
0.5
MILES
Reserve Forest
Reserve Forest
Santal / Majhi
Santal / Majhi
Kharkai river
Amadapani Village
To Amadapani village
INDEX
Forest
Tube well
River
Road
Pasture Land
Bridge
Cultivated Land
Sacred Grove (Jahira)

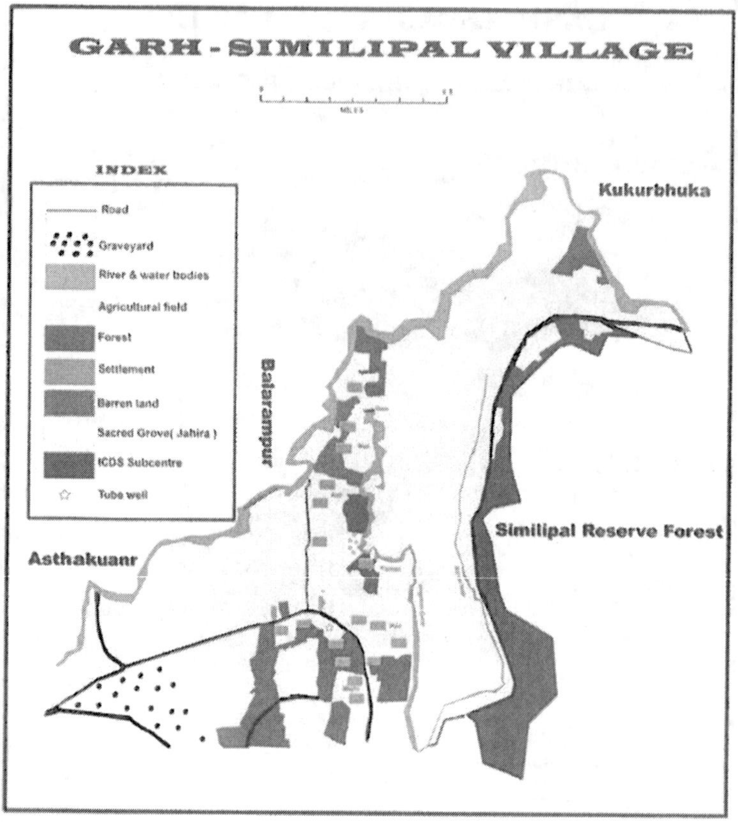
GARH - SIMILIPAL VILLAGE
INDEX
Road
Graveyard
River & water bodies
Agricultural field
Forest
Settlement
Barren land
Sacred Grove(Jahira)
ICDS Subcentre
Tube well
Kukurbhuka
Balarampur
Similipal Reserve Forest
Asthakuanr

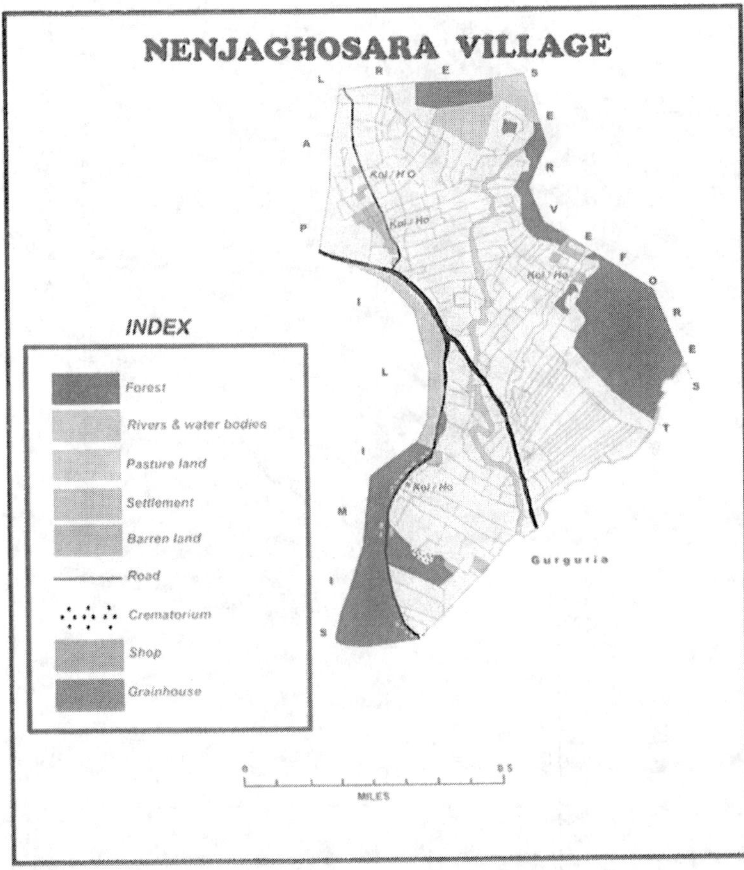
NENJAGHOSARA VILLAGE
Kol / HO
Kol / Ho
Kol / Ho
Kol / Ho
Gurguria
INDEX
Forest
Rivers & water bodies
Pasture land
Settlement
Barren land
Road
Crematorium
Shop
Grainhouse
0
0.5
MILES

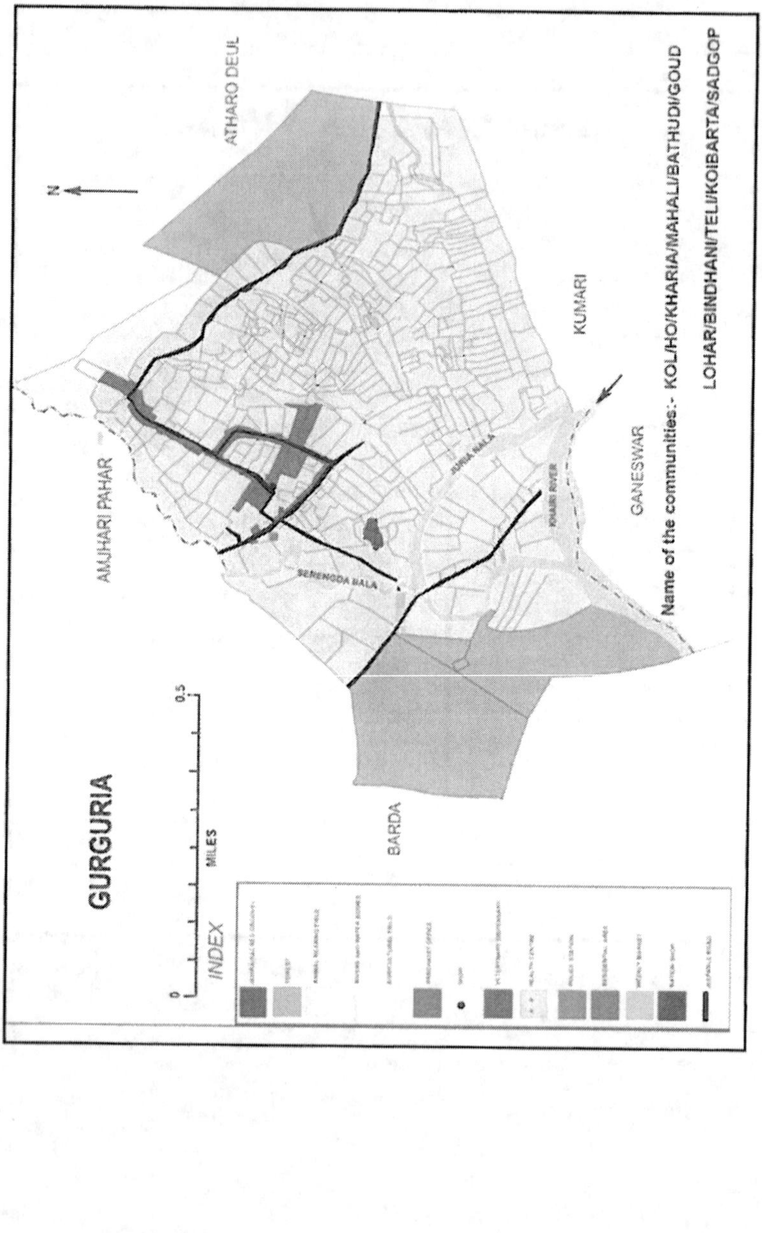
GURGURIA
INDEX
0
0.5
MILES
N
ATHARO DEUL
AMJHARI PAHAR
SERENGDA NALA
KHAIRI RIVER
KUMARI
GANESWAR
BARDA
Name of the communities:- KOL/HO/KHARIA/MAHALI/BATHUDI/GOUD
LOHAR/BINDHANI/TELI/KOIBARTA/SADGOP

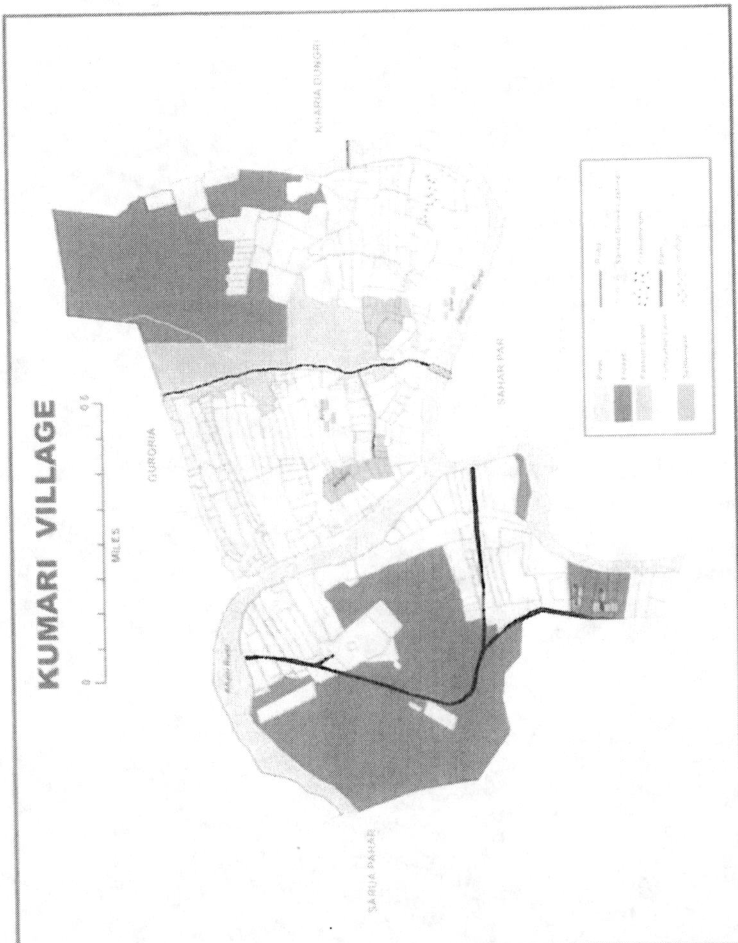
KUMARI VILLAGE
MILES

9

Introducing the Core Area Villages

Samit Ghosal, P.K. Guha and *Tamal Maity*

There are only four villages, namely, Kabatghai, Jamunagarh, Jenabil and Bakua and a colony of forest labours at Barakamra, in the 845.75 sq. km. of the Core Area of Similipal Biosphere Reserve. In the recent past as per the policy of relocation of the core area villages 20 families from Kabatghai had been shifted to a place called Banabasa near Gate No 1 at the vicinity of Jasipur. From Jenabil 12 Kharia and 13 Bathudi families had been shifted to Udala and they have been given 3 acre of well irrigated land.

The four villages of the core area are well connected with good all weather roads. To reach inside the core area one has to cross at least four gates manned by the forest guards. Tourists are not allowed to enter the core area and they are stopped at the last entrance of the core area, however they are allowed to go in the buffer area.

Population Structure in Four Decades

In the following table the population figures of four decades of the four core area villages as per the census records are shown.

Sl. No.	*Village*	*Area (in acre)*	*Census Years*	*Male*	*Female*	*Total Population*
1	Kabatghai	216	1961	54	56	107
			1971	64	64	128
			1981	46	47	93
			2001	55	60	115

(contd...)

(contd...)

S. No.	*Village*	*Area (in acre)*	*Census Years*	*Male*	*Female*	*Total Population*
2	Jamunagarh	345	1961	21	20	41
			1971	-	-	-*
			1981	38	27	65
			1991	37	42	79
			2001	45	51	96
3	Jenabil	345	1961	22	17	39
			1971	36	35	71
			1981	96	94	190
			1991	86	77	163
			2001	74	71	145
4	Bakua	146	1961	21	20	41
			1971	16	14	30
			1981	26	20	46
			1991	28	28	56
			2001	36	44	80

*Shown uninhabited in the census year 1971. Census Figure

Community wise Population in Core Villages

Sl. No.	*Village*	*Communities*	*Nos. of Families*	*Male*	*Female*	*Total*
1	Kabatghai	Ho Santal	4 13	42	38	80
2	Jamunagarh	Ho	18	39	47	86
3	Jenabil	Ho Bathudi	28 1	93	91	184
4	Bakua	Ho	21	44	53	97
	Total		84	218	229	447

Source: Field Data- Nov-Dec:2006

Beside the above at Upper Bara Kamra, there is a camp for forest labour. In that settlement about 13 Kharia families are living who solely depending upon forest department for their livelihood as they do not have any land for agriculture. However, they extensively collect and sell forest produce, which has been formed a substantial part of their livelihood.

Age Group Wise Population in the Core Villages

Age Group	*Kabatghai*	*Jamunagarh*	*Jenabil*	*Bakua*	*Total*
0-4	8	15	36	14	73
5-9	13	15	27	8	63
10-14	8	10	13	9	40
15-19	9	8	9	12	38
20-24	6	6	17	5	34
25-29	7	3	18	6	34
30-34	7	6	11	5	29
35-39	7	6	13	5	31
40-44	6	4	7	9	26
45-49	3	3	10	6	22
50-54	4	5	7	6	22
55-59	1	-	6	5	12
60-64	1	5	5	3	14
65+	-	-	5	4	9

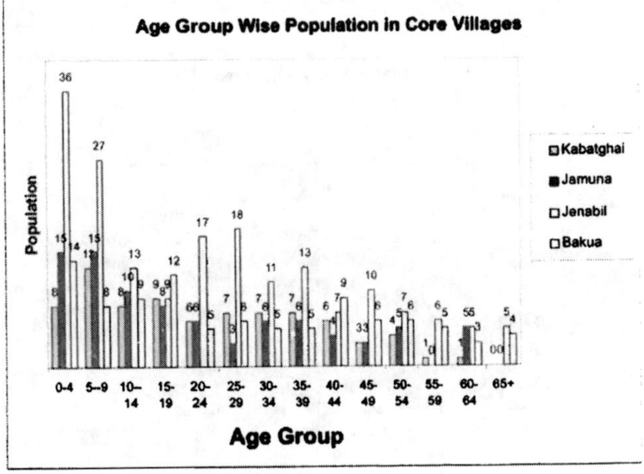

From the above table it is seen that in general maximum number of population is in the age group of up to nine years and also in the range of 20 to 30 years and thereafter a gradual decline of population with the increase of the age groups which is quite normal due to old age.

Family Types in Core Villages

Family Types	*Kabat Ghai*	*Jamunagarh*	*Jenabil*	*Bakua*	*Total*
Nuclear	11	16	11	14	52
Joint	6	2	15	7	30
Broken			2		2
Total Family	17	18	28	21	84

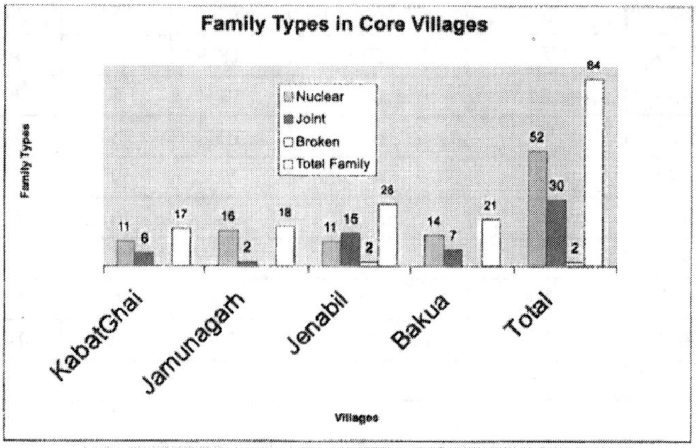

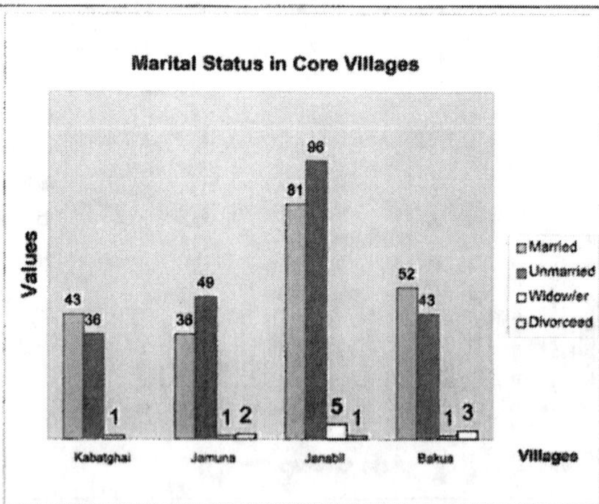

Marital Status in Core Village

Sl. No.	*Village*	*Married*		*Total*	*Unmarried*		*Total*	*Widow*	*Widower*	*Divorce*		*Total*
		M	*F*		*M*	*F*				*M*	*F*	
1	Kabatghai	22	21	43	20	16	36	1	-	-	-	80
2	Jamunagarh	18	18	36	20	29	49		1			86
3	Jenabil	41	40	81	50	46	96	4	1	1	1	184
4	Bakua	26	26	52	17	26	43	1			1	97
	Total	107	105	212	107	117	224	6	2	2	1	447

Village Wise Literacy in Core Village

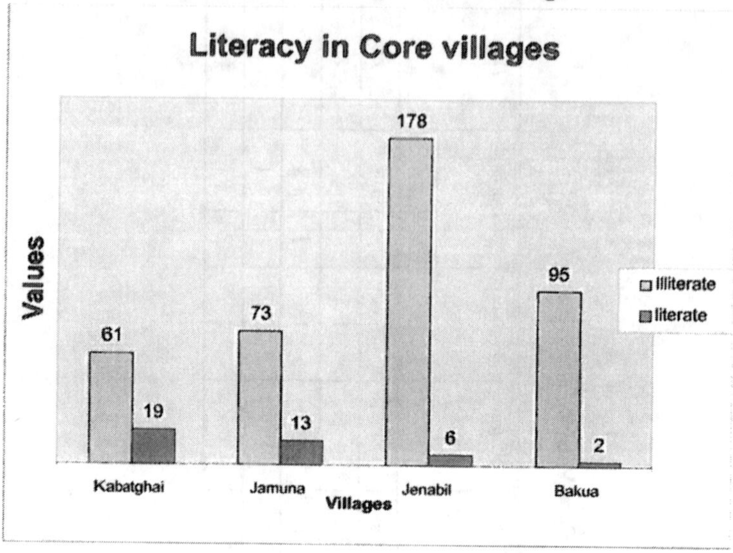

Livestock

Livestock population is an integral part of village agrarian economy. In a forest village the number of livestock is the direct reflection of the exogenic pressure on its neighbouring forests. The following are the livestock population in the four core area villages.

Village	*Cows*	*Oxen*	*Goats*	*Buffaloes*
Kabatghai	46	42	80	5
Jamunagarh	118	20	219	x
Jenabil	200	24	288	33
Bakua	109	52	242	x
Total	473	138	829	38

From the table it is seen that Jenabil has the highest number of bovine population. Jenabil has a vast pasture/grazing ground near the village, which was earlier occupied by the Kharia and Bathudi families but now turned into a pasture. Beside all the villages have a little amount of private forest to support the livestock. However, nonetheless they take the cattle to the forest for grazing within a radius of 5 kilometer only and are not allowed to take them into the deep forest.

It is observed that Govt. vehicle mainly of the forest department are plying inside the core area even during the night which should be stopped immediately as its disturb the animal movement in the night. The forest department has been trying to relocate the four core village for quite some time now but without any result. It is found that the core area villages have a very sound agricultural economy and well irrigated field which made them less dependent on the forest resources. Until and unless they are given land of similar quality outside the core area there is a very little chance of their leaving the villages. The compensation money, whatever is the amount does not mean any thing to them as they do not know how to use the money and they are very much afraid that the money would change hands as soon as they receive it.

10

Indigenous Management of Land and Forest Resources in the Core Area Villages

Samit Ghosal

BIOSPHERE RESERVE

The programme of Biosphere Reserve was initiated under the 'Man and Biosphere' (MAB) programme by UNESCO in 1971. Biosphere Reserves are areas of terrestrial and coastal ecosystem promoting the conservation of biodiversity with its sustainable use. Biosphere Reserves are designed locally recommended nationally and recognised internationally but remain under the sovereign jurisdiction of the states where they are located.

Biosphere reserve fulfills three basic functions which are complimenting and mutually reinforcing:-

(a) **Conservation Function** : Conservation of biological resources and biodiversity.

(b) **Developmental Function** : Promote environmentally compatible economic and human development which is socio-culturally and ecologically sustainable.

(c) **Logistic Function** : Support environmental research, monitoring human activities and imparting education and training.

Biosphere reserves are organized into three interrelated zones, known as the Core Zone, Buffer Zone and Transition Zone. Core area is the legally protected area where biodiversity is preserved. In the buffer area sustainable use of natural resources are permitted. Transition zone is the area where all the 'stake holders' work together to maintain the balance between conservation and economic development based on the cultural values of the local communities.

SIMILIPAL BIOSPHERE RESERVE

A Tiger Reserve was created in 1973 with an area of 303 sq. km. as the core area in the Similipal Reserve Forest which was set aside as National Park and later the area was expanded to 845.75 sq. km. with a buffer area of 1904.25 sq. km. In 1980 the core area of the tiger reserve was transformed into a National Park and the buffer area of the tiger reserve was declared as Sanctuary. The Government of Orissa imposed moratorium on tree felling in the Similipal since 1982. Similipal was declared as the 8th Biosphere Reserve of India with a total area of 4374 sq. km, of which 845.75 sq. km. was the core area which was also the core area of the tiger reserve, and another 2128.25 sq. km. was the buffer area which was the conglomeration of Tiger Reserve Buffer area of 1904.25, Nato Reserve Forest of 77.00 sq. km. and Satkosia reserve forest of 147 sq. km. A radius of 10 km from the outer limit of buffer area, covering an area of 1400 sq. km. is considered as the transition zone of the Similipal Biosphere Reserve. There are four villages in the core zone, 61 villages in the buffer area and 1200 villages in the transition zone.

OBJECTIVES OF THE STUDY

Nature of exploitation of the land and forest resources by the people living in the Core Area of Similipal Biosphere Reserve is directly related to the sustainable maintenance of the bio-diversity of the Biosphere Reserve Ecosystem. Material needs of the people living in the Core Area have to be addressed by sustainable use of natural resources. People living within the confinement of forest ecology have evolved strategies to support their subsistence economy by sustainable and optimum use of land and forest resources at their disposal. Anthropologists' interest in the study of Indigenous Knowledge (IK) has been growing in importance in development policy and practice and in academic research, especially after the Rio conference in 1992. Therefore, the indigenous ways of managing the natural resources by the communities living in the Biosphere Reserve's Core Area is an important issue. The human dimensions of the Biosphere Reserves make them special from other protected areas. Interaction with local communities is essential to keep intact the ecological knowledge and cultural values of the communities living in the area. We have to conserve the natural resource of the Biosphere Reserve vis-à-vis meeting the material needs of the local communities through the sustainable use of natural resources. The indigenous knowledge and techniques of the people in the field of subsistence agriculture and forest exploitation can make valuable contribution to the modern and contemporary knowledge. Core Areas in a BR can also be seen as the areas where the traditional knowledge of the tribal people can be preserved and enhanced. IK is a sum total of the customs, myths and ritual which help the tribal community to live in harmony with their environment and replenish vital and valuable elements dormant in their culture. Much of this knowledge form an important contribution to research and development in the areas of pharmacy, agriculture, aromatic and cosmetic products.

Following are the parameters of indigenous knowledge:

1. Unique to a given community
2. Based on cumulative experience
3. Often tested over centuries of use
4. Adapted to local culture and environment
5. Dynamic and changing with conditions
6. Transmitted through indigenous communication channels

The present issue of indigenous knowledge in land and forest management, which are present among the tribal people living in the area and are used by them in the subsistence agriculture practised by them as well as the nature of exploitation of forest wealth. This needs to be urgently addressed because the bio-diversity present in the Biosphere Reserve is a big repository of the human's traditional knowledge base. The indigenous knowledge and techniques of the people in the field of subsistence agriculture and forest exploitation can make valuable contribution to the modern and contemporary knowledge. Biosphere Reserves are seen to be the areas where the traditional knowledge of the tribal people can be preserved and enhanced. Since human activities have been implicated in the majority of the problems related to plant and animal biodiversity it would be rational to put research effort to human activities and attitudes, which are responsible for such problems. Generally speaking biodiversity involves undomesticated plants and animal species in which biologist, ecologist and foresters are more interested but for an anthropologist cultural value of biodiversity is more important which involves the diversity of domesticated plants and animals and associated value system in peoples' cognitive domain.

METHODOLOGY

Taking village as an unit of study all the four core villages in the Similipal Biosphere Reserve has been studied intensively by using standard anthropological tools and techniques. The nature and extent of exploitation of natural resources by the two tribal communities, namely, Santal and Ho, living in the area has been studied. The four core area villages are situated deep inside the forests and their level and nature of exploitation of forest have a direct correlation with the population pressure of the villages. An attempt has been made to understand the effective areas of forest exploitation based on the population density. An inventory of the collected items from forest has been made to see whether its orientation is towards subsistence or profit.

Participatory research approach and tools were used for information collection and analysis including:

- Historical overview and time line
- Social mapping
- Resource mapping
- Trend analysis
- Focused group discussion

The areas of enquiries was centered around the following parameters

1. Agriculture – operation and practices
2. Food and cereal preservation
3. Conservation and regeneration of soil
4. Protection and conservation of local /village forests and sacred forest (sacred grooves) based on local ecology and related cultural dimension
5. Land and Forest Management
6. Economic importance of plants/trees and their conservation and protection
7. Trees/plants and crops that can grow well together and their uses
8. Crop rotation
9. Marketing of local/forest/indigenous produces

10. Local peoples' awareness of marketing Indigenous knowledge/ Intellectual Property Right (IPR)
11. Extent of disturbance to Biosphere reserve by the local people
12. Nature of Agricultural Biodiversity and plant varieties
13. Forest Biodiversity.

VILLAGES UNDER STUDY

The four villages situated in the core area of the Similipal Biosphere Reserve, namely, Kabatgnai, Jamuna Garh, Jenabil and Bakua have been studied intensively against the extensive backdrop of the Biosphere Reserve mandate and concept of forest ecology. Village Kabatghai is inhabited by both Santal and Kol communities, rest of the three villages are almost exclusively populated by the Hos or Kol people.

DEMOGRAPHIC STRUCTURE OF THE VILLAGES

Table 1. Population Structure (Census: 2001)

Sl. No	*Villages*	*Persons*	*Nos of Family*	*Male*	*Female*	*Sex Ratio*	*Family Size*
1	Kabatghai	115	18	55	60	1090	6.38
2	Jamunagarh	96	10	45	51	1133	9.6
3	Jenabil	145	25	74	71	959	5.8
4	Bakua	80	20	36	44	1222	4
	Total	**436**	**73**	**210**	**226**	**1076**	**5.97**

Table 2. Population Structure (Field Data : 2006)

Sl No	*Villages*	*Persons*	*Nos of Family*	*Male*	*Female*	*Sex Ratio*	*Family Size*
1	Kabatghai	80	17	42	38	904	4.70
2	Jamunagarh	86	18	39	47	1205	4.77
3	Jenabil	184	28	93	91	978	6.57
4	Bakua	97	21	44	53	1204	4.61
	Total	**447**	**84**	**218**	**229**	**1050**	**5.32**

From the table it is observed that the sex ration is maximum, i.e. females per thousand males, in Jamunagarh and least in Kabatghai, where as the family size remain same more or less in all the four villages of the core area. It is pertinent to mention here that Kabatghai is the most exposed village and Jamunagarh is the least exposed village.

MANAGEMENT OF NATURAL RESOURCES

People living in the core area villages of the SBR manage their natural resources in a sustainable way by adapting to the local conditions. Topography, soil types and other geomorphologic features are different in each village. Land and forest are the two natural resources available to them and they manage it by using their indigenous knowledge as imbedded in their culture. As the policy of relocating the villages inside the STR, annexation of village cultivable land to forest has started.

LAND RESOURCES

Table 3. Village Wise Availability of Land (in acre)

Sl. No.	*Name of Villages*	*Agricultural Land*	*Per cent*	*Fallow Land*	*Per cent*	*Total Land*
1	Kabat Ghai	153.57	67.72	73.22	32.28	226.79
2	Jamunagarh	86.85	25.18	258.07	74.82	344.92
3	Jenabil	184.30	53.55	159.91	46.45	344.21
4	Bakua	80.47	55.42	64.75	44.58	145.22
	Total	**505.19**	**47.60**	**555.95**	**52.40**	**1061.14**

From the above table it is seen that the ratio of agricultural land to fallow land is highest in Kabatghai and it is lowest in Jamunagarh. In fact all the four core villages have adopted modern technique of irrigated cultivation which has reduced their dependency on forest product for subsistence and livelihood vis-à-vis the maintenance of low population growth. Kabatghai has increased the percentage of land under agricultural not for the subsistence but for the selling of the surplus production adopting a peasant economy due to their proximity to market place of Jasipur.

The fallow lands are used mainly for the grazing of cattle as pasture and for growing of commercially valuable timbers as asset building measures. As we know that people of Jenabil and Jamunagarh have transformed their agricultural land into private sal forests. A few Bathudi and Kharia families were living in Jenabil about eight years ago, they did not terraced their land and were cultivation ashudhan,millet,maize and kandamul in their unterraced land. So when they left the village due to the policy of relocation the land possessed by them looks like meadows or grass land instead of agricultural field.

FAMILY WISE LAND HOLDING

Table 4. Village Wise Availability of Land (in acre)

Sl. No	*Name of Villages*	*Number of Family*	*Persons*	*Total Land (In Acre)*	*Land Holding Per Family (In Acre)*	*Per Capita Land (In acre)*
1	Kabat Ghai	17	115	226.79	13.34	1.97
2	Jamunagarh	18	96	344.92	19.16	3.59
3	Jenabil	28	145	344.21	12.29	2.37
4	Bakua	21	80	145.22	6.91	1.81
	Total	**84**	**436**	**1061.14**	**12.63**	**2.43**

From table 3 it is seen that both family wise land holding and per capita land holding are more in Jamunagarh and Jenabil villages, one interesting feature is this, that though per family land holding (13.34) in Kabatghai is much more than the per family land holding of Bakua (6.91), there is not much significant differences in the per capita land holding of these two villages, which are 1.97 and 1.81, respectively. This is due to the presence of less population in Bakua than the Kabatghai, which are 115 and 80, respectively.

CROPPING PATTERN

Yield or productivity of the land is an important phenomenon to understand and evaluate a subsistence agricultural economic system. Amount of production of a particular crop in a village is very important to measure the productivity of the land. Each village produces different quantity of crops. Some produce more rice than others and some produce more wheat. The difference in the production of same crop in various villages (two or more) accounts for the variation in soil types, technology involved, manure/ fertilizer used and some geomorphologic factors along with the necessity of that crop in that area. A vital aspect of food production in the villages under study is that there is large difference in the quantity of food crops raised by different villages. As a result different villages tend to produce more or less of a certain kind of crop. It is shown in the next table.

Table 5. Village Wise production of Crops

Villages	*Crops Produced (in kg)*						
	Paddy	*Wheat*	*Maize (Maka)*	*Millet (Kodu)*	*Rashi*	*Mustard (Sarshu)*	*Kand-amul*
Kabatghai	40600	0	8950	15	7750	10760	0
Jamuna-garh	72300	2250	11100	6100	3150	2300	900
Jenabil	67000	1500	16100	9440	1262	1592	14200
Bakua	54700	0	400	300	2850	2050	3750
Total	**234600**	**3750**	**36550**	**15855**	**15012**	**16702**	**18850**

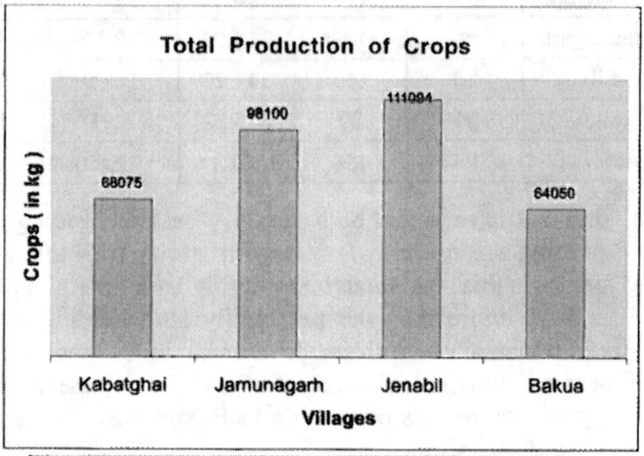

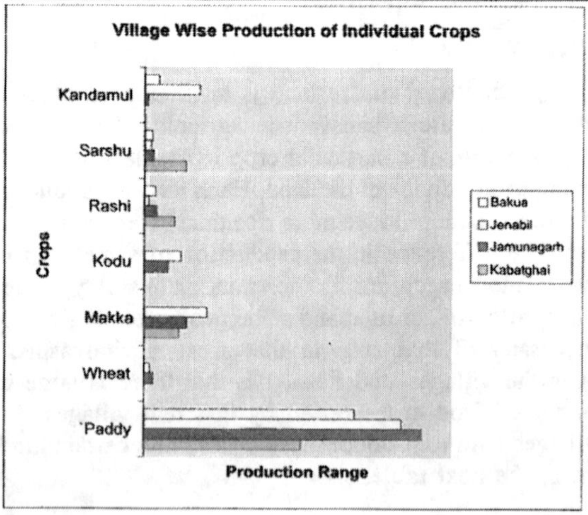

From the above tables it is seen that total production of crops is highest at Jenabil and lowest at Bakua but within this broader limit there has been a preference of the production of individual crops amongst the four studied villages. Paddy is grown profusely in all the four villages, but wheat is produced only at Jamunagarh and Jenabil, where as Kabatghai and Bakua do not produce any wheat. Amongst the two wheat producing village more wheat is produced at Jamunagarh because at Jenabil due to extreme cold and frost wheat crop is damaged. This harse climatic condition of Jenabil doest not affects the production of millet (kodon) because millet can grow in a wide range of climate. But Bakua produce less millet due to a different reason altogether, it is learnt that Bakua used to grow decent amount of wheat even a few years back in the unterraced lands but later on they converted their slopes into terraced fields and start cultivation of paddy. All the villages produce both rashi and sarshu in equal quantity. So far kandamul, a type of yam, is concerned it is grown profusely at Jenabil, some quantity at Jamunagargh, a little at Bakua and nothing at Kabatghai.

Table 6. Per Capita Crop Production and Per Capita Land Holding

Village	*Per Capita Land Holding (In acre)*	*Per Capita Production of Crops (In Quintal)*
Kabatghai	1.97	5.91
Jamunagarh	3.59	10.21
Jenabil	2.37	7.66
Bakua	1.81	8.00

From the above table it is seen that Jamunagarh reflects highest figure in both per capita land holding and per capita crop production, where as Bakua though has less per capita land holding than Kabatghai and Jenabil but produce more crop per capita than Kabatghai and Jenabil. This is because of the fact that both per capita land holding and per capita production of crop are directly linked with the population number of the village and here Bakua stands in an advantage due to its low population.

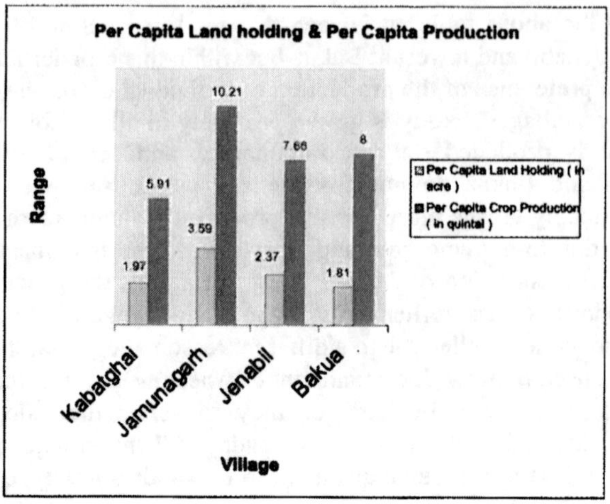

AGRICULTURE

Agricultural operation, selection of crops for cultivation, crop rotation and all other parameters are used to obtain optimum output in effective and efficient resource management.

AGRICULTURAL CALENDAR AND MONTHS

Agricultural season in this area starts from the month of *Maghe* and the last month of the local calendar year is *Poush.*

LOCAL CALENDAR

Local Months	*English Equivalents*
Maghe	Middle of January to Middle of February
Ba/phagun	Middle of February to Middle of March
Chaiti	Middle of March to Middle of April
Baisakh	Middle of April to Middle of May
Josto	Middle of May to Middle of June
Ashara	Middle of June to Middle of July
Sandh	Middle of July to Middle of August
Bhador	Middle of August to Middle of September
Dasai/ kuar	Middle of September to Middle of October
Bandhna/ kartik	Middle of October to Middle of November
Magshir	Middle of November to Middle of December
Poosh	Middle of December to Middle of January

LAND AND SOIL TYPE

From the very beginning all the families living in the core area villages of the Similipal Biosphere Reserve had been allotted 8 acre of land. House tax is Rs. 3/- per decimal and agricultural land tax is Rs. 7/- per decimal. There are three categories of land found in the core area villages of the Similipal Biosphere Reserve, Beda land (low land) Dhipa land (high land) and Guda land (situated near the house) where maize, kandamul and mustards (sarshu) are cultivated both in monsoon and in winter, these are well manured lands with the animal dung available near the house. They most judiciously manage the cropping pattern of the land situated near the house called Guda jamin. Dhipa is a high land where only one crop can grow in monsoon and one crop in the winter, paddy (ashu dhan) is grown in monsoon and sarshu and/ or rashi and millet are cultivated in winter. Beda land where only two times paddy is grown. Guda land is the flat land without terrace. But when population grows and to support the growing population they need more crop and adopted better technology and start terracing.

CULTIVATION OF CROPS

Paddy (*Dhan*)

The following six types of paddy seeds are available in the area

1. Sorno dhan – it is golden in colour
2. Kadua dhan – it is black
3. Lolat dhan- it is white in colour
4. Gudumoni dhan
5. Mugdi
6. Borha sal
7. Guda dhan
8. Suru dhan
9. Nala dhan

Highest quantity of paddy is produced in Jamunagarh (72,300) followed by Jenabil (67,000), Bakua (54,700) and Kabatghai (40,600). Lolat and Kadua are the two oldest paddy seeds, next is Borha sal. Sorno is the new variety, Sorno seeds and Lolat seeds are provided by the block office. Rests of the paddy seeds are available in the villages. Lolat paddy is totally depending on rain. Guda dhan is a type of paddy which grows in two or three month where as other types of paddy takes 5 to 6 months.

In villages of the core area, mostly two types of paddy seeds are grown, Sorno and Borha sal. Sorno is supplied by block office and it requires more fertilizer but Borha sal variety of paddy seed does not require any fertilizer being a local traditional variety. They prefer Borha sal because use of fertilizer is not allowed inside the Biosphere Reserve/ Similipal Tiger Reserve. Bora sal is cultivated at the higher fields which are rain fed, where as Sorno is cultivated at lower fields near the river bed where lift irrigation is possible. Paddy is grown two times in a year. One in monsoon and one in summer, monsoon paddy is sown in Ashar and harvested in Magshir, summer paddy is sown in Phagun and harvested in Jesto. Suru dhan is grown in low land, it's a monsoon crop, Nala dhan is grown near nala or water source in low lying land and Dhipo dhan is grown in high land with less water. They cultivate paddy both by transplantation and direct broadcasting method, in the low lying river side plots (jolo), they practise transplantation method but in high fileds (dhipo) they do it by direct broadcasting. In the lower fields they cultivate paddy two times a year, one in summer, from February to April and the other one is the normal monsoon crop, from July / August to November / December. In the high land (dhipo) they cultivate one single monsoon crop of paddy and let it fallow during winter for regeneration of the soil. But in the land where they grow double crop of paddy they use more manure to retain the productivity of the land. Preparation of field for summer paddy cultivation starts from Chait month, Ploughing is done in Maghe, sowing is done for summer paddy in June both at high and low fields. Suru dhan is the best in quality. It is grown in Beda land, where water is retained. In Beda land two types of paddy are cultivated, Suru dhan and Lolat dhan. Suru dhan is sown in the month of ashar and lolat dhan is in the month of phagun. Suru dhan is sown in Ashar and harvested in the month of Margoshir. Lolat dhan is sown in Phagun and harvested in Ashar. Normally they cultivate Ashudhan at Dipo jamin in monsoon by direct droadcasting method where irrigation is not possible and the land is dependent on rain water. Lolat dhan is cultivated in the low land by transplantation method where irrigation is possible both in monsoon and in summer. For sowing in transplantation method 10 kg of paddy is required per 1 acre of land and the production is 20 quintal but for sowing in broadcasting method 15 kg of paddy is required per 1 acre of land and the production is 15 quintal per acre. They cultivate in the same land and in the same time at least 4 varieties of paddy seeds, as a measure against crop failure some what equivalent to crop insurance. They change and rotate the seeds in the field so they get more production. They never cultivate one type of paddy in the same plot of land in two successive/consecutive years.

Maize (*Maka*)

Highest quantity of maize is produced in Jenabil (16,100 kg) followed by Jamunagarh (11,100), Kabatghai (8,950) and Bakua (400). In one acre of land 15 kg seed is required, and the production will be two and half quintal. They cultivate maize in Jaisto and harvest in Bhadro. Maize is cultivated by making holes in the soils and putting individual seeds in the whole one at a time. No manure is used for maize. Maka is cultivated in Beda land like Sarshu. Maka is cultivated in the same time of Ashu dhan but in different field. Maka is sown in Jaisto and harvested in Bhado. Sarshu is cultivated in the same land of maka after the harvesting of maka. Sarshu is harvested in the month of poush/magh. Maka is sold @ Rs. 7/- kg. and 3 kg of maka seed is required for 1 acre of land and the production is 50 quintal including the stuff. Maize is cultivated in the fields near the house like mustard because crop tending and manuring is required. Before sowing of maize they put the maize seeds in a water bowl, bad seeds come to the surface but good ones sink to the bottom and they use these seeds only. They cultivate maize in the same land where they grow Sarshu /Rai, this also increase the fertility of the soil.

Millet (*Kodo/Kodu*)

Highest quantity of millet is produced in Jenabil (9,440) followed by Jamunagarh (6,100), Bakua (300) and Kabatghai (15). Kodu and Ashu dhan is cultivated in the same land away from the house as both do not require any manure. Millet is sown in ashar and harvested in Maghshir along with paddy. Millet and Ashu dhan are grown in the same plot of land but in different area. Kodo is sold @ Rs.10/- per kg. at the local market. For 1 acre of land 4 kg of millet seeds are required and production is 15 quintal. There are two types of millet, one is big and the other one is small. They also rotate it but they rotate it in a different way. They cultivate both the varieties simultaneously in the same field at 50:50 ratio and in the next year rotate the area under the big variety with the small one keeping the ratio intact. This supposedly increase the fertility of the soil and the production. There are three types of millet seeds, rabal kodo, dasai kodo and aghan kodo. Rabal kodo is dwarf one, it is grown from Ashar to Bhado, dasai kodo is a big plant grown from Ashar to Ashwin, aghan kodo is a creeper, grown from Ashar to Aghan. This way they will get kodo ready in the field for harvesting for a long period so that they can get it when ever they need it. Kodo is sold @ Rs. 4 per kg. Kodo is sown in ashar/sravan and harvested in Bhado (rabel kodo) and Ashiwn (dasia kodo) and in Aghan (aghan kodo). Gundhli is a small variety of millet. It is boiled and eaten as a monsoon crop. The benefit of cultivating multiple varieties of millet at a given point of time is that, they get a standing crop of millet in the field for a prolonged time.

Wheat

Wheat is grown in Dhipa and Beda land. Wheat is sown in direct broadcasting method after harvesting of Nala dhan, i.e., which is grown near nala or water source, in the month of Maghsir and wheat is harvested in the month of Chait/Baisak. Wheat is never cultivated in the land where Ashu dhan is grown. Some chemical fertilizer is used only for wheat in spite of the ban on the use of chemical fertilizer inside the park. Wheat is sold @ Rs. 9/- per kg. For cultivation of wheat in 1 acre of land 20 kg seed is required to get 5 quintal of wheat.

Mastard (*Turi*) and Rashi (*Til*)

Sarshu is cultivated near the homestead and it does not require any fertilizer except cow dung. But Rashi is cultivated at the far away field where they use fertilizer like urea and potash. Mastard is sold @ Rs.15/- per kg and Rashi (til) is sold @ Rs.13/- per kg. Rashi and Sarshu are grown in the same time but in different filed. For 1 acre of land 3 kg mustard seed is required and the production is 10 quintal. For 1 acre of land 4 kg Rashi seed is required and production is 8 quintal. There are two types of mustard, one is Rai sarshu (gach sarshu), it is a small plant and grows upto cheast high and the other one is Lutari sarshu, a herbal creeper type and is only 1 feet in length. Lutari or creeper sarshu is grown up a little early than the plant or Gach sarshu. They practice crop rotation by alternately cultivating plant Sarshu in one year and creeper sarshu in another year. Rashi is sown before Sarshu. Rashi is sown from Bhador to Aghan, and Sarshu is from Ashin to Poosh. For self consumption of oil, they mix and use Rashi and Sarshu together at the ratio of 60:40, i.e., Sarshu 60% and Rashi 40%. Rashi needs more water but mastard needs more manure. So they grow mastard in the field near the house where they can manure it and grow Rashi in the field away from the house where water is available in plenty.

Kondamul (*Sanga*)

Kondamul is some kind yam (tuber). It is sold @ Rs. 6/- per kg. They cultivate kandamul in Ashar and started selling it from Poosh. Kandamul is grown in Dhipo or high land. They cultivate kandamul in the same land as millet. They believe that if kandamul is cultivated the land becomes fertile next year for the cultivation of millet.

Weight and Measure

Pura is the unit of local measurement, which is more or less 400 kg, i.e. 4 quintal. Land is measured in terms of Manr; 1 acre is one and half Manr.

Crops in terms of water requirement—

Most to least

1. Paddy
2. Maize
3. Rashi
4. Mastard
5. Kodo

Crops in term of requirement of manure

1. Paddy
2. Mastard
3. Maize
4. Rashi and Kodo –no manure

Crops in terms of animal attack

Pig – Paddy and Maize, Elephant – Paddy, Monkey – Paddy

Crops in terms of food for animals

Rashi- sambar

Alu, Kodo, Paddy – Pig wild boar

Paddy, Kodo, Maka – Elephant

Kandamul – Pig, Porcupine

Elephants are driven away by beating drums and by lighting fire. Elephants come mainly for the paddy.

MANAGEMENT OF FOREST RESOURCES

Each village in the core area of the Similipal Biosphere Reserve has its definite and demarcated forest area for exploitation, which is a radius of 10 kms. In this area only cattle (Cow or Goat) of that village can graze. Cattle of other villages are not permitted in this area. Since there is no such restriction on the movement of humans, men and women can roam around for collection of firewood and other forest produces. For collection peoples go out of their home as early as 4.00 a.m. in the morning to the forest of other villages' vicinity also. Collection parties are constituted in groups with persons of same sex. Males and females form separate groups. Only the husbands and wives can go together in the forest.

The villages do not buy fodder from the market. They take their cattle for grazing in the forest. Goats eat only leaves but cows eat grass which creates problem of finding new pasture for the cows. Goats and cows can graze in the same place since their eating habits are different. Goats go to higher land and eat leaves whereas cows graze in lower land in the meadows and eats grass. The existence of privately owned forest is a unique feature of the core area of the Similipal Biosphere reserve. Here, the people after modernized their agriculture into developed terrace cultivation with two crops in a year with irrigation facilities, have been able to generate surplus agricultural production for their population and converted some agricultural fields into sal forest for their own consumption and sell also. Core area village in fact do not heavily depend on forest produces for their livelihood due to a good agricultural production, they only collect forest produces mainly for self use, however some landless labour from the Kharia community settled at the Upper Bara Kamra labour camp collect and sell forest produces for their livelihood. Human interference in the forest are mainly in the form of fuel wood collection, collection of NTP or MFP, lopping and grazing. Capacity building of the tribal population living in the forest area is the best solution for forest conservation. This is evident from the fact that after the many fold increase of agricultural production in the core villages' dependency on forest for livelihood has been decreased many fold. Forest dependency is more in Jenabil than Bakua, Jamunagarh and Kabatghai are least depend on forest as far as MFP is concerned.

Hunting

Hunting (during the colonial period mostly by the native Rajas as part of their recreation) was made illegal in 1972. But hunting for cultural compulsions is still continued in a sporadic manner in the Similipal forest. This social hunting is totally devoid of any economic reasons and widely known as Akhanda Shikar. This practice was initiated by the early Rajas of Mayurbhanj to establish their hunting right over the Similipal forest and as a good gesture towards the subject the king allowed the people also to do hunting for one day only and termed as 'Akhanda Shikar'. This was traditionally held on the day of pana sankranti in the month of April/May. Taking advantage of this cultural sanction people had started mass scale hunting operation in the month of April and May causing indiscriminate killing of wild animals and thus which once had been a one day affair became a act of concern by the forest department. At present the nature and extent of this ritual hunting becoming less due to regular counselling by the forest officials, NGO's and people's participation. It may be mention here that Kharias do not participate in Akhanda shikar.

Poaching

Unlike hunting poaching is not a localized problem and poses much more threat to the wild life being external in nature and often with inter state affinity. Similipal had a sizeable tusker population and poacher as far as from Karnataka has an interest in Similipal.

Eco Development Committee (EDC)

As a measure of forest conservation with the help of people's participation government has started eco development committee or EDC in the villages under the eco development programme in Orissa. This initiative has been projected to cover 65 villages within the Tiger Reserve and 1200 villages with in a periphery of 10 km. from the Tiger Reserve which are situated with in Similipal Biosphere Reserve. Eco development Programmes attempts to involve local people in forest conservation because it is well understood that without the participation of local people it is impossible for a forest guard to manage about 40 sq. km. of forest patch allotted to him.

Sabuj Bahini: In the core area of the Similipal Biosphere Reserve sabuj bahini is the protector of wild life. Their members are employed by the forest department but from the villages only for their supposedly good knowledge of the local forest terrain. They also prevent and control forest fire. The members of the sabuj bahini petrol the near by forest and look out for the poachers. They are equipped with a GPS hand set and report immediately to the head quarters if they find any sign of intruders and any pug mark of tiger. Sabuj bahini staffs are paid only Rs. 1,500/- p.m. from the forest department and but at present demanding Rs. 3,000/- at least to meet their livelihood.

Self Help Group

In the villages of the buffer area located near the boundary of the core zone numerous self-help groups or SHG have been formed. They mainly develop a alternative livelihood for the people living in the area to minimize the dependency on forest by developing cottage industry based on renewable forest resources like leaves, resins, flowers, pottery etc.

Forest Produces

Fuel wood : It is an important economic activity for the people living in the core area of Similipal Biosphere Reserve. Women collect firewood mainly for the domestic consumption. They collect only dry and dead fallen tree, twigs and branches from the ground. A head load of fire wood is consisting of 40 kg. As per some local calculation one family of five people requires 5 kg of fuel wood, therefore seventeen families in the Kabatghai village will require 85 kg of fuel wood per day.

Table 7. Annual Requirement of Fire Wood

Village	*Number of Family*	*Average Fuel wood consumption per day (kg)*	*Average Fuel wood consumption per year (in quintal)*
Kabatghai	17	85	310.25
Jamunagarh	18	90	328.50
Jenabil	28	140	511.00
Bakua	21	105	383.25

From the above table it is evident that the two villages which consumes more firewood are located deep inside the core area where forest cover is available in plenty.

Kendu Leaves (Diasporas Melanoxylon) : Women collect kendu leaves from the forest in the month of April – July. The leaves are sold to the local traders who collect leaves from the village itself. In the core village due to the ban on free movement by the outsiders traders can not visit and at times the people has to go at least upto Gudgudia in the buffer village. These leaves are used to manufacture traditional cigarettes known as bidi.

Siali Leaves (Bauchinia Vahili) : The peak period for the collection is immediately after the monsoon when there is a flush of new leaves. This is an important source of livelihood for the poor and marginalized farmers in the area where sal leaves are not available. There are numerous use of siali leaves in fact more than the sal leaves-

1. Thatching material
2. Protection to rain as umbrella
3. Wrapping material
4. Cup
5. Plates

Sal (Shorea Robusta) – leaves is an important NTP and used as a source of income to many forest dwellers. Sal leaf plates and cups has a great demand in the market for its biodegradability. Most of the sacred grooves or Zahira in the Similipal Biosphere has a predominant population of sal. This shows that how important it has been considered by the people.

Sal Seeds – Sal seed or decorticated kernel of sal fruit is an important non timber forest produce of Similipal. It is used in solvent extraction plants for the manufacture of fats and de-oiled cakes. Fats in turn are used in soap making and in vanaspati. The collection of sal seeds starts in the month of May and continues till the end of June. After the collection of kendu leaf and mohua flower is over, this is the period when no other forest produce is available and agricultural operations are yet to start. The fruits are spread on the ground and dried for two days. Then the seeds are crushed and the wings and the coats are removed manually. This decorticated seeds are then ready for sell.

Hill Broom Grass (Thysanoloena Maxima) – It is one of the most significant MRP of Similipal. This renewable natural resource belongs to the family of grasses is in great demand for its inflorescence which is used to make brooms.

Beside these main forest produces mushroom and different tubers are available in plenty in the forest. There are two varieties of mushroom (chattu), small in size and dark in colour, it grown on the ground and considered to be the best. Another variety is called Bidoni mushroom, it is big in size and grows on the white ant hills only. People of Jenabil do not collect mushrooms much as it is not lucrative for them because they had to take it out of the core area traveling a long distance and at times bribing the forest guards also.

Honey is another forest produce exclusively collected by the Kharia community. There are two seasons for honey collection, one is in Baisak / Josto and another is in the month of Kuar. The honey collected in the month of Kuar is slightly sweeter than the honey collected in the month of Baisak / Josto. But in case of heavy rainfall the honey collected in Kuar month could be greatly affected. There are also two types of honey bees, one is called Lelle and the other is called Hurumsuku. The hives of Lelle bees are bigger in size but the hives of Hurumsuku bees are smaller in size. Lelle hives are found both in hills and trees, but the Kharias could only collect honey from the hives located at hills and no other tribes have the expertise of collecting honey from the hill slopes. For tree hives they put fire on the ground and carry the fire with them while climbing the tree. After cutting the hive they put down the hive gently with the help of a rope, press it with hand and use a clean cloth as sieve to collect the honey. Hurumsuka bees build their hives first by digging a horizontal hole in the tree trunk and extent it vertically up and deep inside the tree trunk to build the hive. This type of hives is not seen from the outside and the hives are identified only by the movement of the bees. For the collection of Hurumsuka honey they take out the hive by vertically cutting a portion of the tree. They always collect the honey during the darker period or new moon phases of the lunar cycle so that the bees can not see them in light.

The wax is sold @ Rs. 100 per kg. and honey is sold @ Rs. 50 per kg. at the Guddudia market.

Sal Resin or dhuno is another important forest produce in Similipal. Resin is available from the sal tree only in the month of Bhodo, Kuar and Kartik. Not all Sal tree produce dhuno but only the trees infected with sal borer produce dhuno. Sal resin or dhuno is sold @ Rs. 50 per kg at the local market but its value increases manifold in the city.

For body oil they use Kusum seed oil which is collected in the month of Ashara and Sravan.

However, generally speaking villages of the core area in the Similipal Biosphere reserve do not depend much on the forest produce due to very good production of agricultural crops and they don't have a ready market nearby to sell their forest produces for which they had to make a long distance travel.

Sacred Grooves

In all the core villages there is at least one sacred groove – a patch of forest land preserved as a gene pool for maintaining biodiversity. In this Jahira / Jair numerous species of plants are found but out side it in the neighboring forests no such varieties are observed, which is predominantly Sal forest. In the forests near the village areas lack of fruit bearing trees are observed. Fruit bearing trees and species other than Sal is found in deep forest.

Disturbance Index

1. To quantitatively evaluate the actual disturbances caused by human intervention in the forest areas near the core villages.
2. To statistically assess the human disturbances caused by the villages in core area 20 quadrates (10 m × 10 m) was randomly selected in 4 sides of the villages.
3. The DI is computed by adopting a linear combination of the defined parameters on the basis of probabilistic weight ages.
4. Circumference at breast height (cbh 1.37 m from the ground) was adopted for all the trees.
5. Shrubs and saplings were taken separately as these have no timber value and therefore are not likely to be directly intervened by humans expect by the cattle.
6. Normal standing trees as well as the standing dead, dead fallen, living fallen, Lopped/damaged and clear cut at trunk were taken into consideration.
7. D.I. Disturbance Index was calculated as the percentage of damaged trees to the total number trees per 2000 sq. mt. (i.e. 20 quadrates of 10 m × 10 m).
8. Irrespective of the value of the index, data shows that the numbers of clear cut trees at trunk level and living fallen are very few, this negate the human role as the frequency of dead trees could be attributed to natural phenomenon.

A chart showing the disturbance index of the four core area villages are shown below:

Village	*Normal Trees (N)*	*Damaged Trees (N)*	*Total (N + D)*	*D.I. (%)*	*Disturbance Level**
Kabatghai	143	41	184	22.28	Medium Disturbance
Jamunagarh	150	30	180	16.66	No Disturbance
Jenabil	158	24	182	13.18	No Disturbance
Bakua	134	42	176	23.86	Medium Disturbance

* As per Knight, 1975.

Bio Diversity Index

1. In ecology a **diversity index** is a *statistic* which is intended to measure the biodiversity of an ecosystem.
2. Shannon's diversity index has also been calculated for the core villages to see the diversity of the plant species (trees).
3. Calculation is based on the numbers plant/tree species available in a given quadrate to the total tree population.
4. There has been a phenomenal difference observed between the sacred grooves and in general forest areas.

A chart showing the Shannon's (1963) diversity index of the four core area villages are shown below :

Village	*Neighbouring Forest*	*Sacred Groove*
Kabatghai	0.116	1.342
Jamunagarh	0.099	1.207
Jenabil	0.075	1.322
Bakua	0.072	1.486

Therefore, it is presumed that long back Similipal forest in core area had good diversity of plants/trees, which is still intact inside the sacred groove due to religious restrictions whereas outside the sacred grooves diversity was not maintained.

Summary and Observation

The following five features are identified as livelihood problem for the people living in the core area of the Similipal Biosphere Reserve.

1. Increase of population
2. Limitation of land resources
3. Loss of soil fertility
4. Less occupational mobility
5. Restriction on the use of forest resources
6. Emotional attachment to their habitat

There has been a great deal of micro-agricultural variation present in the core area villages due to local adaptation in crop selection and production. Some villages produce some types of crops which others do not; some villages tend to grow more of a certain types of crops than others. Jamunagarh produce wheat whereas wheat is not grown in

Jenabil due to cold. Cultivation of millet has been stopped in Bakua whereas other three villages continue to produce millet. Cultivation of a crop highly depends on land and soil types.

There are a few types of paddy, maize and millet seeds available. These seeds are selected according to the soil types, slope and water retaining capacity of the land. They choose to cultivate a particular variety of seed for a particular crop based on its suitability according to the nature of land and soil and other climatic condition. In a single village one can find three types of paddy, two types of maize and two types of millet are in use for different categories of land.

Since human activities have been implicated in the majority of the problems related to plant and animal biodiversity it would be rational to put research effort to human activities and attitudes that are responsible for such problems. Generally speaking biodiversity involves undomesticated plants and animal species in which biologist, ecologist and foresters are more interested but for an anthropologist cultural value of biodiversity is more important which involves the diversity of domesticated plants and animals and associated value system in peoples' cognitive domain. The study finds that the communities living in the four villages inside the core area of the Similipal Biosphere Reserve have built unique terraced agricultural fields along the village elevation line so that water can be channelised across the contour of the village. They also lift river water to irrigate the field. They judiciously synergized the traditional method of crop rotation, soil preparation and seed selection along with other modern method of agriculture to get a good production. Good productions of crops have lessened their dependency on forests which indirectly helps the growth of forest in the core areas. The core area villages mainly collect forest produces for their own use and consumption because marketing of these products are difficult due to the distance of these villages from the market centre.

The biodiversity of the general forest in the core area is poor which is predominantly a Sal forest where as in the sacred forest/groove (Jahira) the biodiversity of the plants/tress are quite well. As mentioned earlier the peoples residing in the core area depend not much on forest as they have very good agriculture along with impoverished irrigation system. Issue of relocating these four core area villages as part of Biosphere Reserve Policy has been addressed for quite some time now but without any results because the land offered to them is much inferior to the land they held at present inside the core area and grossly unacceptable to the people.

11

Land Use Changes: A Study in the Core Villages

P.K. Guha

Introduction

Government of India declared Similipal as Biosphere Reserve in 1994. The reserve derives its name from the magnificent *Simul* (Silk Cotton Tree). It has been aptly described as Salmali Saila (the hill of *Simul*) in the poem 'Usha' written by the famous Oriya poet, Radhanath Ray (*Tiger Reserve status report, 2001 issued by Ministry of Environment and Forest, Bhubaneswar, Orissa*). The Biosphere Reserve is located on the north eastern part of Orissa in Mayurbhanj district, adjoining to the states of West Bengal and Jharkhand. It lies between 31°88′ to 21°27′ North latitude and 86°04′ to 87°17′ East longitude. Similipal gives a picture of rugged terrain, dense forest and wild life species. Administratively, Similipal Biosphere Reserve covers three taluks, namely, Baripada,Panchir and Baghmati of the Mayurbhanj district of Orissa.This Biosphere Reserve has three zones namely *Core zone, Buffer zone and Transition zone* spread over 5324 sq. km.out of which 845 sq. km. falls under the Core zone.

Physical Environment

The Similipal Biosphere Reserve is situated in the central part of Mayurbhanj district sloping towards east and west is covered by *Sal* forest. The *Meghasini* hills (seat of clouds) which rises to a height of 3824 ft. is situated in the extreme south. The other important hills are the *Dhudruchampa* (3310 ft.), *Gorumahasani* (2964 ft.), *Badampahar* (2730 ft.), *Chahala* (2541ft.) and *Balidiha* (2044 ft.). There are several

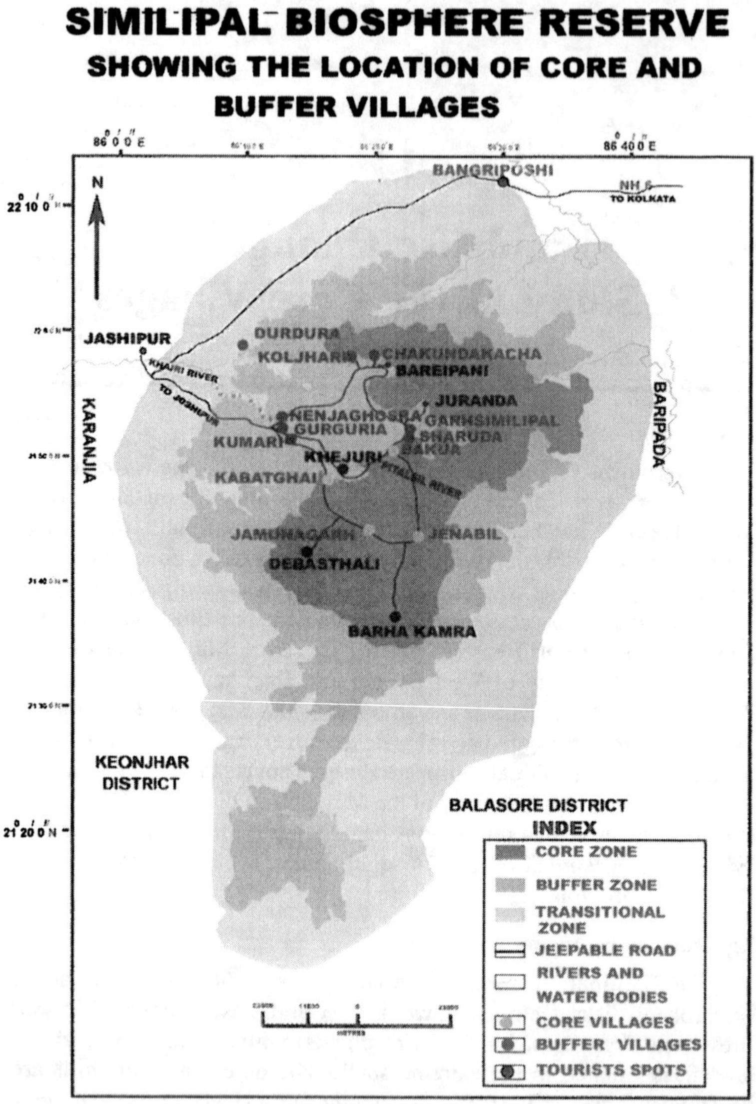
SIMILIPAL BIOSPHERE RESERVE
SHOWING THE LOCATION OF CORE AND BUFFER VILLAGES
86 0 0 E
86 40 0 E
22 10 0
21 20 0 N
N
BANGRIPOSHI
NH 6
TO KOLKATA
JASHIPUR
KHAIRI RIVER
TO JOSHIPUR
KARANJIA
DURDURA
KOLJHARI
CHAKUNDAKACHA
BAREIPANI
JURANDA
NENJAGHOSRA
GARHSIMILIPAL
GURGURIA
SHARUDA
KUMARI
BAKUA
KHEJURI
KABATGHAI
JAMUNAGARH
JENABIL
DEBASTHALI
BARHA KAMRA
BARIPADA
KEONJHAR DISTRICT
BALASORE DISTRICT
INDEX
CORE ZONE
BUFFER ZONE
TRANSITIONAL ZONE
JEEPABLE ROAD
RIVERS AND WATER BODIES
CORE VILLAGES
BUFFER VILLAGES
TOURISTS SPOTS

Map 1

streams flowing in all directions, which ultimately meet the Bay of Bengal. The major perennial streams are the *Budhabalanga, Palpala, East Deo, Nekedanacha (Salandi), West Deo, Khairi, Bhandan, Khadkei and Jamira*. The rivers are not navigable in any season but give rise to floods during rainy season.

Soil

The most usual classification of the soil of this area is based on its position or level. The area is mainly undulating except along the few rivers and it comprises ridges, slope and depressions.

According to farmers cognition, there are five types of soil namely *(1) Balimati, (2) Dorasa mati, (3) Matial or Chikita mati, (4) Tilak mati* and *(5) Lal mati.*

Bali mati contains more than 2/3rd of sand and is very poor in fertility. *Chikita mati* is a strong clay which retains more water as the drainage system in this area is not good and not favourable for cultivation. *Dorasa mati* is a kind of mixed soil of sand and clay. It retains moisture and easy to cultivate. *Tilak mati* is grey in colour and is sandy. It is of lateritic origin.

From the general observation, the soil of this region may broadly be classified into two groups the Red soil and Lateritic soil.

1. Red soil : The typical red soil mostly Aus paddy, lesser millets, sabai grass and other minor crops are grown.

2. Lateritic soil: Two types of lateritic soil are found in this area. These are lateritic morum and lateritic rock. For agriculture purposes, the soil has no productive value. The main two classifications of agricultural land are – *Jal* (low cultivated land) and *Daha* (high cultivated land).

The *Jal* lands are those that are found in the depression but the *Daha* lands are found in the upland where the water is not stored.

Vegetation

The virgin semi-evergreen forest dominates the reserve. The growth is thick and impenetrable. The reserve consists of different types of forests such as Northern Tropical Semi Evergreen Forests, Northern Tropical Moist Deciduous Forests, Dry Deciduous Hill Forests, High-level Sal Forest, Grassland and Savannah.

There are 1076 species of plants of which 92 are orchids having similarities to the Northern and Southern Indian orchids. Seven percent of the Indian species of flowering plants and eight percent of orchids of India are found in Similipal. Northern Tropical Semi Evergreen Forests extend over an area of about 80 sq. km. Northern Tropical Moist Deciduous Forests extend over an area of about 1540 sq. km. It is found all over Similipal except the moist valleys on the Southern and Eastern aspects of the hills. Similipal is rich in Orchidaceous flora 92 species of orchids has been identified.

Fauna

There are 42 species of major mammals, 242 species of birds and 30 species of reptiles recorded so far in the Reserve. Tiger, Panther, Elephant, Gaur, and four horned antelope etc. are found in different levels of abundance. Ratel, Pangolin, Giant squirrel, Sambar and Cheetal are among the other few worth mentioning.

Tiger (*Panthera tigris*) including melanistic forms, Leopard (*Panthera pardus*), Elephant (*Elephas maximus*), Sambar (*Cervus unicolor*), Barking Deer (*Muntiacus munjak*), Bison (*Bos gaurus*), Forest cat (*Felis chaus*), Wild boar (*Sus scrofa*), Giant Squirrel (*Ratufa indica*) and Common Langur (*Presbytis entellus*) are found in Similipal Biosphere Reserve.

Climate

Oppressive high summer, high humidity and well distributed rainfall during monsoon characterized the climatic situation of this Biosphere Reserve. The temperature rises up to 32.5 degree centigrade during summer (March to May) and it goes down to 5 degree centigrade during winter (December to February). The annual average rainfall is 200 cm spreading over the whole Biosphere Reserve.

Land use

The value of land depends on its use. The use depends on the following three factors:

1. Accessibility: It includes both time and money costs associated with traveling, as an immobile factor.

2. Space: The space factor depends on the need for land in accordance to its size and quality.

3. Environment: This includes physical and cultural land characteristics.

Objectives

The land use is an important part of cultural landscape which depends on environment, both physical and cultural. The former includes highland, low land and water bodies; and later includes settlement pattern, public utility places and cultural places. So both the elements of environment, directly or indirectly, have been deflected through the land utilization.

The main objective of the present study is to show the change in land use of four villages of core zone of the Biosphere Reserve which are Bakua, Jenabil, Jamunagarh and Kobatghai. The field work was done in the month of November and December, 2006.

Methodology

To understand the land use change over the years of the above mentioned four villages of core zone of Similipal Biosphere Reserve, data have been collected on each plot of these villages. First, the informations were collected on each plot from 1934 - 35 land records from Land Revenue Office, Baripada. The detail of land characteristics of 1934 - 35 of each number of plot of each village have been noted down from the record books along with cadastral maps of each village. The information on land characteristics have been collected on (Bakua 147 nos. of plots, Jenabil 92 nos. of plots, Jamunagarh 31 nos. of plots and Kobatghai 184 nos. of plots) 454 nos. of plots of the above mentioned four villges. While dealing with the data several units of cadastral map like quantum of land, different elements of physical and cultural environments, Government and non-Government land, etc. have been taken into consideration. Accordingly different colors and symbols have been put on different numbers of plot. For an understanding of the land use maps of these four villages, a standardize pattern of colours has been used like blue for water, yellow for arable land, red for residential areas, green for forest and so on. From the data analysis and from the cadastral maps of 1934-35, tables of four villages have been prepared.

Bakua village

Bakua village is one of the four in habitat villages in the core zone of Similipal Biosphere Reserve and its extension recorded between 86°21′ 10.2″E to 86°21′42.4″E longitude and 21°48′40.1″N to 21°49′17.5″N latitude. The village is located at a distance of 65 kilometers east from Jashipur. This village is surrounded by dense forest of mainly the Sal trees. In the north Cherabil forest is located, the east is bounded by Badhargora forest, Ghardhu forest is found in south and the west of this village is bounded by Cheroananhi forest. The average height of this village is more than 900 m. The central part of this village is slightly lower than its surrounding areas. The highest (988.8 m.) point is found at the northwest corner of the village The Kikkirondu river is flowing south to north in the village and the villagers use this river water for irrigation in agriculture, mainly the paddy. The summer temperature rises to 36 degree Celsius but in winter it goes down to 7 degree Celsius. The average rainfall is 24 mm. and the average rainy days are 24 during rainy season. Two Jeepable roads on the west and the southern side of this village connects to the town of Jashipur, Garhsimilipal and Jenabil village. The village has 145 acres of land.

According to the household census conducted during 2006, the village has 21families of Ho tribe having 97 individuals, of which 44 are males and 53 are females. The sex ratio 1204 female per thousand males and the family size are 4.

The economy of the people mainly depends on agriculture - mainly the rice for which they use cattle. According to 2003 census, 99 cattles are found in this village. Besides cattles, 57 no. of goats are found. They sell out the well-grown goats at the nearby market i.e., Gurguria market. 2003 census also gives an idea of poultry.

Table 1. Land use characteristics of Bakua village in percentage (1934)

Sl. No.	*Land characteristics*	*Area in acres*	*Percentage*
1	High cultivated land	5.39	3.71
2	Low cultivated land	75.96	52.31
3	Forest	34.26	23.59
4	Pasture land	11.o7	7.62
5	Fallow land	14.96	10.30
6	Water bodies	0.78	0.54
7	Residential areas	1.61	1.11
8	Sacred groves	0.76	0.52
9	Jeepable road	0.44	0.30
	Total	145.23	100.00

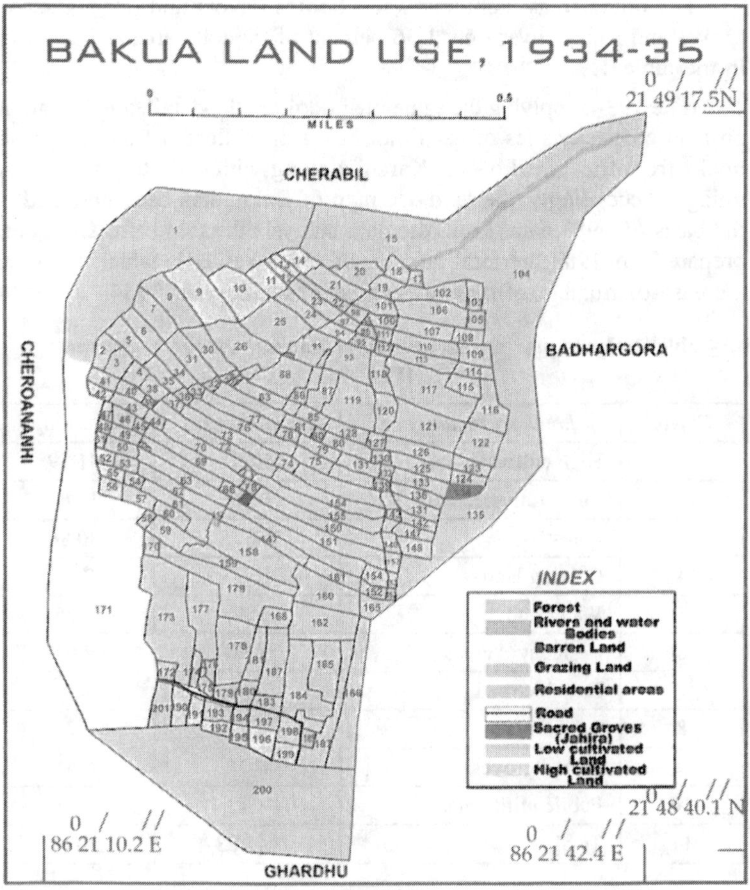

Map 2

The low cultivated land is followed by forest area, which is only 23.59 percent. The map No. 2 shows that the barren land is located in the western part of this village which is only 10.30 percent. Grazing lands follows the barren land and it covers only 7.62 percent area where the cattles are grazed. The grazing land is followed by high-cultivated land which covers only 3.71 percent area. In this cultivated land mainly maize and mustard are grown and is located in the northwest corner of this village. High-cultivated land is followed by residential areas covering only 1.11 percent area of this village. The figure shows that in the year 1934 this village had very thin population.

The other areas including water bodies (rivers and ponds), sacred groves and roads cover only 1.36 percent of the total area as has shown in the table 1.

Like 1934, applying the same methodology the detailed informations on land characteristics of each plot has been collected from the record books from the Tahsil office, Karanjia along with cadastral map of this village. Accordingly, the land use map of Bakua has been prepared on the basis of 1983 data. From the data analysis the table No. 2 has been prepared. In 1983 the total number of plots was 201, which shows an increase of number of plots as compared to the year 1934.

Table 2. Land use characteristics of Bakua village in percentage (1982-83)

Sl No.	*Land characteristics*	*Area in acres*	*Percentage*
1	High cultivated land	59.82	41.19
2	Low cultivated land	19.76	13.61
3	Forest	44.75	30.86
4	Grazing land	10.93	7.52
5	Water bodies	1.24	0.86
6	Residential areas	2.96	2.04
7	Jeepable road	3.58	2.47
8	Sacred place	0.31	0.22
9	Sacred groves	0.75	0.52
10	Public utility place	8.13	0.71
11	Total	145.23	100

The table 2 shows that 41.19 percent areas have been covered under high-cultivated land. It shows that the increase is more than fourteen times as compared to the data available in 1934. The proper causes were not known. On the other hand, the low cultivated land has been decreased from 52.31percent in 1934 to13.61 percent in 1983. It shows that the maximum amount of low cultivated land was converted to high-cultivated land. Though the proper explanation could not available from the land revenue office but the villagers informed that previously the rainfall was high so there was no problem for irrigation even in the dry season and they could grow rice in this land. But presently due to the shortage of natural water in this land the rice is grown only at the

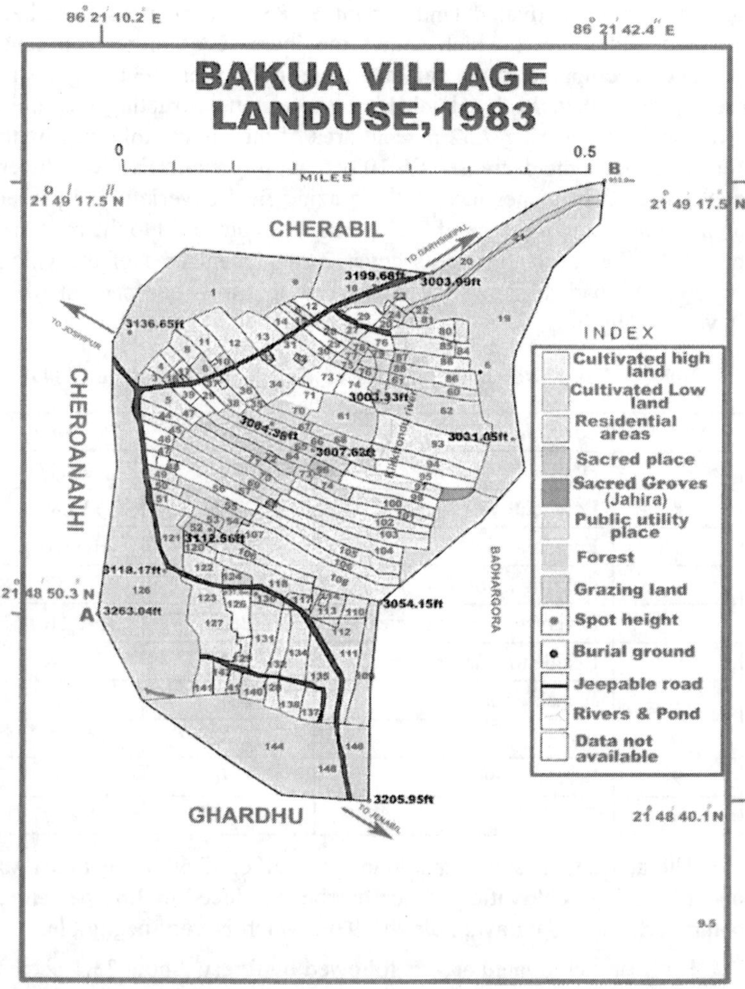
BAKUA VILLAGE
LANDUSE,1983
CHERABIL
CHEROANANHI
GHARDHU
BADHARGORA
INDEX
Cultivated high land
Cultivated Low land
Residential areas
Sacred place
Sacred Groves (Jahira)
Public utility place
Forest
Grazing land
Spot height
Burial ground
Jeepable road
Rivers & Pond
Data not available
MILES

Map 3.

time of rainy season where proper water is available. But in other areas, they cultivate mainly mustard, wheat, maize and til for which very less amount of water is required. According to the villagers in low cultivated land mainly the rice are grown but mustard, wheat, maize and til are grown in high cultivated land. About 30.86 percent areas have been covered under forest which shows the increase of more than seven percent as compared to the data available in the year 1934. It gives the clear picture that the local inhabitants are not destructing the forest. Grazing field covering 7.52 percent area of the village followed by the forest. It shows the decrease of 0.10 percent of grazing field which very negligible. Road comes next to the grazing field covering 2.47 percent area. It shows the increase of 2.17 percent as compared to the road area of 1934. The residential areas cover 2.04 percent area of the village follows the road. It shows the increase is less than one percent which is very negligible.

Table. 3 Land use characteristics of Bakua in percentage (2006)

Sl. No.	*Land characteristics*	*Area in acres*	*Percentage*
1	High cultivated land	54.56	37.57
2	Low cultivated land	20.04	13.80
3	Forest	48.07	33.1
4	Grazing land	10.92	7.52
5	Rivers and water bodies	1.25	0.86
6	Residential areas	5.72	3.94
7	Jeepable road	3.59	2.47
8	Sacred place	0.32	0.22
9	Sacred groves	0.76	0.52
10	Total	145.23	100.00

The above table shows that in the year 2006, 37.64 percent area was under high land cultivation, which has been reduced to 3.62 percent as compared to the data available in 1934, which is very negligible.

High land cultivated area is followed by forest. About 33.1 percent area was under forest, it shows that increase of 2.24 percent of forest which gives an idea that the forest area is not badly affected by villagers. The percentage of the forest area has been increased because the families who used to stay in plot number 18 on the north, shifted to plot number 111 and 112 on the southern part of the village to get

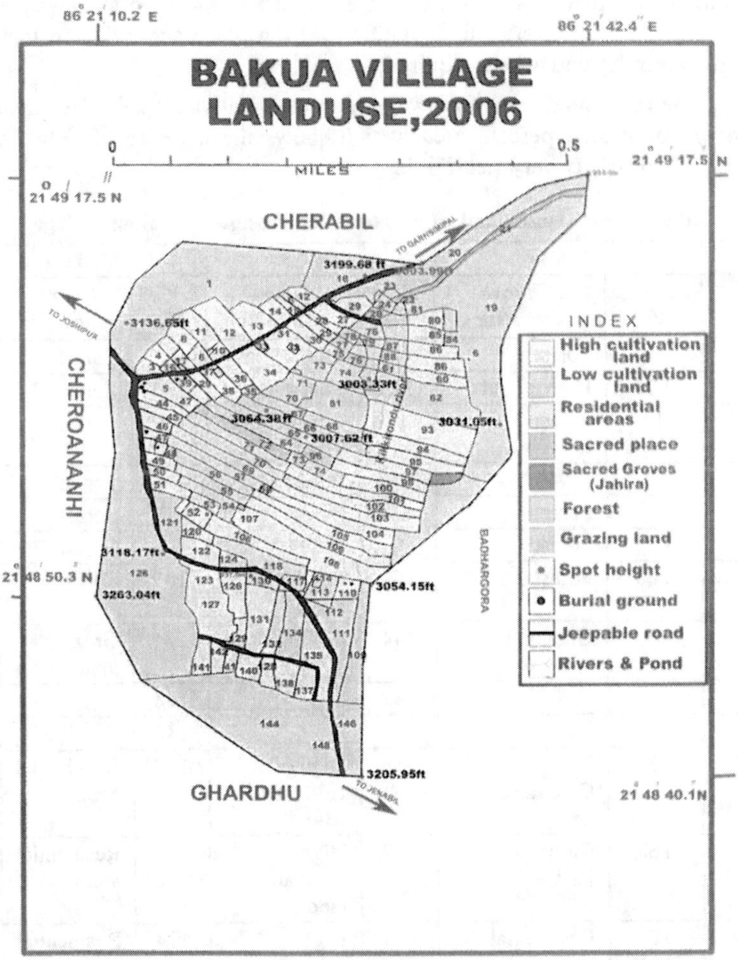

BAKUA VILLAGE
LANDUSE,2006
CHERABIL
CHEROANANHI
GHARDHU
BADHARGORA
INDEX
High cultivation land
Low cultivation land
Residential areas
Sacred place
Sacred Groves (Jahira)
Forest
Grazing land
Spot height
Burial ground
Jeepable road
Rivers & Pond

Map 4.

the agricultural land nearer to their residence. Presently this plot is fully covered by Sal trees. In 1983 plot number 109 shown as public utility place but the author's visit to this plot gave the idea of forest. After enquiring it is learnt that the villagers stopped this plot of using as public utility place long back because they are using some other vacant small plot (not fixed) as their public meeting place according to their convenient as and when required.

The residential areas follow the high land cultivated area. The areas cover about 3.94 percent area, which shows the increase of only 1.9 percent, which is very negligible.

Table 4. Plot wise land characteristics change in Bakua village

	1934-35		*1982-83*		*2006*	
	Plot No.	*Chara-cteristics*	*Plot No.*	*Chara-cteristics*	*Plot No.*	*Chara-cteristics*
1.	1	Forest	1	Forest	1	Forest
2.	171	Barren land	126	Forest	126	Forest
3.	15	Forest	18	Residential areas	18	Forest
			20	Forest	20	Forest
4.	104	Forest	19	Forest	19	Forest
	135	Forest		Forest		Forest
5.	166	Forest	109	Residential areas	109	Forest
6.	200	Grazing land	144	Grazing land	144	Grazing land
			148	"	148	"
			146	"	146	"
7.	176	Low Cultivated land	118	High cultivated land	118	Residential areas
8.	168	Cultivated land	117	High cultivated land	117	Residential areas
9.	187 183 180 179 175 174	Residential areas " " " "	134 132 140 131 129	High cultivated land " "	134 132	Residential areas

(contd...)

LAND USE CHANGE OF BAKUA VILLAGE OVER THE DECADES

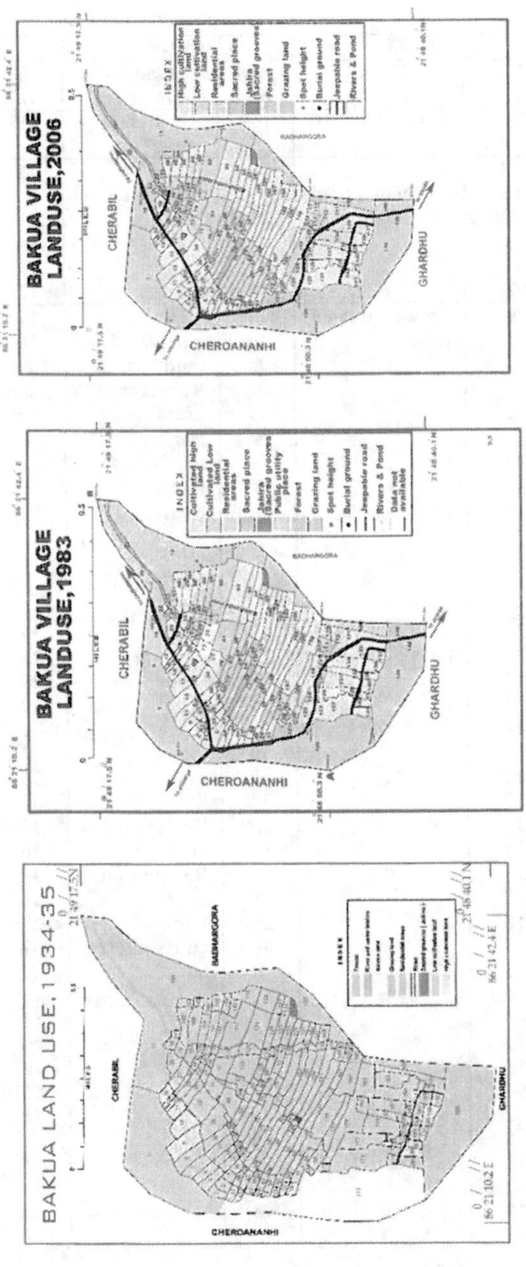

Map 5.

(contd...)

	1934-35		*1982-83*		*2006*	
	Plot No.	*Chara-cteristics*	*Plot No.*	*Chara-cteristics*	*Plot No.*	*Chara-cteristics*
10	166	Forest	109	Residential areas	109	Forest
11	14 13 12 11 10 9	High cultivated land	12 6 15 14 13 12	High cultivated land	12 6 15 14 13 12	High cultivated land
12	8 7 6 5 2 3 41 42	Low cultivated land	11 10 8 6 17 4 16 3	High cultivated land Residential areas High cultivated Land Residential areas High cultivated land	11 10 8 6 17 4 16 3	High cultivated land
13	20 21 22 23 24 25 26 30 31 34 35 38 40 39 43 46 45 49 50 54 57 56 55 52 48 47	Low cultivated land High cultivated land	27 28 29 30 31 32 33 34 36 36 37 38 29 39 5 47 44 46 49 50 51 56 55 52 48 47	High cultivated land Residential area Residential area High cultivated land	27 28 29 30 31 32 33 34 36 36 37 38 29 39 5 47 44 46 49 50 51	High cultivated land Residential area Residential area High cultivated land

(contd...)

	1934-35		1982-83		2006	
	Plot No.	*Chara-cteristics*	*Plot No.*	*Chara-cteristics*	*Plot No.*	*Chara-cteristics*
14	102 106 107 110 113 117 121 126 125 133 136 131 142 147 148 140 157 163 165 160 158 159 147 151 150 155 154 75 131 79 131 130 127	Low cultivated land	80 84 85 86 87 93 94 97 100 101 102 103 104 105 106 108 107 106 108 108 107 106 107 106 105 NA NA NA NA NA 65 97 95 94	High cultivated land	80 84 85 86 87 93 94 97 100 101 102 103 104 105 106 108 107 106 108 108 107 106 107 106 105 NA NA NA NA NA 65 97 95 94	High cultivated land
	134	Sacred groves	134	Sacred groves	134	Sacred groves
15	75	Low cultivated land	96	Residential areas	96	Residential areas

The table 4 shows that how the land characteristics of each plot has been changed. Since 1934, the plot number 1(one) remains as forest till 2006. In 1934, plot no. 171 was under barren land but in 1982 the same plot (126 no.) has been converted to forest. In 1934 plot no. 15 is referred as forest but in1982 the same plot bearing no.18 and 20 was converted to residential and forest respectively but in 2006 plot no. 18 was converted to forest because the families who used to stay in plot no. 18 shifted to plot no. 111 and 112 on the southern part of the village as the agricultural field would be nearer to their residence. Presently the plot no. 18 is fully covered by Sal trees. Plot no. 166 was under forest in the year 1934 but in 1982 the same plot bearing no. 109 shown as public utility place but in 2006 my visit to this plot shows that the area was under forest cover. After enquiring it is learnt that the villagers have stopped using this plot as public utility place long back and presently they are using some other vacant small plot (not fixed) as their public meeting place in accordance as when required. In 1934, the plot no. 176 was under low cultivated land but in1982 the same plot bearing no. 118 has been converted as high-cultivated land but the same plot has been converted to residential area in the year 2006 because as this plot is located on high land so the villagers have chosen this place for their residential use. In the year 1934, the plot no. 187, 183, 180,179, 17&, 175, is shown as residential areas but in the year 1982 all the plots shown as high land cultivated areas bearing no.134, 132,140,131 and 129 from the field investigation it has been observed that these people have shifted to plot no. 6 and 10 to get the road facilities. In 1934, the plot no. 26 and 30 have been shown as low cultivated areas but in 1982 &2006 the two parts of these two plots have been shown as residential areas bearing plot no. 33 and 34 respectively. The plot no. 7 which was low cultivated land in 1934 has been changed to residential areas in 1982 bearing plot no.10, the causes of this change has already been mentioned earlier. Plot no. 75 was referred as low cultivated land in the year 1934 but this plot was converted as residential areas in the year 1982 bearing plot no. 96. On author's visit to this plot in the year 2006 also shows the existence of residence.

Discussion: From the above analysis it is observed that the forest area has been increased to 9.52 percent since 1934-35, which is a very good signal to the administrators, forest management officials and the researchers that the forest resource is well maintained and the villagers are not destructing the pristine forest of this biosphere reserve. The residential areas have been increased to only 2.83 percent since 1934, which is not also very alarming.

Jenabil village

Jenabil is located on the south of Bakua village and it extends between 86°21′12.8″E to 86°21′43″E longitude and 21°43′41.6″N to 21°44′06.1″N latitude. The village is located at a distance of 65 km. south-east from Jashipur. This village is surrounded by dense forest, Rohali Buru forest is located in the north of this village and the east is bounded by Duarsuni forest. Hatsai forest is found on the south of village and the west is bounded by Kurtapur forest. The average height of this village is more than 850 m. The highest point is found on the north of this village (1044 m.) where the forest guesthouse is located. This village has gentle slope from north to south and from west to east. The Khairi river is flowing from north - east to west of this village and it meets with Jenabil river at the west and after meeting with Jenabil river at the west, both the rivers taking the name of Jenabil flows towards south. The villagers use the water of these two rivers for cultivation and for drinking purpose. The summer temperature rises to 36 degree Celsius but in winter it goes down to 4 to 5 degree Celsius. Average rainfall is 24 mm. during rainy season and the average rainy days are 24.Two Jeepable roads are found on the north, west and south connecting Khunta, Kabatghai and Jamunagarh. According to the household census out of 145 individuals, 74 are males and 71 are females, sex ratio is 959 females per thousand males having 25 nos. of family. The average family size is 5.8. According to 2003 census 119 cattles are found in this village. Besides cattles, 143 no. of goats are also counted. They sell out the well-grown goats at the nearby market i.e., Gurguria market.

Table 5. Land use characteristics of Jenabil village in percentage (1934-35)

Land characteristic	*Area in acres*	*Percentage*
Forest	110.05	31.97
Lowland cultivated areas	160.76	46.7
Rivers and water bodies	37.30	10.84
Inspection Bung low	7.64	2.22
Fallow and Stony land	13.18	3.83
Hills	15.28	4.44
Total	344.21	100.00

The table 5 shows that 46.7 percent land was under low land cultivation. From the map it is observed that in the year 1934 there was no high land cultivated areas in this village. Low land cultivated area is followed by forest which covers only 31. 97 percent area of this village. The rivers and water bodies follow forest which cover 10.84 percent of the total area. The river and water bodies are followed by hills, covering 4.44 percent area of the village. Only 3.83 percent area is covered by fallow and stony land, the only administrative building (forest inspection bunglow) is found in this village covering only 2.22 percent area of the village. The above informatons collected from Land Revenue office, Baripada along with cadastral map of Jenabil.

According to the record supplied by the land revenue office, Baripada it is observed that in 1934 there was no residential areas in Jenabil village, but the cultivated land. The people in the historical past could not settle down here because this village is located in the remote

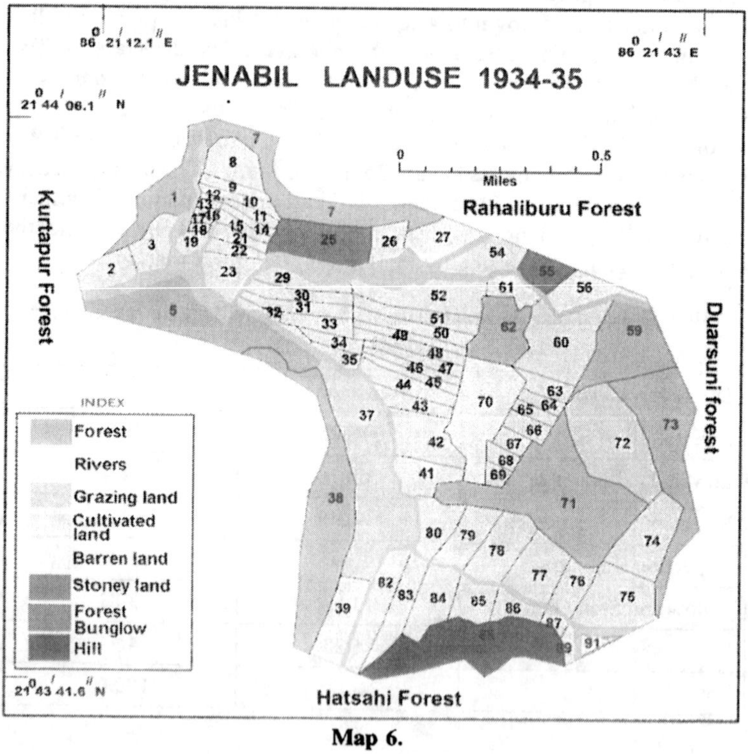

Map 6.

of core zone surrounded by dense forest, also there was no proper communication facility. So people from Jamunagarh came to cultivate the land during daytime.

Like 1934, applying the same methodology the detailed informatons on land characteristics of each plot has been collected from the record books from the Tehsil office, Karanjia along with cadastral map of Jenabil village. Accordingly, the land use map was prepared on the basis of 1982-83 data. From the data analysis the table No. 6 has been prepared. In 1983 the total number of plots were 310, which shows an increase of number of plots as compared to the year 1934.

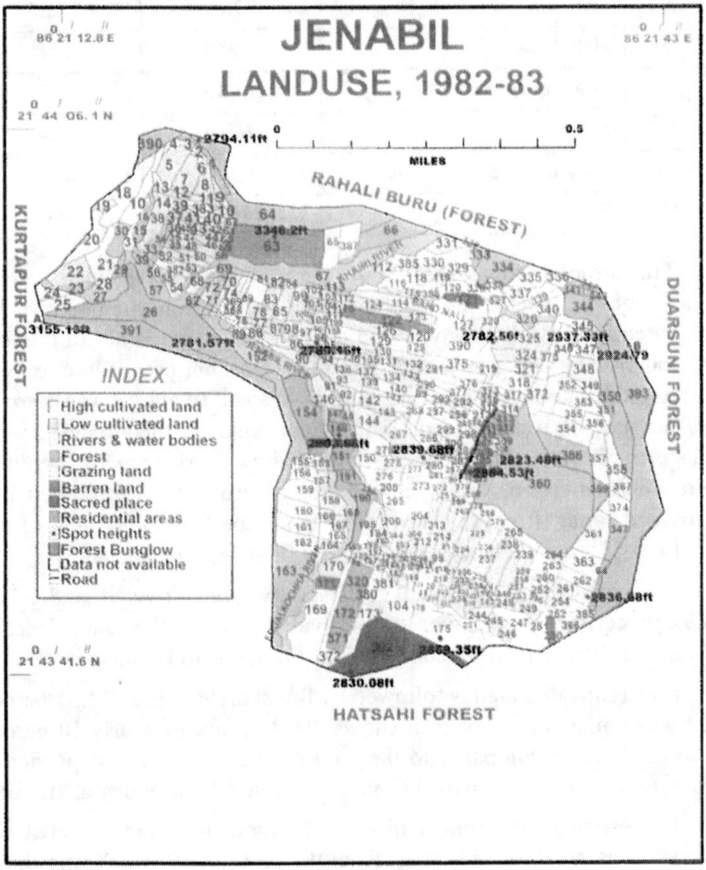

Map 7.

Table 6. Land use characteristics of Jenabil village in percentage (1982-83)

Sl No.	*Land Characteristics*	*Area in acres*	*Percentage*
1	Lowland cultivated areas	99.56	28.92
2	Highland cultivated areas	97.27	28.26
3	Water bodies	36.86	10.71
4	Residential areas	8.55	2.48
5	Fallow land	0.96	0.28
6	Forest	77.62	22.55
7	Hill	1.99	0.58
8	Grazing land	14.35	4.17
9	Stony area	6.33	1.84
10	Sacred place	0.72	0.21
11	Total	344.21	100

The table 6 shows that 28.26 percent areas have been covered under high-cultivated land. It shows that the maximum amount of low cultivated land has been converted to high-cultivated land. In this land previously rice was grown during rainy season but presently due to the shortage of natural water the mustard, maize and til are grown where the less amount of water is required. It may also be mentioned that the villagers call low cultivated land where mainly the rice are grown but the land where mustard, wheat, maize and til are grown are called high cultivated land. It may be mentioned that a part of forest area of plot no. 1and plot no. 71 are being used as high cultivated land.

The high-cultivated land is followed by low cultivated land. About 28.92 percent area was under low cultivated areas. It shows the decrease of low cultivated land as compared to the data of 1934.

Low cultivated land is followed by forest areas. About 22.55 percent area was under forest cover. It shows the decrease of nearly 10 percent of forest land as compared to the data of 1934. This has happened as this area was converted to the high-cultivated land and grazing land.

The grazing land follow the forest. About 4.17 percent area was converted as grazing land- apart from the plot no. 59 which was shown as forest in 1934, where mainly the cattles are fed.

Residential areas follow grazing land. In 1934 there was no sign of residence, but the data of 1982-83 shown that 2.48 percent area was under residence.

But Water bodies, Fallow land, Hill, Stony area and sacred place covered other 13.46 percent area of this village.

The existing land use has been prepared of this village on the basis of map supplied by land revenue office, Karanjia and plot- to- plot survey in the year 2006.

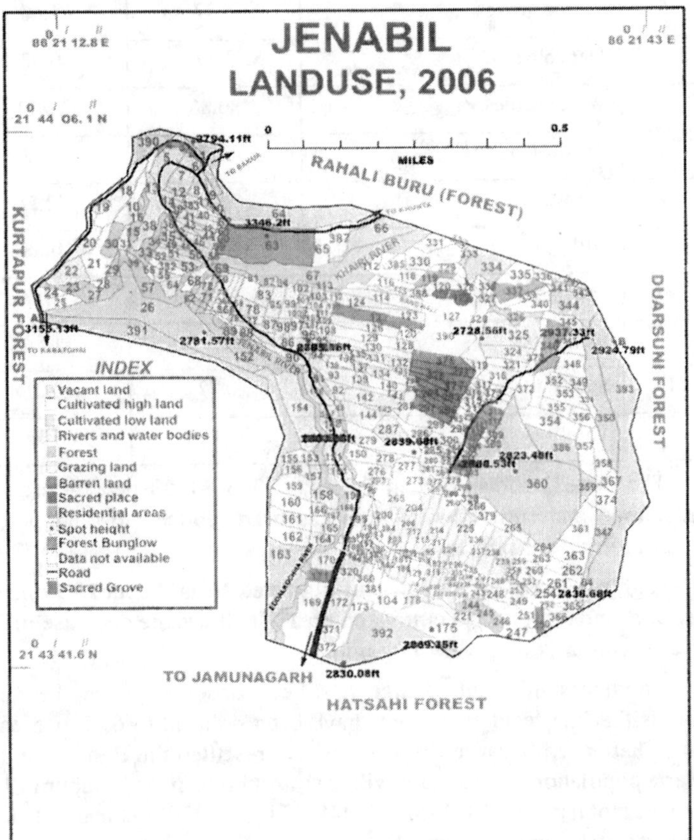

Map 8.

Table 7. Land use characteristics of Jenabil village in percentage (2006)

Sl. No.	*Land characteristics*	*Area in acres*	*Percentage*
1	Low land cultivated areas	96.66	28.08
2	Highland cultivated areas	87.05	25.29
3	Residential areas	6.41	1.86
4	Vacant land	6.74	1.96
5	Sacred groves	5.37	1.56
6	Jeepable road	3.17	0.92
7	Water bodies	36.86	10.71
8	Fallow land	0.96	0.28
9	Forest	77.62	22.55
10	Hill	1.99	0.58
11	Grazing land	14.36	4.17
12	Stony area	6.3	1.83
13	Sacred place	0.72	0.21
14	Total	344.21	100

The table 7 shows that in the year 2006, 28.08 percent area was under low land cultivation, which has been reduced to 0.84 percent, which is very negligible.

High land areas follow lowland cultivated areas. About 25.29 percent area was under high land cultivated areas. It shows the decrease of 2.99 percent which is also very negligible.

Another significant change has been observed from the above table is that the residential areas have been reduced to 0.62 percent. It has so happened because the Government resettled the Bathudi and the Kharia population from Jenabil village to Ambodia near Jashipur in 1993 from the plot number 154, 146, 147, 148, 151, 191, 320, 171 and 371. These plots are still lying vacant.

From the above table, it has also been observed the all weather road has been increased to 0.92 percent. But other land use categories

LAND USE CHANGE OF JENABIL VILLAGE OVER THE DECADES

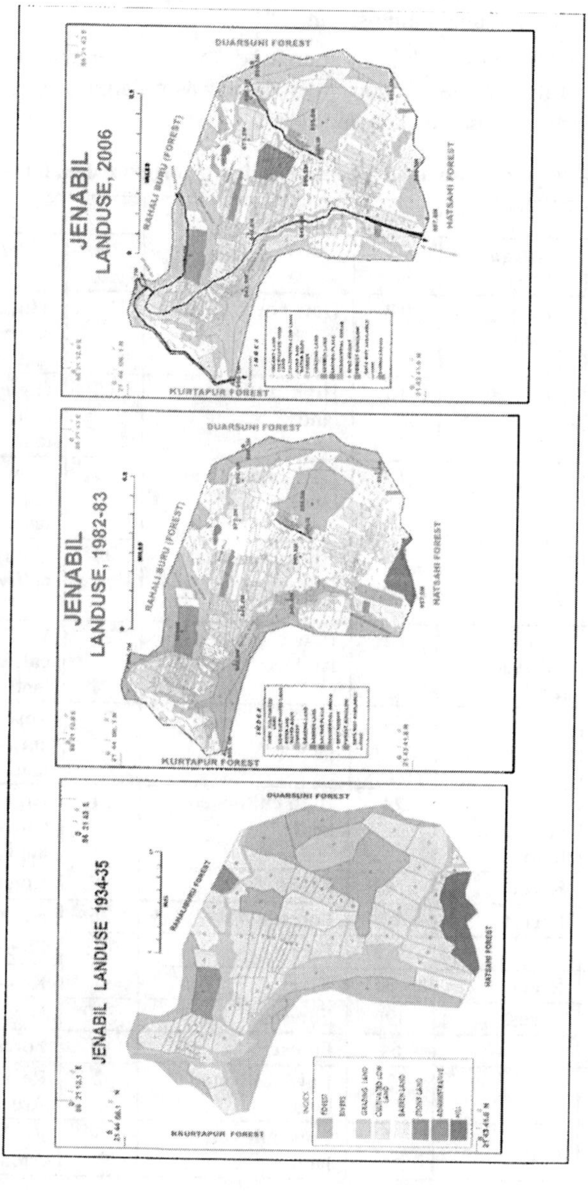

Map 9.

like Water bodies, Fallow land, Forest, Hill, Grazing land, Stony area and Sacred place remained almost same.

Sacred groves had been increased to 1.56 percent. It may be mentioned that in the previous year there was no sign of this category and the causes are not known.

Table 8. Plot wise land characteristics change in Jenabil village

1934		*1982-83*		*2006*	
Plot no.	*Category*	*Plot no.*	*Category*	*Plot no.*	*Category*
1	Forest	10	Data not available	10	High cultivated land
		18	High cultivated land	18	High cultivated land
		19	Low cultivated land	19	Low cultivated land
		20	High cultivated land	20	High cultivated land
2	Low cultivated land	22	Low cultivated land	22	Low cultivated land
		24	Low cultivated land	24	Low cultivated land
3	Low cultivated land	21	High cultivated land	21	High cultivated land
4	River	4	River	4	River
5	Forest	26	Forest	26	Forest
		391	Forest	391	Forest
6	River	6	River	6	River
7	Forest	390	Forest	390	Forest
		64	Forest	64	Forest
		2	Low cultivated land	2	Residential Areas
		3	Low cultivated land	3	Residential Areas

(contd...)

(contd...)

	1934		*1982-83*		*2006*
Plot no.	***Category***	***Plot no.***	***Category***	***Plot no.***	***Category***
		4	Low cultivated land	4	Residential Areas
		1	Forest	1	Forest
		9	Low cultivated land	9	Low cultivated land
8	Low cultivated land	5	Low cultivated land	5	Low cultivated land
		6	Data not available	6	Data not available
		7	Low cultivated land	7	Low cultivated land
		8	Low cultivated land	8	Low cultivated land
9	Low cultivated land	11	Low cultivated land	11	Low cultivated land
		12	Low cultivated land	12	Low cultivated land
10	Low cultivated land	10	Low cultivated land	10	Low cultivated land
		383	Data not available	383	Data not available
		13	Low cultivated land	13	Low cultivated land
11	Low cultivated land	62	Low cultivated land	62	Low cultivated land
		40	Low cultivated land	40	Low cultivated land
12	Low cultivated land	14	Low cultivated land	14	Low cultivated land

(contd...)

1934		*1982-83*		*2006*	
Plot no.	*Category*	*Plot no.*	*Category*	*Plot no.*	*Category*
13	Low cultivated land	38	Low cultivated land	38	Low cultivated land
14	Low cultivated land	61	Low cultivated land	61	Low cultivated land
15	Low cultivated land	42	Low cultivated land	42	Low cultivated land
		41	Low cultivated land	41	Low cultivated land
16	Low cultivated land	38	Low cultivated land	38	Low cultivated land
17	Low cultivated land	38	Low cultivated land	38	Low cultivated land
18	Low cultivated land	38	Low cultivated land	38	Low cultivated land
19	Low cultivated land	38	Low cultivated land	38	Low cultivated land
20	River	20	River	20	River
21	Low cultivated land	60	Low cultivated land	60	Low cultivated land
		48	Low cultivated land	48	Low cultivated land
		45	Low cultivated land	45	Low cultivated land
		43	Low cultivated land	43	Low cultivated land
		9	Low cultivated land	9	Low cultivated land

(contd...)

1934		*1982-83*		*2006*	
Plot no.	*Category*	*Plot no.*	*Category*	*Plot no.*	*Category*
		46	Low cultivated land	46	Low cultivated land
		47	Low cultivated land	47	Low cultivated land
22	Low cultivated land	58	Low cultivated land	58	Low cultivated land
		35	Low cultivated land	35	Low cultivated land
		49	Low cultivated land	49	Low cultivated land
23	Low cultivated land	50	Low cultivated land	50	Low cultivated land
		51	Low cultivated land	51	Low cultivated land
		52	Low cultivated land	52	Low cultivated land
		53	Low cultivated land	53	Low cultivated land
		382	Data not available	382	Data not available
		54	Low cultivated land	54	Low cultivated land
		55	Low cultivated land	55	Low cultivated land
		56	Data not available	56	Data not available
24	River	24	River	24	River
25	Inspection bungalow	63	Inspection bungalow	63	Inspection bungalow

(contd...)

	1934		*1982-83*		*2006*
Plot no.	*Category*	*Plot no.*	*Category*	*Plot no.*	*Category*
26	Low cultivated land	387	High cultivated land	387	High cultivated land
		65	Data not available	65	Data not available
27	Barren land	66	Forest	66	Forest
28	River	28	River	28	River
29 30 31 33	Low cultivated land	69	Low cultivated land	69	Low cultivated land
		68	Low cultivated land	68	Low cultivated land
		81	Low cultivated land	81	Low cultivated land
		82	Low cultivated land	82	Low cultivated land
		84	Low cultivated land	84	Low cultivated land
		73	Low cultivated land	73	Low cultivated land
		71	Low cultivated land	71	Low cultivated land
		70	Low cultivated land	70	Low cultivated land
		74	Low cultivated land	74	Low cultivated land
		369	Data not available	369	Data not available
		368	Data not available	368	Data not available
		80	Data not available	80	Data not available

(contd...)

	1934		*1982-83*		*2006*
Plot no.	*Category*	*Plot no.*	*Category*	*Plot no.*	*Category*
		83	Low cultivated land	83	Low cultivated land
		85	Low cultivated land	85	Low cultivated land
		99	Low cultivated land	99	Low cultivated land
		98	Low cultivated land	98	Low cultivated land
		78	Low cultivated land	78	Low cultivated land
		69	Low cultivated land	69	Low cultivated land
		77	Low cultivated land	77	Low cultivated land
32	Low cultivated land	88	Low cultivated land	88	Low cultivated land
		62	Low cultivated land	62	Low cultivated land
34	Low cultivated land	86	Low cultivated land	86	Low cultivated land
35	Low cultivated land	90	Low cultivated land	90	Low cultivated land
36	River	36	River	36	River
37	grassland	154	Residential areas	154	Vacant Land
		146	Low cultivated land	146	Vacant Land
		147	Residential areas	147	Vacant Land
		148	Residential areas	148	Vacant Land

(contd...)

1934		1982-83		2006	
Plot no.	Category	Plot no.	Category	Plot no.	Category
		151	Residential areas	151	Vacant Land
		191	Residential areas	191	Vacant Land
		196	Residential areas	196	Vacant Land
		190	Residential areas	190	Vacant Land
		153	Low cultivated land	153	Low cultivated land
		157	Low cultivated land	157	Low cultivated land
		158	Low cultivated land	158	Low cultivated land
		167	Low cultivated land	167	Low cultivated land
		166	Low cultivated land	166	Low cultivated land
		165	Data not available	165	Data not available
		164	Low cultivated land	164	Low cultivated land
		170	Low cultivated land	170	Low cultivated land
38	Forest	163	Forest	163	Forest
		162	High cultivated land	162	High cultivated land
		161	High cultivated land	161	High cultivated land
		160	High cultivated land	160	High cultivated land
		159	High cultivated land	159	High cultivated land

(contd...)

	1934		*1982-83*		*2006*
Plot no.	*Category*	*Plot no.*	*Category*	*Plot no.*	*Category*
		156	High cultivated land	156	High cultivated land
		155	High cultivated land	155	High cultivated land
39	Barren land	169	Barren land	169	Barren land
40	River	40	River	40	River
41	Low cultivated land	150	Data not available	150	Data not available
		278	High cultivated land	278	High cultivated land
		276	High cultivated land	276	High cultivated land
		277	High cultivated land	277	High cultivated land
42	Low cultivated land	279	High cultivated land	279	High cultivated land
		287	Low cultivated land	287	Low cultivated land
		143	Low cultivated land	143	Low cultivated land
		144	Low cultivated land	144	Low cultivated land
43	Low cultivated land	142	Low cultivated land	142	Low cultivated land
		192	Low cultivated land	192	Low cultivated land

(contd...)

1934		1982-83		2006	
Plot no.	*Category*	*Plot no.*	*Category*	*Plot no.*	*Category*
44	Low cultivated land	121	Low cultivated land	121	Low cultivated land
		139	Low cultivated land	139	Low cultivated land
		140	High cultivated land	140	High cultivated land
		93	Low cultivated land	93	Low cultivated land
		138	Low cultivated land	138	Low cultivated land
45	Low cultivated land	134	Low cultivated land	134	Low cultivated land
		137	Low cultivated land	137	Low cultivated land
		138	Low cultivated land	138	Low cultivated land
		94	Low cultivated land	94	Low cultivated land
46	Low cultivated land	133	Low cultivated land	133	Low cultivated land
		136	Low cultivated land	136	Low cultivated land
47	Low cultivated land	132	Low cultivated land	132	Low cultivated land
		131	Low cultivated land	131	Low cultivated land

(contd...)

	1934		*1982-83*		*2006*
Plot no.	*Category*	*Plot no.*	*Category*	*Plot no.*	*Category*
		135	Low cultivated land	135	Low cultivated land
		95	Low cultivated land	95	Low cultivated land
48	Low cultivated land	128	Low cultivated land	128	Low cultivated land
		130	Low cultivated land	130	Low cultivated land
		129	Data not available	129	Data not available
49	Low cultivated land	120	Low cultivated land	120	Low cultivated land
		126	Data not available	126	Data not available
50	Low cultivated land	122	Low cultivated land	122	Low cultivated land
51	Low cultivated land	163	Low cultivated land	163	Low cultivated land
		124	Low cultivated land	124	Low cultivated land
52	Low cultivated land	114	Low cultivated land	114	Low cultivated land
		115	Low cultivated land	115	Low cultivated land
		388	Low cultivated land	388	Low cultivated land
53	River	53	River	53	River

(contd...)

	1934		1982-83		2006
Plot no.	***Category***	***Plot no.***	***Category***	***Plot no.***	***Category***
54	Low cultivated land	331	High cultivated land	331	High cultivated land
		330	High cultivated land	330	High cultivated land
		385	High cultivated land	385	High cultivated land
		112	Data not available	112	Data not available
		118	Data not available	118	Data not available
		119	Data not available	119	Residential Areas
		332	High cultivated land	332	High cultivated land
		329	High cultivated land	329	High cultivated land
55	Stone	333	Forest	333	Forest
		334	Forest	334	Forest
56	Barren land	337	High cultivated land	337	Residential Areas
		335	High cultivated land	335	High cultivated land
		336	High cultivated land	336	High cultivated land
57	Barren land	347	Forest	347	Forest
58	River	58	River	58	River
59	Forest	373	High cultivated land	373	High cultivated land
		346	Grazing land	346	Grazing land
		347	Grazing land	347	Grazing land
		348	Grazing land	348	Grazing land

(contd...)

	1934		*1982-83*		*2006*
Plot no.	*Category*	*Plot no.*	*Category*	*Plot no.*	*Category*
		350	Forest	350	Forest
		393	Forest	393	Forest
60	Low cultivated land	327	Low cultivated land	327	Low cultivated land
		320	High cultivated land	320	High cultivated land
		325	Low cultivated land	325	Low cultivated land
		324	Low cultivated land	324	Low cultivated land
		321	High cultivated land	321	High cultivated land
61	Barren land	120	High cultivated land	120	High cultivated land
		121	Sacred place	121	Sacred place
62	Forest	127	Data not available	127	Data not available
		390	Data not available	390	Data not available
63	Low cultivated land	318	High cultivated land	318	High cultivated land
		319	High cultivated land	319	High cultivated land
64	Low cultivated land	317	High cultivated land	317	High cultivated land
		373	High cultivated land	373	High cultivated land
65	Low cultivated land	314	High cultivated land	314	High cultivated land

(contd...)

1934		1982-83		2006	
Plot no.	*Category*	*Plot no.*	*Category*	*Plot no.*	*Category*
		316	Low cultivated land	316	Low cultivated land
		315	Data not available	315	Data not available
		313	Residential areas	313	Residential areas
66	Low cultivated land	312	Residential areas	312	Residential areas
		213	Data not available	213	Data not available
		311	Residential areas	311	Residential areas
		294	High cultivated land	294	High cultivated land
		295	High cultivated land	295	High cultivated land
		310	Residential areas	310	Residential areas
		309	High cultivated land	309	High cultivated land
		308	High cultivated land	308	High cultivated land
		307	Residential areas	307	Residential areas
		298	High cultivated land	298	High cultivated land
67	Low cultivated land	306	Residential areas	306	Residential areas
68	Low cultivated land	305	Residential areas	305	Residential areas

(contd...)

	1934		*1982-83*		*2006*
Plot no.	*Category*	*Plot no.*	*Category*	*Plot no.*	*Category*
		284	High cultivated land	284	High cultivated land
		304	Residential areas	304	Residential areas
		283	High cultivated land	283	High cultivated land
69	Low cultivated land	303	Residential areas	303	Residential areas
		302	Residential areas	302	Residential areas
		287	Data not available	287	Data not available
70	Barren land	375	High cultivated land	375	Sacred groves
		291	High cultivated land	291	Sacred groves
		376	Data not available	376	Sacred groves
		377	Data not available	377	Sacred groves
		292	High cultivated land	292	Sacred groves
		290	Low cultivated land	290	Low cultivated land
		294	High cultivated land	294	High cultivated land
		296	High cultivated land	296	Sacred groves
		89	Low cultivated land	89	Low cultivated land
		288	High cultivated land	288	High cultivated land

(contd...)

1934		1982-83		2006	
Plot no.	*Category*	*Plot no.*	*Category*	*Plot no.*	*Category*
		297	High cultivated land	297	Sacred groves
		299	High cultivated land	299	High cultivated land
		286	High cultivated land	286	High cultivated land
		285	High cultivated land	285	High cultivated land
		280	High cultivated land	280	High cultivated land
		300	High cultivated land	300	High cultivated land
71	Forest	360	Forest	360	Forest
		268	Data not available	268	Data not available
		270	High cultivated land	270	High cultivated land
		261	Data not available	261	Data not available
		202	High cultivated land	202	High cultivated land
		273	High cultivated land	273	High cultivated land
		277	High cultivated land	277	High cultivated land
		281	High cultivated land	281	High cultivated land
		269	High cultivated land	269	High cultivated land

(contd...)

	1934		1982-83		2006
Plot no.	*Category*	*Plot no.*	*Category*	*Plot no.*	*Category*
		301	Road	301	Road
72	grassland	372	High cultivated land	372	High cultivated land
		352	High cultivated land	352	High cultivated land
		353	High cultivated land	353	High cultivated land
		355	Data not available	355	Data not available
		356	High cultivated land	356	High cultivated land
		354	High cultivated land	354	High cultivated land
		386	Forest	386	Forest
		357	High cultivated land	357	High cultivated land
		358	High cultivated land	358	High cultivated land
73	Forest	Part of 343	Grazing land	Part of 343	Grazing land
		349	Grazing land	349	Grazing land
		52	Grazing land	52	Grazing land
		350	Forest	350	Forest
		393	Forest	393	Forest
74	Low cultivated land	367	Data not available	367	Data not available
		374	Data not available	374	Data not available

(contd...)

1934		1982-83		2006	
Plot no.	*Category*	*Plot no.*	*Category*	*Plot no.*	*Category*
		361	High cultivated land	361	High cultivated land
		347	Forest	347	Forest
75	Low cultivated land	363	High cultivated land	363	High cultivated land
		262	High cultivated land	262	High cultivated land
		261	High cultivated land	261	High cultivated land
		254	High cultivated land	254	High cultivated land
		64	Forest	64	Forest
76	Low cultivated land	264	High cultivated land	264	High cultivated land
		263	Data not available	263	Data not available
		260	High cultivated land	260	High cultivated land
		252	High cultivated land	252	High cultivated land
		256	High cultivated land	256	High cultivated land
77	Low cultivated land	265	High cultivated land	265	High cultivated land
		238	Low cultivated land	238	Low cultivated land
		237	Low cultivated land	237	Low cultivated land

(contd...)

	1934		*1982-83*		*2006*
Plot no.	*Category*	*Plot no.*	*Category*	*Plot no.*	*Category*
		212	Low cultivated land	212	Low cultivated land
		206	Low cultivated land	206	Low cultivated land
		95	Low cultivated land	95	Low cultivated land
81	River	81	River	81	River
82	Low cultivated land	320	Residential areas	320	Vacant land
89	Barren land	105	Grazing land	105	Grazing land
90	River	105	Grazing land	105	Grazing land
91	Barren land	105	Grazing land	105	Grazing land
92	Low cultivated land	191	Residential areas	191	Residential areas
		196	Residential areas	196	Residential areas
		194	High cultivated land	194	High cultivated land
		195	High cultivated land	195	High cultivated land
		190	Residential areas	190	Residential areas
		189	Data not available	189	Data not available
		170	Low cultivated land	170	Low cultivated land
		171	Barren land	171	Vacant land

The table 8 shows that how the land characteristics of each plot has changed over the years. The plot no. 1 was under forest cover in 1934 but in 1982 this plot has been converted to low and high cultivated land and broken into four plots and it remains as low and high cultivated land in 2006. The plot no. 7 was under forest cover but in 1982 this has been broken into six plots of which three remained as forest and rest three plots were converted to low cultivated land but in 2006 it has been observed that the plots covering forest remained as forest but the other three plots (plot no. 2, 3 and 4) were converted to residential areas from low cultivated land. In 1934 plot no. 37 was under grass. But it is observed from 1982 data that this plot had been broken into sixteen plots, out of these plots (plot no. 147, 148, 151, 190, 191, 196 and 154) seven plots were converted to residential areas. Among other plots eight plots (plot no. 146, 153, 157, 158, 167, 166, 164 and 170) were converted to low cultivated land. But in 2006 it has been seen that all the residential areas are lying as vacant land because the resident of these plots had been shifted to Ambodia by the government and it has been mentioned earlier. In 1934, plot no. 38 was under forest but in 1982 this plot had been broken into seven plots out these seven plots, only plot no. 163 remained as forest but other six plots (plot no. 162, 161, 160, 159, 156 and 155) converted to high cultivated land, where mainly maize, til, wheat are grown for which less amount of water is required. In 2006, no changes in these plots. In 1934 the plot no. 41 shown as low cultivated land but in 1982 this plot has been broken into four plots out of which three plots had been converted to high cultivated land it has happened due to non availability of proper water facility. In 1934, plot no. 54 was under low cultivated land but in 1982 this plot has been broken into eight plots and the plot no. 331, 330, 385, 332 and 329 had been converted to high cultivated land but no information's available on plot no. 112, 118 and119. No significant change has been observed on all these plots in 2006. In 1934, plot no. 65 was under low cultivated land but in 1982 this has been broken into four plots. Out of these plots, plot no. 313 was converted to residential area and other remained same. In 2006 no significant change has been observed. In 1934 plot no. 66 was under low cultivated land. But in 1982, this plot had been broken into ten nos. of plots out of which four plots (plot no. 312, 311, 310 and 307) had been changed as residential areas but other five plots (plot no. 294, 295, 308, 298 and 309) had been converted to high cultivated land but the data were not available on pot no. 213. In 2006 no significant change had been observed on these plots. In 1934 the plot no. 67 was under low cultivated land but in 1982 the plot no. had been broken into two plots bearing no. 305 and 306 and had been converted to residential areas. In 2006 these two plots remain as residential plots. In 1934 the plot no. 68 was under low cultivated land but in 1982 this plot had been broken into four plots out of which two plots (plot no. 305 and 304) had been converted to residential areas but the other two plots (plot no. 284 and 283) were converted to high cultivated land. In 2006 no significant change has been observed on these plots. In 1934 plot no. 69 was under low cultivated land but in 1982 this plot was changed into three nos. of plots out of which two plots bearing no. 302 and 303 were converted to residential areas. In 2006 no significant change is observed on these plots. In 1934 the plot no. 70 is shown as barren land but in 1982 this plot was broken in 16 plots and almost all these plots were converted to high land cultivated land. But in 2006 plot no. 375, 291, 376, 377, 292, 296 and 297 were converted to sacred groves from high cultivated land. In 1934 plot no. 71 was under forest cover but in 1982 this plot were broken into ten nos. of plots covering high land cultivated land, road and forest. No change had been observed in 2006 on these plots. In 1934 plot no. 72 was under grass cover but in 1982 and 2006 this plot were broken into nine plots covering high cultivated land and forest. In 1934 the plot no.73 was under forest but in 1982 and 2006 this plot had broken into five number of plots covering grazing land and forest. In 1934 the plot nos. 74 and 75 were under low cultivated land but in 1982 and 2006 these plots had been broken into four and five number of plots respectively covering high cultivated land and forest. In 1934 plot no. 82 was under low cultivated land but in 1982 this plot had been broken into four plots, out of which three plots bearing no. 320, 171 and 371 were under residential areas, but in 2006 these three plots were converted to vacant land since the villagers mainly the Kharia were shifted to Ambodia by the government. In 1934 the plot number 90 was under rivulet but in 1982 and also in 2006 it has been observed that this land had been converted to grazing land, as due to the siltation the width of the river had been filled up resulting in the formation of grazing land. In 1934, the plot no. 92 was under low cultivated land but in 1982 and in 2006 it has been observed that this plot had been broken into eight number of plots. Out of which the plot number 190, 191 and 196 were converted to residential use but other remained as cultivated land.

Discussion: From the above analysis it has been observed that in this village ten percent forest has been reduced and this area has been utilised as high cultivated land and grazing land. Another significant change has been observed from the above table is that the residential areas have been reduced to 0.62 percent. It has happened because the Government resettled the Bathuri and the Kharia population from Jenabil village to Ambodia near Jashipur in 1993 from the plot number 154,146,147,148,151,191,320,171and 371.These plots is still lying vacant.

Jamunagarh

Jamunagarh is located on the south of Jenabil village and is extended between 86°19′ 3.73″E to 86°20′ 23.43″E longitude and 21°42′ 17.13″N to 21°42′ 50.23″N latitude. The village is located at a distance of 60 km. south from Jashipur. This village is surrounded by dense forest. On the north of this village Jonoburu hill is located, the east is bounded by Thakurani Sal hill, Baradanda *maiden* (field) is found on the south of village and the west is bounded by Jonoburu hill. This village has gentle slope from north- east to south west. The Jamuna river is flowing from north-east to north-west of this village and another river Daldali is flowing to the north-east side of this village. The villagers use the water of these two rivers for cultivation and for drinking purpose. The summer temperature rises to 36 degree Celsius but in winter it goes down to 4 to 5 degree Celsius. The average rainfall is 24 mm. during rainy season and the average rainy days are 24. One Jeepable road is found from north-east to south-west connecting Jenabil and Kobatghai. According to the household census out of 96 individuals 45 are males and 51 are females, sex ratio is 1133 having 10 nos. of family. The family size is 9.6.

The economy of the people of this village mainly depends on agriculture - the rice. To cultivate the land they use cattle. According to 2003 census 66 cattles are found in this village. Besides cattles, 64 no. of goats are found. They sell out the well-grown goats at the nearby market i.e., Gurguria market. 2003 census also shows 171 chickens and ducks per house hold.

Table 9. Land use characteristics of Jamunagarh village in percentage (1934)

Sl. No.	*Characteristics*	*Area in acres*	*Percentage*
1	Forest	110.17	31.94
2	Barren land	215.16	62.38
3	Jahira	0.45	0.13
4	Jeepable road	4.66	1.35
5	River and water bodies	14.48	4.2
6	Total	344.92	100

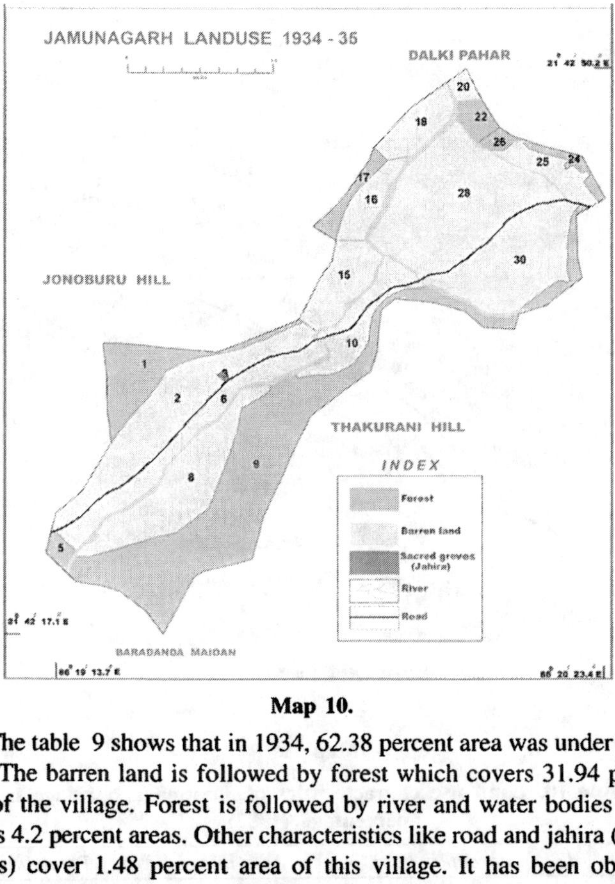

Map 10.

The table 9 shows that in 1934, 62.38 percent area was under barren land. The barren land is followed by forest which covers 31.94 percent area of the village. Forest is followed by river and water bodies which covers 4.2 percent areas. Other characteristics like road and jahira (sacred groves) cover 1.48 percent area of this village. It has been observed from the data and from the map that there was no sign of cultivated land nor residential areas. It means that up to 1934, people could not enter in this area because of non-availability of proper communication and fear of wild animals like elephants and tigers. Information were collected on 31 nos. of plots.

Like 1934, applying the same methodology the detailed informatons on land characteristics of each plot has been collected from the record books from the Tehsil office, Karanjia along with cadastral map of this village. Accordingly, the land use map of Jamunagarh was prepared on the basis of 1983 data. From the data analysis the following table has been prepared. In 1983 the total number of plots was 162, which shows an increase of number of plots as compared to the year 1934.

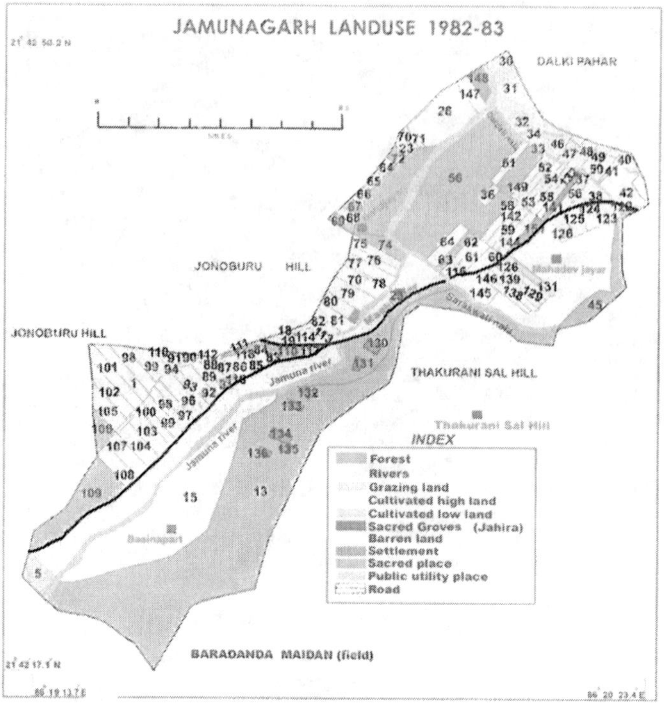

Map 11.

Table 10. Land use characteristics of Jamunagarh village in percentage (1982)

Sl. No.	*Land characteristics*	*Area in acres*	*Percentage*
1	High cultivated land	103.87	30.12
2	Low cultivated land	0.23	0.07
3	Barren land	81.85	23.73
4	Forest	108.57	31.47
5	Water bodies	13.87	4.02
6	Residential areas	9.93	2.88
7	Sacred place	0.07	0.02
8	Jeepable road	5.17	1.50
9	Public utility place	3.70	1.07
10	Pasture land	17.66	5.12
	Total	344.92	100

The table 10 shows that 31.47 percent of area was under forest cover. It shows the decrease of nearly 0.47 percent as compared to the data of 1934 which is very negligible. Forest area is followed by high cultivated land and it covers about 30.12 percent area of this village. In 1934 there was no sign of cultivated land. But the people of this village after the settlement converted the barren lands to high cultivated land. The above table also shows that 2.88 percent area was under residence. It means that the people came to settle down here after 1934.The above table also shows that the barren land had been reduced to 38.65and roughly this part had been converted to high cultivated land, residential areas and pasture land. Pasture land covers 5.12 percent area of this village where the cattles move freely. The other land characteristics of this village (water bodies, Low cultivated land, Sacred place, Road and Public utility place) cover about 11.8 percent area of this village.

The existing (2006) land use has been prepared for this village on the basis of map supplied by land revenue office, Karanjia and plot-to-plot survey in the year 2006.

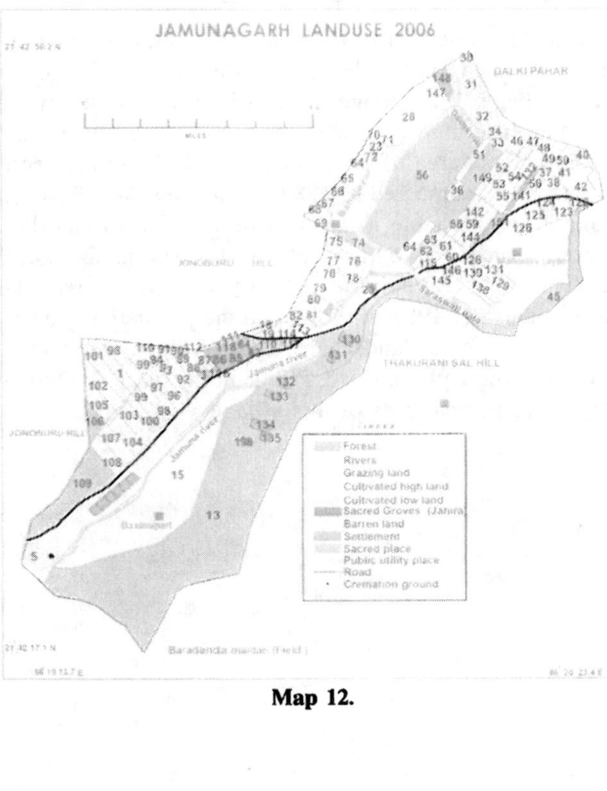

Map 12.

Table 11: Land use characteristics of Jamunagarh village in percentage (2006)

	Land Characteristics	*Areas in acres*	*Percentage*
1	High cultivated land	115.79	33.57
2	Residential areas	14.38	4.17
3	Barren land	65.88	19.1
4	Forest	105.82	30.68
5	water bodies	13.87	4.02
6	Low cultivated land	2.59	0.75
7	Sacred place	0.07	0.02
8	Jeepable road	5.17	1.5
9	public place	3.69	1.07
10	Pasture land	17.66	5.12
11	Total	344.92	100

The table 11 shows that in the year 2006, 33.57 percent area was under high cultivated land which shows the increase of 3.45 percent area of this village. The villagers are using the barren land as high cultivated land. The high cultivated land is followed by forest and it covers 30.68 percent area of this village which shows the decrease of 1.19 percent area as compared to 1982 and for this the villagers are not responsible. From this table it has also been observed that the barren land has been reduced to 4.63 percent and this part had been converted to high cultivated land. The residential areas had been increased to 1.29 percent as compared to 1982, it shows that the population growth is less in this village. The other land characteristics like Water bodies, Low cultivated land, Sacred place, Road, Public utility place and Pasture land cover only 11.12 percent area of this village.

LAND USE CHANGE OF JAMUNAGARH VILLAGE OVER THE DECADES

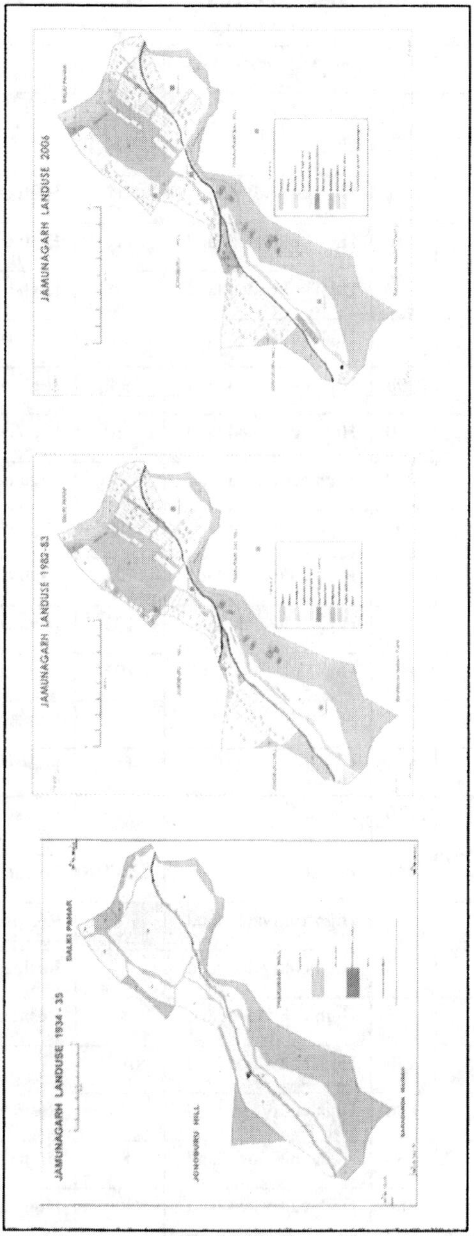

Map 13.

Table 12: Plot wise land characteristics change in Jamunagarh village

	1934		*1982*		*2006*
Plot No.	*Land Charact-teristics*	*Plot No.*	*Land chara-cteristics*	*Plot No.*	*Land Characteristics*
1	Forest	1	High cultivated land	1	High cultivated land
		112	High cultivated land	112	High cultivated land
		90	High cultivated land	90	High cultivated land
		91	High cultivated land	91	High cultivated land
		94	High cultivated land	94	High cultivated land
		110	High cultivated land	110	High cultivated land
		99	High cultivated land	99	High cultivated land
		98	High cultivated land	98	High cultivated land
		101	High cultivated land	101	High cultivated land
		102	High cultivated land	102	High cultivated land
		105	High cultivated land	105	High cultivated land
		106	Forest	106	Forest
		111	Forest	111	Forest
2	Barren land	110	Forest	110	Forest
		117	Forest	117	Forest
		83	High cultivated land	83	Residential areas
		84	High cultivated land	84	Residential areas
		85	High cultivated land	85	Residential areas
		178	High cultivated land	NA	
		86	High cultivated land	86	Residential areas
		87	High cultivated land	87	High cultivated land
		88	High cultivated land	88	High cultivated land

(contd...)

	1934		*1982*		*2006*
Plot No.	*Land Charact-teristics*	*Plot No.*	*Land chara-cteristics*	*Plot No.*	*Land Characteristics*
		89	High cultivated land	89	High cultivated land
		92	High cultivated land	92	High cultivated land
		93	High cultivated land	93	High cultivated land
		96	High cultivated land	96	High cultivated land
		97	High cultivated land	97	High cultivated land
		98	High cultivated land	98	High cultivated land
		99	High cultivated land	99	High cultivated land
		100	High cultivated land	100	High cultivated land
		103	High cultivated land	103	High cultivated land
		104	High cultivated land	104	High cultivated land
		107	High cultivated land	107	High cultivated land
		108	High cultivated land	108	High cultivated land
		109	Forest	109	Forest
		118	Forest	118	Forest
3	Sacred Grove	3	Sacred Grove	3	Sacred Grove
4	Road	4	Road	4	Road
5	Forest	5	Public utility place	5	Public utility place
6	Barren land	15	Barren land	15	Barren land
7	River	7	River	7	River
8	Barren land	15	Barren land and sacred grove	15	High cultivated land and residential areas
9	Forest	13	Forest	13	Forest
		132	Residential areas	132	Residential areas
		133	Residential areas	133	Residential areas

(contd...)

1934		1982		2006	
Plot No.	*Land Charact-teristics*	*Plot No.*	*Land chara-cteristics*	*Plot No.*	*Land Characteristics*
		134	Residential areas	134	Residential areas
		135	Residential areas	135	Residential areas
		136	Residential areas	136	Residential areas
10	Barren land	130	Residential areas	130	Residential areas
		131	Residential areas	131	Residential areas
		NIL	Forest	NIL	Forest
11	River	11	River	11	River
12	Road	12	Road	12	Road
13	Barren land	23	Barren land and sacred grove	23	Barren land and sacred grove
14	River	14	River	14	River
15	Barren land	75	High cultivated land	75	Highland
		76	High cultivated land	76	High cultivated land
		77	High cultivated land	77	High cultivated land
		78	High cultivated land	78	High cultivated land
		70	High cultivated land	70	High cultivated land
		79	High cultivated land	79	High cultivated land
		80	High cultivated land	80	High cultivated land and residential areas
		81	High cultivated land	81	High cultivated land
		82	High cultivated land	82	High cultivated land
		113	High cultivated land	113	High cultivated land
		114	High cultivated land	114	High cultivated land
		19	High cultivated land	19	High cultivated land
		18	High cultivated land	18	High cultivated land

(contd...)

	1934		*1982*		*2006*
Plot No.	*Land Charact-teristics*	*Plot No.*	*Land chara-cteristics*	*Plot No.*	*Land Characteristics*
		NIL	Residential areas	NIL	Residential areas
16	Barren land	16	Forest and sacred grove	16	Forest and sacred grove
17	Forest	23	Barren land	23	High cultivated land
		72	Forest	72	High cultivated land
		64	High cultivated land	64	High cultivated land
		65	High cultivated land	65	High cultivated land
		66	High cultivated land	66	High cultivated land
		67	Forest	67	High cultivated land
		68	Low cultivated areas	68	Low cultivated areas
		69	Forest	69	High cultivated land
18	Barren land	148	High land cultivated land	148	High land cultivated land
		147	Vacant land	147	Vacant land
		28	Barren land	28	Barren land
		71	High cultivated land	71	High cultivated land
		70	High cultivated land	70	High cultivated land
19	River	19	River	19	River
20	Barren land	30	Public utility place	30	Public utility place
		31	Public utility place	31	Public utility place
21	River	21	River	21	River
22	Forest	32	Grazing land	32	Grazing land
		33	Grazing land	33	Grazing land
23	River	23	River	23	River
24	Forest	40	Barren land	40	Barren land

(contd...)

	1934		1982		2006
Plot No.	*Land Charact-teristics*	*Plot No.*	*Land chara-cteristics*	*Plot No.*	*Land Characteristics*
25	Barren land	46	Forest	46	Forest
		47	Forest	47	Forest
		48	High cultivated land	48	High cultivated land
		49	High cultivated land	49	High cultivated land
		50	High cultivated land	50	High cultivated land
		41	Forest	41	Forest
		42	Barren land	42	Barren land
26	Forest	34	Grazing land	34	Grazing land
27	River	27	River	27	River
28	Barren land	56	Forest	56	Forest
		51	Barren land	51	Barren land
		36	High cultivated land	36	High cultivated land
		64	High cultivated land	64	High cultivated land
		63	Forest	63	Forest
		62	High cultivated land	62	High cultivated land
		61	High cultivated land	61	High cultivated land
		115	High cultivated land	115	High cultivated land
		60	High cultivated land	60	High cultivated land
		151	High cultivated land	151	High cultivated land
		144	High cultivated land	144	High cultivated land
		39	High cultivated land	39	High cultivated land
		142	High cultivated land	142	High cultivated land

(contd...)

	1934		*1982*		*2006*
Plot No.	*Land Charact-teristics*	*Plot No.*	*Land chara-cteristics*	*Plot No.*	*Land Characteristics*
		58	High cultivated land	58	High cultivated land
		38	High cultivated land	38	High cultivated land
		37	High cultivated land	37	High cultivated land
		58	High cultivated land	58	High cultivated land
		141	High cultivated land	141	High cultivated land
		132	High cultivated land	132	High cultivated land
		55	High cultivated land	55	High cultivated land
		54	High cultivated land	54	High cultivated land
		53	High cultivated land	53	High cultivated land
		52	High cultivated land	52	High cultivated land
		149	High cultivated land	149	High cultivated land
29	River	29	River	29	River
30	Barren land	122	Barren land and sacred place	122	Barren land and sacred place
		123	High cultivated land	123	High cultivated land
		124	High cultivated land	124	High cultivated land
		125	High cultivated land	125	High cultivated land
		126	High cultivated land	126	High cultivated land
		131	Low cultivated land	131	Low cultivated land
		126	High cultivated land	126	High cultivated land
		129	High cultivated land	129	High cultivated land
		139	High cultivated land	139	High cultivated land
		138	High cultivated land	138	High cultivated land
		146	High cultivated land	146	High cultivated land
		145	High cultivated land	145	High cultivated land
31	Forest	31	Forest	31	Forest

The table 12 shows that the change of land characteristics of each plot. Plot no. 1 was under forest cover in 1934 but in 1982 this plot had been broken into 13 plots. Out of these plots, 11 plots (bearing no. 1, 112, 90, 91, 94, 110, 99, 98, 101, 102 and 105) had been converted to high cultivated land but other two plots bearing no. 106 and 111 remained as forest. No significant change has been observed in 2006 on these plots. In 1934 plot no. 2 was under barren land but in 1982 this plot was broken into 23 nos. of plots out of which four plots (plot no. 110, 117, 109 and 118) were converted to forest but other plots bearing no. 83, 84, 85, 178, 86, 87, 88, 89, 92, 93, 96, 97, 98, 99, 100, 103, 104, 107 and 108 were converted to high cultivated land. The plot wise land characteristics change has not been observed on these plots in 2006. In 1934 the plot no. 5 was under forest but this plot had been converted to public utility place in 1982 and remained as same in 2006. The villagers selected this place for their social function. In 1934 the plot no. 8 was under barren land but in1982 within this barren land a piece of land was formed as sacred grove. In 2006 six families have been settled down on this barren land. In 1934, the plot no. 9 was under forest but in 1982 this plot had been broken into six plots out of which the plot no. 132, 133, 134, 135 and 136 had been converted to residential areas but the major portion remained as forest. In 2006, no major change is seen on these plots. In 1934, the plot no. 10 was under barren land but in 1982 this was broken into three pieces out of which the plot no. 130 and 131 was converted to residential areas but major part remained as barren land. In 2006, no significant change has been seen on this plot. In 1934, the plot no. 15 was under barren land but in 1982 this plot had been broken into 14 plots out of which 13 plots bearing no. 75, 76, 77, 78, 70, 79, 80, 81, 82, 113, 114, 18 and 19 had been converted as high cultivated land but rest one was under residential areas. In 1934 plot no. 17 was under forest but in 1982 this plot had been broken into 8 nos. of plots out of which plot no. 23 converted to barren land. Plot no. 72, 67 and 69 remained as forest, but the plot no. 64, 65 and 66 converted to high cultivated land and plot no. 68 was converted to low cultivated land. But in 2006 it is observed from the plot to plot survey that all the plots were converted to high cultivated land except plot no. 68 which remained as low cultivated land. From the above table it has been observed that in 1934 the plot no. 18 was under barren land but in 1982 this plot has been broken into 5 plots out of these plots, three plots bearing no. 148,70 and 71 converted to high cultivated land, plot no. 147 remained as vacant land and the plot no. 28 remained as barren land. In 2006 all the land characteristics of this plot remained same as in 1982. In 1934 the plot no. 22 was under forest area but in 1982 and in 2006 it has been observed that this plot was broken into two plots bearing no. 32 and 33 and converted to grazing land. In 1934 the plot no. 24 was under forest but in 1982 and in 2006 this plot was converted to barren land. In 1934 the plot no. 25 was under barren land but in 1982 this plot was broken into seven plots out of which three plots bearing nos. 41, 46 and 47 converted to forest, the plot nos. 48,49 and 50 converted to high cultivated land and plot no. 42 remained as barren land. No significant change has been observed on this plot in 2006. In 1934 the plot no. 28 was under barren land but in 1982 this plot has been broken into 24 plots, out of which two plots bearing no.56 and 63 were converted to forest, one plot (51) remained as barren land but the other plots (36, 64, 62, 61, 151, 144, 39, 142, 58, 141, 132, 55, 54, 53, 52 and 149) had been converted to high cultivated land. In 2006 the land characteristics of this plot remained same. In 1934 the plot no. 30 was under barren land but this plot had been broken into 12 plots. Out of these plots only one plot remained as barren land (plot no. 122) but other plots (123, 124, 125, 126, 131, 126, 129, 139, 138, 146 and145) had been converted to high and low cultivated land.

Kobatghai

Kobatghai is located on the south of Bakua village and its extension is between 86°15′34.2″E to 86°16′21.4″E longitude and 21°47′15.0″N to 21°48′1″N latitude. The village is located at a distance of 60 km. south from Jashipur. This village is surrounded by dense forest. In the north of this village Baula Pahar forest is located, the east is bounded by Jamuna Pahar forest. Debai Dungri forest is found on the south of village and the west is bounded by Atna Bera forest. This village is elongated in shape and only river the Jamuna is flowing on the north of this village from south-east to north-west. The villagers use the water of these two rivers for cultivation and for drinking purpose. The summer temperature rises to 36 degree Celsius but in winter it goes down to 4 to 5 degree Celsius. Average rainfall is 24 mm. during rainy season and the average rainy days are 24. One Jeepable roads is found from north-west to south-east connecting Joshipur and Jamungarh. According to the household census, out of 115 individuals 55 are males and 60 are females, sex ratio is 1090 having 18 nos. of family. The family size is 6.38.

According to 2003 census 95 cattles and 5 buffaloes are found in this village. Besides cattles and buffaloes, 29 nos. of sheep and 102 nos. of goats are found. They sell out the well-grown goats at the nearby market i.e., Gurguria market. 2003 census also shows 165 chickens and ducks per house hold.

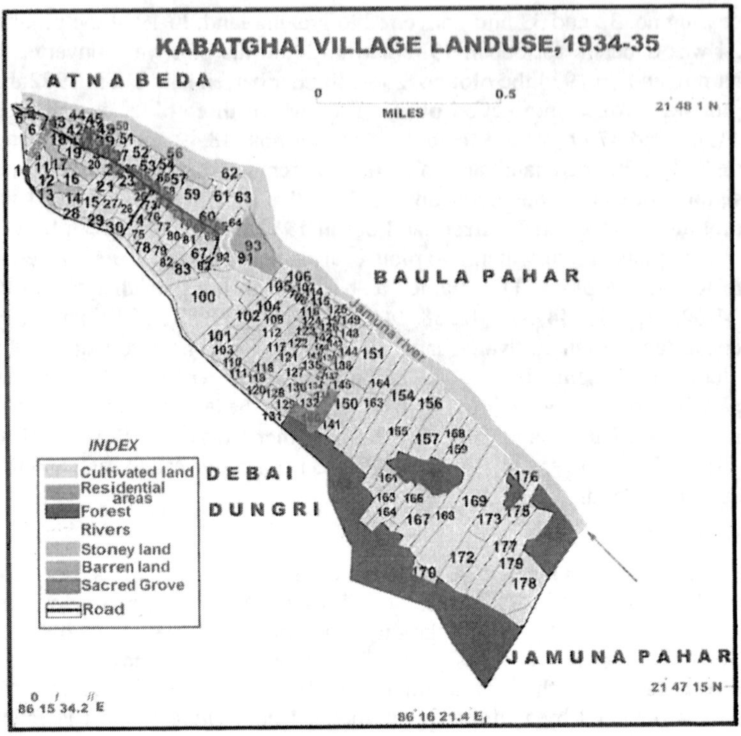

Map 14.

Table 13. Land use characteristics of Kobatghai village in percentage (1934)

Sl No.	*Land characteristics*	*Area in acres*	*Percentage*
1	Water bodies	12.63	5.57
2	Low cultivated land	157.08	69.26
3	Forest	40.55	17.88
4	Residential areas	4.32	1.9
5	Stony area	4.92	2.17
6	Jeepable road	3.03	1.34
7	Sacred grove	4.26	1.88
8	Total	226.79	100

The table 13 shows that 69.26 percent areas were under low cultivated land. This low cultivated land has been divided into four categories namely: *Jaraoul, Jarasayam, Jaradayam* and *Ashu*. Among all these categories Jaraoul is the best where *Surudhan* (the best quality rice) is grown. From the map it is observed that almost all the areas of this village are covered by low land cultivation. Low cultivated land is followed by forest which covers 17.88 percent areas. Forest is followed by water bodies which covers 5.57 percent area of this village. From the above table it has been observed that only 1.9 percent area was under tribal settlement. The rest of the area (5.39 percent) of this village was covered by Stony land, Road and *Jahira* (Sacred Grove). Like 1934, applying the same methodology the detailed informatons on land characteristics of each plot has been collected from the record books from the Tehsil office, Karanjia along with cadastral map of this village. Accordingly the land use map of Kobatghai has been prepared on the basis of 1982 data. From the data analysis the following table has been prepared. In 1982 the total number of plots was 214, which shows an increase of number of plots as compared to the year 1934.

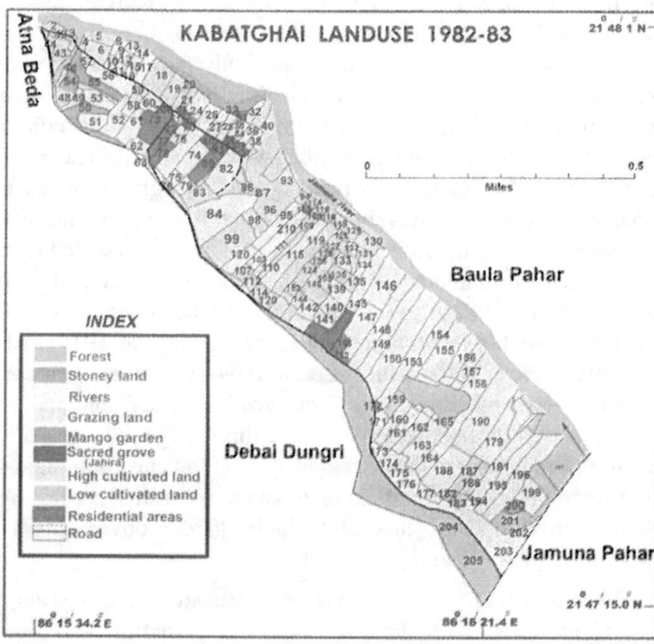

Map 15.

Table 14. Land use characteristics of Kobatghai village in percentage (1982)

Sl. No.	*Land characteristics*	*Area in acres*	*Percentage*
1	Water bodies	12.63	5.57
2	Forest	24.02	10.59
3	Low cultivated land	65.89	29.05
4	High cultivated land	98.38	43.38
5	Residential areas	5.76	2.54
6	Stony area	1.63	0.72
7	Jeepable road	5.03	2.22
8	Pasture land	8.71	3.84
9	Sacred groves	4.26	1.88
10	Mango garden	0.48	0.21
11	Total	226.79	100

The table 14 shows that 43.38 percent area was under high land cultivated area. It may be mentioned that in 1934 there was no such category of land in this village. In these lands they cultivate mainly mustard, wheat, maize and til for which very less amount of water is required. It may also be mentioned that the villagers called low cultivated land where mainly the rice are grown but where mustard, wheat, maize and til are grown they called this type of land as high cultivated land. This change in land use has happened due to non availability of irrigated water. In 1982 the low cultivated area had been reduced to more than 40 percent as compared to 1934 and had been changed to high cultivated land. In 1982, the forest area had been reduced to 7.29 percent as compared to information collected for the year 1934. In 1982 the residential areas had been increased to 0.64 percent as compared to 1934, which is very negligible. In 1982, the area in road and communication had been increased to 0.88 percent area of this village, which is a good sign that in this remote area the development in road and communication is getting improved. Other land characteristics like Water bodies, Stony area, Sacred grove, Pasture land and Mango garden cover only 12.22 percent area of this village.

The existing (2006) land use has been prepared of this village on the basis of map supplied by land revenue office, Karanjia and plot- to-plot survey in the year 2006.

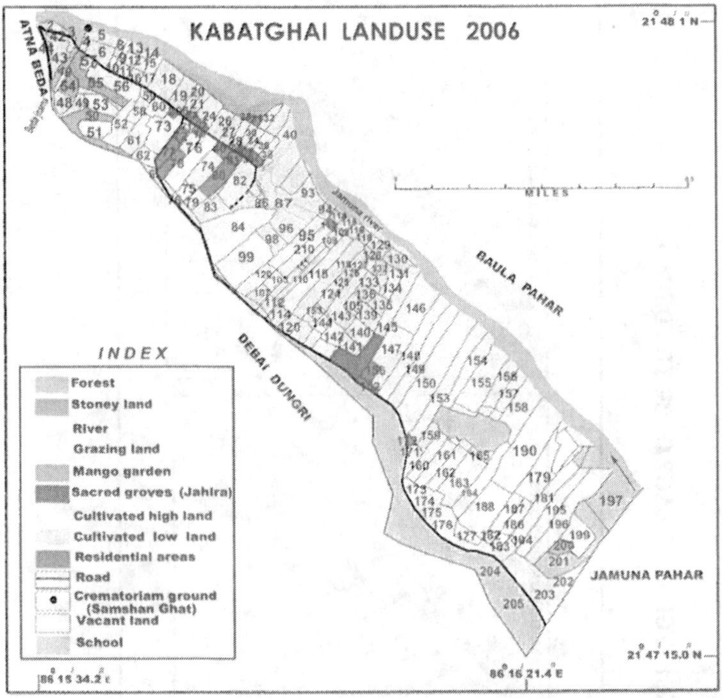

Map 16.

Table 15. Land use characteristics of Kobatghai village in percentage (2006)

Sl.No.	*Land characteristics*	*Area in acres*	*Percentge*
1.	Water bodies	12.62	5.57
2.	Forest	23.79	10.49
3.	Low cultivated land	60.90	26.84
4.	High cultivated land	79.36	34.99
5.	Residential areas	5.44	2.4
6.	Stony area	1.63	0.72
7.	Jeepable road	5.03	2.22
8.	Pasture land	8.70	3.84
9.	Sacred groves	4.08	1.8
10.	Mango garden	0.50	0.22
11.	Vacant land	24.74	10.91
12.	Total	226.79	100

LAND USE CHANGE OF KOBATGHAI VILLAGE OVER THE DECADE

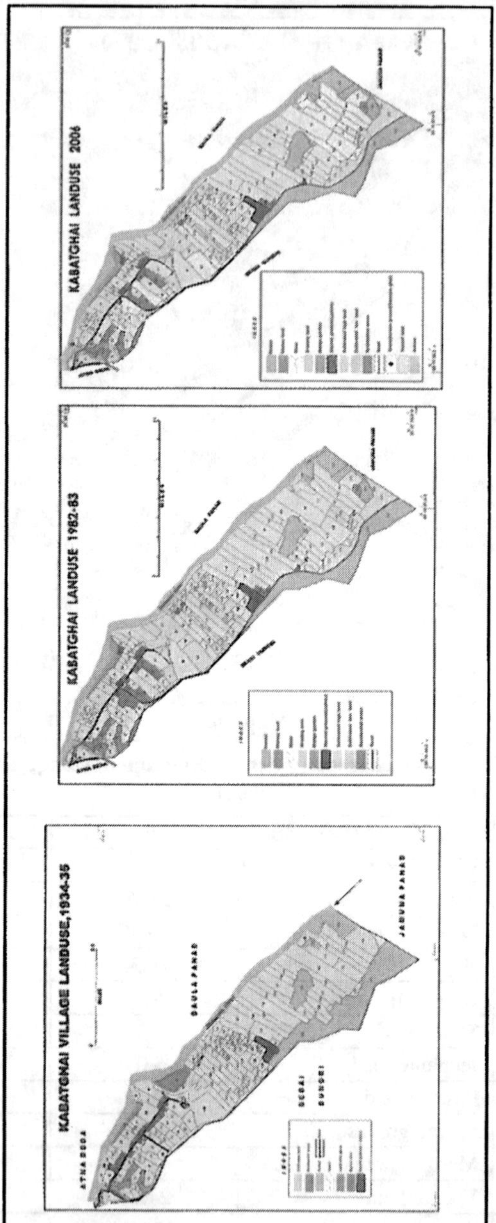

Map 17.

From the table 15 it has been observed that the forest area had been reduced from 17.88 percent in 1934 to10.59 percent in 1982 to 10.49 percent in 2006 in Kobatghai village of core region of Similipal Biosphere Reserve which is very alarming. But for this decrease of forest the tribal people who are residing in this village are not solely responsible. Another significant change has been observed that the residential area had been reduced to 0.14 percent, it has happened, because from the plot no. 73, Govt. had resettled twenty Kharia families near Joshipur in the year 1970, from that period this plot is lying vacant. Low and High cultivated land had been reduced to three percent and nearly nine percent respectively. Vacant land had been increased to10.91 percent because the people of this village had left cultivation in the bordering area of the forest because the animal like elephants destroy the ripe agricultural product resulting the increase of vacant land. Other land characteristics like Stony area, Road, Sacred groves and Mango garden remain almost same.

Table 16. Plot wise land characteristics change in Kobatghai village

1934-35		*1982-83*		*2006*	
Plot no.	*Category*	*Plot no.*	*Category*	*Plot no.*	*Category*
1	River	1	River	1	River
2	Low cultivated land	2	Low cultivated land	2	Low cultivated land
3	Low cultivated land	3	Low cultivated land	3	Low cultivated land
4	Low cultivated land	4	Low cultivated land	4	Low cultivated land
5	Low cultivated land	5	Low cultivated land	5	Low cultivated land
6	Low cultivated land	6	Low cultivated land	6	Low cultivated land
7	Low cultivated land	7	Low cultivated land	7	Low cultivated land

(contd...)

	1934-35		*1982-83*		*2006*
Plot no.	*Category*	*Plot no.*	*Category*	*Plot no.*	*Category*
8	River	8	River	8	River
9	Low cultivated land	9	Low cultivated land	9	Low cultivated land
10	Forest	10	Forest	10	Forest
11	Low cultivated land	54	Low cultivated land	54	Low cultivated land
12	Low cultivated land	12	Low cultivated land	12	Low cultivated land
13	Stony area	50	Stony area	50	Stony area
14	Low cultivated land	14	High cultivated land	14	High cultivated land
15	Low cultivated land	15	High cultivated land	15	High cultivated land
16	Low cultivated land	16	High cultivated land	16	High cultivated land
17	Stony area	17	Stony area	17	Stony area
18	Low cultivated land	57	School	57	School
19	Low cultivated land	56	High cultivated land	56	High cultivated land
20	Low cultivated land	56	High cultivated land	56	High cultivated land
21	Low cultivated land	21	Low cultivated land	21	Low cultivated land
22	Residential areas	22	Residential areas	22	Residential areas
23	Low cultivated land	23	Low cultivated land	23	Low cultivated land

(contd...)

	1934-35		*1982-83*		*2006*
Plot no.	*Category*	*Plot no.*	*Category*	*Plot no.*	*Category*
24	Residential areas	206	Residential areas	206	Residential areas
25	Low cultivated land	25	Residential areas	25	Residential areas
26	Low cultivated land	73	Residential areas	73	Vacant land
27	Low cultivated land	27	Low cultivated land	27	Low cultivated land
28	Low cultivated land	28	Low cultivated land	28	Low cultivated land
29	Low cultivated land	29	Low cultivated land	29	Low cultivated land
30	Low cultivated land	30	High cultivated land	30	High cultivated land
31	Road	31	Road	31	Road
32	Road	32	Road	32	Road
33	Residential areas	33	Residential areas	33	Residential areas
34	Residential areas	34	Residential areas	34	Residential areas
35	Residential areas	19	Low cultivated land	19	Low cultivated land
36	Residential areas	20	Low cultivated land	20	Low cultivated land
37	Residential areas	17	High cultivated land	17	High cultivated land
38	Residential areas	16	High cultivated land	16	High cultivated land
39	Residential areas	11	High cultivated land	11	High cultivated land

(contd...)

1934-35		1982-83		2006	
Plot no.	*Category*	*Plot no.*	*Category*	*Plot no.*	*Category*
40	Residential areas	8	High cultivated land	8	High cultivated land
41	Residential areas	8	High cultivated land	8	High cultivated land
42	Low cultivated land	42	Low cultivated land	42	Low cultivated land
43	Low cultivated land	43	Low cultivated land	43	Low cultivated land
44	Low cultivated land	44	Low cultivated land	44	Low cultivated land
45	Low cultivated land	8	Low cultivated land	8	Low cultivated land
46	Residential areas	46	Residential areas	46	Residential areas
47	Residential areas	47	Residential areas	47	Residential areas
48	Low cultivated land	48	Low cultivated land	48	Low cultivated land
49	Low cultivated land	12	High cultivated land	12	High cultivated land
		13	High cultivated land	13	High cultivated land
50	Low cultivated land	14	High cultivated land	14	High cultivated land
		15	High cultivated land	15	High cultivated land

(contd...)

	1934-35		*1982-83*		*2006*
Plot no.	*Category*	*Plot no.*	*Category*	*Plot no.*	*Category*
51	Low cultivated land	17	High cultivated land	17	High cultivated land
52	Low cultivated land	52	Low cultivated land	52	Low cultivated land
53	Low cultivated land	53	Low cultivated land	53	Low cultivated land
54	Low cultivated land	24	High cultivated land	24	High cultivated land
55	Low cultivated land	24	High cultivated land	24	High cultivated land
56	Barren and Stony area	56	Mango garden	56	Mango garden
		33	Residential areas	33	Residential areas
		31	Residential areas	31	Residential areas
57	Low cultivated land	26	High cultivated land	26	High cultivated land
58	Residential areas	58	Residential areas	58	Residential areas
59	Low cultivated land	27	High cultivated land	27	High cultivated land
		28	High cultivated land	28	High cultivated land
		30	High cultivated land	30	High cultivated land
60	Residential areas	28	Residential areas	28	Residential areas
		35	Residential areas	35	Residential areas

(contd...)

1934-35		1982-83		2006	
Plot no.	*Category*	*Plot no.*	*Category*	*Plot no.*	*Category*
61	Low cultivated land	34	Low cultivated land	34	Low cultivated land
		36	Low cultivated land	36	Low cultivated land
62	Low cultivated land	62	Low cultivated land	62	Low cultivated land
63	Low cultivated land	36	Low cultivated land	36	Low cultivated land
		38	Low cultivated land	38	Low cultivated land
		40	Low cultivated land	40	Low cultivated land
64	Low cultivated land	64	Low cultivated land	64	Low cultivated land
65	Residential areas	37	Residential areas	37	Residential areas
		39	Residential areas	39	Residential areas
66	Road	66	Road	66	Road
67	Low cultivated land	67	Low cultivated land	67	Low cultivated land
68	Residential areas	68	Residential areas	68	Residential areas
69	Residential areas	69	Residential areas	69	Residential areas
70	Residential areas	70	Residential areas	70	Residential areas
71	Residential areas	71	Residential areas	71	Residential areas
72	Residential areas	72	Residential areas	72	Residential areas

(contd...)

	1934-35		*1982-83*		*2006*
Plot No.	*Category*	*Plot no.*	*Category*	*Plot no.*	*Category*
73	Low cultivated land	71	Low cultivated land	71	Low cultivated land
		72	Residential areas	72	Residential areas
74	Low cultivated land	77	Residential areas	77	Residential areas
75	Low cultivated land	75	Low cultivated land	75	Low cultivated land
76	Low cultivated land	76	High cultivated land	76	High cultivated land
77	Low cultivated land	77	High cultivated land	77	High cultivated land
78	Low cultivated land	78	Residential areas	78	Residential areas
79	Low cultivated land	79	Low cultivated land	79	Low cultivated land
80	Low cultivated land	80	Low cultivated land	80	Low cultivated land
81	Low cultivated land	81	Residential areas	81	Residential areas
		80	Residential areas	80	Residential areas
82	Low cultivated land	82	Low cultivated land	82	Low cultivated land
83	Low cultivated land	83	Low cultivated land	83	Low cultivated land
84	Low cultivated land	84	Low cultivated land	84	Low cultivated land
85	Low cultivated land	85	Low cultivated land	85	Low cultivated land

(contd...)

	1934-35		1982-83		2006
Plot no.	*Category*	*Plot no.*	*Category*	*Plot no.*	*Category*
86	Low cultivated land	86	Low cultivated land	86	Low cultivated land
87	Low cultivated land	87	Low cultivated land	87	Low cultivated land
88	Low cultivated land	88	Low cultivated land	88	Low cultivated land
89	Low cultivated land	89	Low cultivated land	89	Low cultivated land
90	Low cultivated land	90	Low cultivated land	90	Low cultivated land
91	Low cultivated land	91	Low cultivated land	91	Low cultivated land
92	Low cultivated land	92	Low cultivated land	92	Low cultivated land
93	Stony area	93	Low cultivated land	93	Low cultivated land
94	Low cultivated land	94	Low cultivated land	94	Low cultivated land
95	Low cultivated land	95	Low cultivated land	95	Low cultivated land
96	Low cultivated land	96	Low cultivated land	96	Low cultivated land
97	Low cultivated land	97	Low cultivated land	97	Low cultivated land
98	Low cultivated land	98	Low cultivated land	98	Low cultivated land

(contd...)

	1934-35		*1982-83*		*2006*
Plot no.	*Category*	*Plot no.*	*Category*	*Plot no.*	*Category*
99	Low cultivated land	99	Low cultivated land	99	Vacant land
100	Low cultivated land	84	High cultivated land	84	High cultivated land
	Low cultivated land	99	High cultivated land	99	High cultivated land
101	Low cultivated land	120	High cultivated land	120	High cultivated land
102	Low cultivated land	95	High cultivated land	95	High cultivated land
103	Low cultivated land	103	Low cultivated land	103	Low cultivated land
104	Low cultivated land	104	Low cultivated land	104	Low cultivated land
105	Low cultivated land	105	Low cultivated land	105	Low cultivated land
106	Low cultivated land	106	Low cultivated land	106	Low cultivated land
107	Low cultivated land	107	Low cultivated land	107	Low cultivated land
108	Low cultivated land	108	Low cultivated land	108	Low cultivated land
109	Low cultivated land	210	High cultivated land	210	High cultivated land
	Low cultivated land	103	High cultivated land	103	High cultivated land

(contd...)

	1934-35		*1982-83*		*2006*
Plot no.	*Category*	*Plot no.*	*Category*	*Plot no.*	*Category*
111	Low cultivated land	111	Low cultivated land	111	Low cultivated land
112	Low cultivated land	112	Low cultivated land	112	Low cultivated land
113	Stony area	113	Stony area	113	Stony area
114	Low cultivated land	114	Low cultivated land	114	Low cultivated land
		113	Residential areas	113	Residential areas
115	Low cultivated land	115	High cultivated land	115	High cultivated land
116	Low cultivated land	116	Low cultivated land	116	Low cultivated land
117	Low cultivated land	115	High land	115	High land
118	Low cultivated land	118	Low cultivated land	118	Low cultivated land
119	Low cultivated land	119	Low cultivated land	119	Low cultivated land
120	Low cultivated land	120	Low cultivated land	120	Low cultivated land
121	Low cultivated land	121	Low cultivated land	121	Low cultivated land
122	Low cultivated land	122	Low cultivated land	122	Low cultivated land
123	Low cultivated land	123	Low cultivated land	123	Low cultivated land

(contd...)

	1934-35		*1982-83*		*2006*
Plot no.	*Category*	*Plot no.*	*Category*	*Plot no.*	*Category*
124	Low cultivated land	124	Low cultivated land	124	Low cultivated land
125	Low cultivated land	125	Low cultivated land	125	Low cultivated land
126	Low cultivated land	126	Low cultivated land	126	Low cultivated land
127	Low cultivated land	143	High land	143	High land
128	Low cultivated land	128	Low cultivated land	128	Low cultivated land
129	Low cultivated land	129	High land	129	High land
130	Low cultivated land	130	Low cultivated land	130	Low cultivated land
131	Forest	131	Forest	131	Forest
132	Low cultivated land	141	High cultivated land	141	High cultivated land
133	Low cultivated land	140	High cultivated land	140	High cultivated land
134	Low cultivated land	134	High cultivated land	134	High cultivated land
		140	High cultivated land	140	High cultivated land
135	Low cultivated land	135	Low cultivated land	135	Low cultivated land
136	Low cultivated land	139	High land	139	High land
137	Low cultivated land	139	High land	139	High land

(contd...)

	1934-35		*1982-83*		*2006*
Plot no.	*Category*	*Plot no.*	*Category*	*Plot no.*	*Category*
138	Low cultivated land	138	Low cultivated land	138	Low cultivated land
139	Low cultivated land	139	Low cultivated land	139	Low cultivated land
140	Low cultivated land	140	Low cultivated land	140	Low cultivated land
141	Low cultivated land	156	residential areas	156	residential areas
142	Low cultivated land	142	Low cultivated land	142	Low cultivated land
143	Low cultivated land	143	Low cultivated land	143	Low cultivated land
144	Low cultivated land	144	Low cultivated land	144	Low cultivated land
145	Low cultivated land	145	High cultivated land	145	High cultivated land
146	Low cultivated land	146	Low cultivated land	146	Low cultivated land
147	Low cultivated land	147	Low cultivated land	147	Low cultivated land
	Low cultivated land	152	Residential areas	152	Residential areas
148	Low cultivated land	148	Low cultivated land	148	Low cultivated land
149	Low cultivated land	149	Low cultivated land	149	Low cultivated land
150	Low cultivated land	147	High cultivated land	147	High cultivated land

(contd...)

	1934-35		*1982-83*		*2006*
Plot no.	*Category*	*Plot no.*	*Category*	*Plot no.*	*Category*
151	Low cultivated land	146	High cultivated land	146	High cultivated land
152	Low cultivated land	152	Low cultivated land	152	Low cultivated land
153	Low cultivated land	153	Low cultivated land	153	Low cultivated land
154	Low cultivated land	154	Low cultivated land	154	Low cultivated land
155	Low cultivated land	153	High land	153	High land
156	Low cultivated land	154	High cultivated land	154	High cultivated land
157	Low cultivated land	155	High cultivated land	155	High cultivated land
158	Low cultivated land	154	High cultivated land	154	High cultivated land
159	Low cultivated land	157	High cultivated land	157	High cultivated land
160	Sacred grove	160	Sacred grove	160	Sacred grove
161	Low cultivated land	161	Low cultivated land	161	Low cultivated land
162	Forest	162	Forest	162	Forest
		173	Grazing land	173	Grazing land
		174	Grazing land	174	Grazing land
		175	Grazing land	175	Grazing land
		176	Grazing land	176	Grazing land

(contd...)

1934-35		1982-83		2006	
Plot no.	*Category*	*Plot no.*	*Category*	*Plot no.*	*Category*
163	Low cultivated land	149	High cultivated land	149	High cultivated land
164	Low cultivated land	160	High cultivated land	160	High cultivated land
165	Low cultivated land	161	High cultivated land	161	High cultivated land
166	Low cultivated land	162	High cultivated land	162	High cultivated land
167	Low cultivated land	165	High cultivated land	165	High cultivated land
	Low cultivated land	163	High cultivated land	163	High cultivated land
168	Low cultivated land	164	High cultivated land	164	Vacant land
169	Low cultivated land	190	High cultivated land	190	Vacant land
	Low cultivated land	188	High cultivated land	188	Vacant land
170	Low cultivated land	177	High cultivated land	177	Vacant land
171	Forest	173	Grazing land	173	Grazing land
		174	Grazing land	174	Grazing land
		175	Grazing land	175	Grazing land
		176	Grazing land	176	Grazing land
		203	Grazing land	203	Grazing land
		202	Grazing land	202	Grazing land

(contd...)

	1934-35		*1982-83*		*2006*
Plot no.	*Category*	*Plot no.*	*Category*	*Plot no.*	*Category*
173	Low cultivated land	179	High cultivated land	179	Vacant land
174	Forest	174	High cultivated land	174	High cultivated land
175	Low cultivated land	181	High cultivated land	181	Vacant land
176	Low cultivated land	195	High cultivated land	195	Vacant land
177	Low cultivated land	177	Low cultivated land	181	Vacant land
178	Low cultivated land	199	High cultivated land	199	Vacant land
	Low cultivated land	200	Mango garden	200	Mango garden
	Low cultivated land	201	Forest	201	Forest
179	Low cultivated land	196	High land	196	Vacant land
180	Forest	180	Forest	180	Forest
181	River	181	River	181	River
182	Low cultivated land	182	Low cultivated land	182	Low cultivated land
183	Low cultivated land	183	Low cultivated land	183	Low cultivated land
184	Low cultivated land	184	Low cultivated land	184	Low cultivated land

The table no. 16 shows that how the land characteristics of each plot has been changed in Kobatghai village. In 1934 the plot no. 14, 15, 16 were under low cultivated land but in 1982 and in 2006 these plots had been converted to high cultivated land. It has happened due to the change of land characteristics to the non availability of proper irrigation water. In 1934 the plot no. 18 was under low cultivated land but in 1982 in this plot a school had been set up and the plot no. had been changed to 57, but in 2006 no change has been found on this plot. In 1934 the plot no. 19 and 20 were under low cultivated land but in 1982 these plots had been converted to high cultivated land and no significant change had been observed in 2006. In 1934 plot no. 25 was under low cultivated land But in 1982 and in 2006 this plot had been converted to residential area. In 1934 the plot no. 26 was under low cultivated land but in 1982 this plot had been converted to residential area bearing the plot no. 73 but in 2006 on visit to this plot it had been observed that this plot is lying vacant. It has happened because from the plot no. 73, Government had resettled twenty Kharia families near Jashipur in the year 1970, from that period this plot is lying vacant. In 1934 the plot no. 35, 36, 37, 38, 39, 40 and 41 were under residential area but in 1982 these plots had been converted to low cultivated land bearing no. 19 and 20. But the other plots bearing no. 17, 16, 11 and 8 were converted to high cultivated land. The people of these plots were shifted to some other places within the village according to their convenient. In 1934, the plot no. 49 and 50 was under low cultivated land but in 1982 these plots were broken into two plots bearing no. 12, 13 and 14, 15 respectively and was converted to high cultivated land. No significant change has been observed in 2006 on these plots. In 1934, the plot no. 51, 52, 53, 54 and 55 were under low cultivated land but all these plots were converted to high cultivated land. In 1934 the plot no. 56 was barren and stony land but in 1982 and in 2006 this plot was broken into three plots bearing no. 56, 31 and 33 and was converted to mango garden and residential areas respectively. In 1934 the plot no. 78 was under low cultivated land but in 1982 and in 2006 this plot was converted to residential areas. In 1934, the plot no.81 was under low cultivated land but in 1982 from the office record it had been observed that this plot was broken into two plots bearing no. 81 and 80 and these plots were converted to residential areas, no significant change has been observed in 2006 on these plots. In 1934 the plot no. 114 was under low cultivated area but in 1982 this was broken into two plots bearing no. 113 and 114 and these plots were converted as residential areas and low cultivated land respectively, in 2006 no change has been observed on this plot. In 1934 the plot no. 162 and 171 were under forest but in 1982 and in 2006 it had been observed that this plot was broken into nine plots bearing no. 162, 173, 174, 175, 176, 202, 203, 204 and 205 out of which the plot no. 162 remained as forest but the other plots were converted to grazing land. It has happened because when the road was constructed many trees were cut and later on this land was converted to grazing land. In 1934 the plot no. 178 was under low cultivated land but in 1982 this plot had been broken into three plots bearing no. 199, 200 and 201. Out of these plots the plot no. 199 remained as low cultivated land but the plot no. 200 was converted to mango garden and plot no. 201 was converted to forest. No significant change had been observed in 2006.

Conclusion: From the above analysis it is observed that in Bakua village the forest area has been increased to 9.52 percent since 1934-35, which is a very good signal to the administrators, forest management officials and the researchers. It appears that the forest resource is well maintained and the villagers are not destructing the pristine forest of this biosphere reserve. The residential areas have been increased to only 2.83 percent since 1934, which is not also very alarming. But in Jenabil village it has been observed that ten percent forest has been reduced and this area is being utilized as high cultivated land and grazing land. Another significant change has been observed that the residential areas have been reduced to 0.62 percent. It has happened because the Government resettled the Bathuri and the Kharia population from Jenabil village to Ambodia near Jashipur in 1993 from the plot number 154, 146, 147, 148, 151, 191, 320, 171 and 371. These plots are still lying vacant. In Jamunagarh village (1934) it has been observed from the data and from the map that there was neither the sign of cultivated land nor the residential areas. It means that up to 1934 the people could not enter in this village because of non-availability of proper communication and fear of wild animals like elephants and tigers. But after analyzing the data it has been observed that in this village the forest area had been reduced to 1.26 percent, cultivated area had been increased to more than 34 percent, barren land had been decreased to 43.28 percent and it may be mentioned that this portion had been converted to cultivated land and the residential areas were increased by 4.17 percent. In Kobatghai village the forest area was very less. According to 1934 only 17.88 percent area was under forest that had been decreased to 7.39 percent. The cultivated area was reduced to 7.43 percent because the villagers are not cultivating the forest bordering area where the animals like elephants come during night time and spoil the agricultural product. The residential areas had been increased to only 0.5 percent which is not alarming. This has happened, because from the plot no. 73, Government had resettled twenty Kharia families near Jashipur in the year 1970 giving them compensation. After studying the four villages in core area of Similipal Biosphere Reserve it has been observed that the forest area has been reduced but for this the villagers are not responsible. It is learnt during the time of field work that many unauthorized people come from outside to cut the trees and killed the animals. It is also learnt from the Director, Tiger Reserve Project that the administration wants to vacate all these four villages for the free faunal movement. Under this backdrop it is therefore suggested to initiate the Social Impact Assessment Study for these villages to be rehabilitated in a new locate, which their Socio-Economic attributes should not be disturbed.

12

Study on Material Culture in the Core Area : An Eco-cultural Dimension

Tamal Maity

Similipal Biosphere Reserve:

Similipal one of the most important Biosphere Reserve, is located on the southern extension of Chhotanagpur plateau or northern extent of the Deccan Plateau in the district of Mayurbhanj, Orissa. The Similipal Biosphere Reserve is divided into two zones (a) Core zone (845.70 sq. km. area) and (b) Buffer zone (1904.30 sq. km. area).

Area of Study:

The Study is undertaken in core area (845.70 sq. km.) of which four villages are situated namely, (1) Kabatghai, (2) Jamunagarh, (3) Jenabil and (4) Bakua. The core area is rich in biodiversity with physical and topographical features. The villages come under Karanjia tahsil of Mayurbhanj district of Orissa state.

Ethnic composition of core area:

The core area villages are dominated with scheduled tribes such as Santal / Majhi, Ho / kol and one family of Bathudi.

Ethnic composition in core area of Similipal Biosphere Reserve

Sl. No.	*Village*	*Comm-unities*	*No. of family*	*No. of popul-ation*	*Total popul-ation*	*Male*	*Female*
1	Kabatghai	Santal Ho	13 4	20 60	80	42	38
2	Jamunagarh	Ho	18	86	86	39	47
3	Jenabil	Ho Bathudi	27 1	183 1	184	93	91
4	Bakua	Ho	21	97	97	44	53
	Total		**84**	**447**	**447**	**218**	**229**

Santal

The Santal of Mayurbhanj district of Orissa are also known as Majhi. They are distributed in the districts of Keonjhar, Balasore and Mayurbhanj. According to province 2001 data census their population is 5,30,776. Though their mother tongue is Santali, they are also conversant with Oriya, the Indo-Aryan language.

The Santal are divided into twelve patrilineal totemic clans (*pari*), namely, (1) Hansda, (2) Murmu, (3) Hembram, (4) Soren, (5) Kinku, (6) Marandi, (7) Tudu, (8) Boske, (9) Besra, (10) Pauria, (11) Charrey and (12) Bedia, which are subdivided into several *Khunt* or subclans.

A good number of them are having own agricultural land and are engaged in cultivation. Santali belong to the Austro-Asiatic family of languages. They speak Santali language at home. Rice is their staple food. They are non-vegetarian. Each village has a grove of *sal* trees known as *jahira /saran*, which is worshipped during various rituals.

The traditional village council consists of headman (*majhi*), the messenger (*dakua*), the priest (nache/*dehiri*). The headman (*majhi)* is assisted by other members in looking after village affairs. Their general rate of literacy is low.

The Santals have cordial relationships with the Hos and outsiders in the village. For agricultural implements made of iron and baskets they dependent on Lohar and Mahli, which they purchased from the market.

The Santal are changing due to the influence of education and communication. There is a significant change appearing in their living standard, dressing, agriculture and other aspects of life.

Ho

The Ho of Mayurbhanj district of Orissa migrated from Singbhum district of Bihar. In Orissa they are distributed in Cuttok, Kalahandi, Dhenkanal, Keonjhar, Mayurbhanj and some other districts. According to 1981 census their number is 44,496 persons. They have a number of exogamous clans *(kili)* such as (1) Alda, (2) Mundu, (3) Samad, (4) Purti, (5) Hembram, (6) Hansda, (7) Pingua etc. Marriage within *kili* is strictly prohibited.

Ho's marriage through negotiation is the common mode of acquiring mates. They are marginal and small farmers. A few of them are landless and are sharecroppers. They also work as agricultural labourers, make ropes from *sabai* grass and collect and sell forest produce and firewood. The Ho follows a religion which is mixture of their own religion and Hinduism. They worship *Sing Bonga*, *Marang-Boga* who appears to be their supreme deity and they identify him with the sun. The Hindu goddess Laxmi is worshipped on every Thursday. Their sacred specialist is the *deheri*, who belong to their own community. Singing and dancing are an integral part of their lives.

They have traditional intercommunity relations with Santal and Bathudi. Their literacy rate is very poor due to economic and social reasons. They rely upon modern medicine when their indigenous medicines fail to work.

Bathudi

The Bathudi are one of the original settlers of Bhanjubhum, which consists of the Mayurbhanj and Keonjhar districts in Orisa. According to 1981 census, their population in Orissa is 1,47,969 (73,241 males and 74,728 females). They are mainly rural community. They speak a dialect of Oriya and use the Oriya script.

The Bathudi are divided into number of totemic clans like *Saluka* (wood husking), *Kushmardia* (wild grass), *Nag* (serpent), *Thariyar* (a kind of wild fruit), Panipuk (water insects), Sanklinyar (conch), Dhanphul (flower of the wild paddy). They practise endogamy and clan exogamy. The marriageable age for girls is between fifteen to eighteen years and for boys between eighteen to twenty four years. Spouses are acquired through negotiation and with mutual consent. The rule of residence after marriage is patrilocal. The Bathudi families are mostly of the nuclear type. The son inherits the parental property equally and the eldest son succeeds his father.

The Bathudi are mostly marginal farmer who also work as labourers for contractors in forest and road projects. A few of them are also employed as teachers and government officials. The Bathudi have their own community council (*jati samaj)* controlled by the hereditary heads like *pradhan, dakua, chowkia,* and *dahiri.* The *jati samaj* deals with intercommunity social affairs.

They profess both Hinduism and the traditional tribal religion and worship *sal* grove (*jahira*) or *gramthan.* Mangla Thakur is their family deity. The Bathudi prefer formal education for their boys and girls. They continue to have faith in herbal and indigenous medicines but keenly adopt family planning measures.

Methodology

During two months fieldwork in core area, census forms have been filled up in every household of four villages and collected data through genealogy, observation ad schedules.

Objective of the Study

Present study intended to document the material culture of the core villages of Similipal Biosphere Reserve in relation to ecology and ethnic groups of the area to establish the impact of ecology in material culture production as part of man ecology nature interaction complex. In this connection following objectives are fixed.

1. To understand the indigenous technique and usage employed in the manufacture and use of various tools and implements and house building techniques of the local people.
2. Documentation and preparation of a total inventory of the implement used in various sphere of life.
3. To understand local resources used pattern in material culture production.
4. The ethnic affiliation and gender orientation of material culture.

MATERIAL CULTURE

Introduction

Material culture is the study of those objects (other than human) and all those artifacts with which man surrounds himself in his attempts to ameliorate his position in his environment (Hutton, 1944). Material culture is the cultural manifestation of material use. It consists of certain process involving visible materials and also the results of such processes.

Material culture is visible to naked eye. It involves the transformation of natural items to a cultural item. The cultural semantics of such items are determined by the ecological context and mode of use ascribed by the community. Every culture allows certain transformation process to define its material use. In general it is a need based culturising process of the natural items available in the surrounding environment to define the means of existence. The transformation process involved in the formation of a cultural item in the respective material culture of the ethic groups.

Importance of Material Culture

1. Material life is an index of the general level of the knowledge and of the intellectual attainment of the people.
2. Ecological factors play a great role in shaping the material cultures of the communities.
3. Man utilizes his natural environment/ natural resources with the technology at his disposal.

Material culture as part of Indigenous knowledge

Material culture surveyed in this study is treated as part of indigenous knowledge. In this connection each item manifests its ecological orientation and the social organization of knowledge behind the production and use.

Ecology and Material culture

Role of ecology in shaping material culture is a domain of cultural Anthropologist's endeavor in understanding the man nature interaction specific to the area and the associated communities. Present study reveals that the ecological surroundings helped the ethnic groups to procure materials and shape them to have an eco-friendly subsistence tool production and an indigenous way of utilizing nature for the betterment of human survival.

Ethnic structure of Material culture

In a study of material culture, it is inevitable to make an enquiry that how the material culture is related to a multi-ethnic village setting as to understand the ethnic boundaries in relation with material culture. Present study observed that as such no ethnic boundary can be marked either in the production or in possession of material culture as far as the area of study is concerned. It is manifested more of a regional characterization beyond the boundaries of villages or ethnic groups as no item is marked as ethnic specific or village specific.

Gender and Material culture

The material culture reflects gender demarcation in use and not in production. It is observed that even a similar pattern found among the gender bound use among all ethnic groups settled in the core area.

Modes of documentation

The material culture is classified and described with the help of sketches and photographs in addition to its measurements and process of production.

CULTURAL PROFILE AND MATERIAL CULTURE OF CORE AREA STUDIED VILLAGES

The village

The four villages in core area are situated in the interior forest surrounded by hills. Villages are generally located at the foot of the hills- at safe distances from water-courses. Low lands are generally avoided. Similipal forest being studded with *sal* trees and four villages are contiguous to *sal* forests. Water has to be fetched from the nearest rivers. Rivers are flowing nearly of their villages (Kabatghai, Jamunagarh, Jenabil and Bakua).

The above table it is observed that Santals are only in Kabatghai village and Bathudi is only one member residing in Jenabil village but Hos are dominated in three villages. Usually a village consists of from 17 to 28 households.

It is observed that a central lane running through the village on both sides of which huts are arranged. There is a noticeable feature that the huts of Kabatghai village do not face the street but are built with their back to it. But in Jamunagarh and Jenabil villages huts are scattered and built nearest to their agricultural field according to their convenience. The *marghatti* (sasans) of Ho community are situated outside of the village and different *kilis* are different places. The nearby forests and paddy fields after harvest are the common grazing fields/ grounds for the village cattle. In these villages we noticed small and cultivated trees nim (*Melia-azadiraehta*), tetul (*Tamarindus indica*) could be seen scattered in the sasans of the different *kilis.*

Construction of Huts by Hos

In the four villages of core area it is observed that two different types of dwelling huts, with two sloped *(danda ghar)* and four sloped roofs *(chhata ghar)*. The incline of the roof is not very steep and it comes down on both the sides to about five feet from the ground level. The structure of the hut they use *sal* tree, *parashu*, *mirshi* (small branches of tree), collected from nearby forest. The roofs are securely tied to the *sal* logs and *mann* (bamboo) with *babai* grass ropes for thatching the structure of the roof. The roofs are covered with the leaf of *shaudi* tree collected from the forest. Sometimes they covered the roof with *basu* (paddy straw). The durability of *basu* is less, so they use *shaudi* and it is also resist heat. The four side of the hut is mud wall and an opening for passage. The side walls are seven to eight feet in height and the back walls from six to seven feet in height. The floor is varying in length from fifteen to twenty feet and in width is ten to twelve feet. Outside of the walls are plastered with a particular variety of clay of whitish colour available in the forest. The walls are ornamented with figures of animals or birds or floral design. The painting of wall is generally done by women folk while construction of the huts takes part by both sexes. These huts are not provided with windows and there is usually one opening – the door. A *panda* (varanda) is made about two feet in depth in front of the entrance where it is only one foot in height. The door is constructed of split bamboo (*mann*) or a local variety of wood securely tied together, both the sides of which are plastered with mud. It consists of a single panel and opens inside of the room.

Inside of the hut is often partitioned by a mud wall, which does not always reach the roof. They usually occupy one compartment; the remaining ones are usually used as granaries or storeroom and kitchen. The kitchen and the storeroom remain dark as no windows are provided. There is no outlet in the kitchen for smoke, so the smoke escapes slowly through the thatched roof.

Cattles and other domestic animals are kept in separate room in *goda/ gowa* (cowshed) just adjacent to their hut. It is observed that separate bedrooms are not provided for each married couple and it is not unusual for them to share the same room with their unmarried sons, daughters, brothers and sisters. Adjoining each hut there is an enclosed plot of land, which is put to various uses.

Construction of Huts by Santals

The house of the Santal is not of varied design and the plan of the main structure is always the same. In some a courtyard can be found in the centre, and rooms on two sides and mud walls on the other two

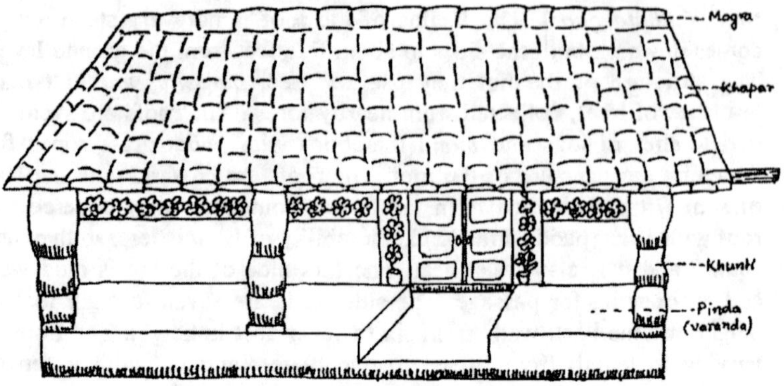

Santal Hut

sides. An average house of the Santals measures 15 to 18 feet in length and 10 to 12 feet in width. Two types of roof is there, one is two slopes and other is four slopes. Four slope roofs they called *chhata ghar* or umbrella house.

In constructing house nine big *sal* posts are required, which are collected from forest. These nine posts are sunk in the ground in three rows, each row consisting of three posts. The middle one they fix a little longer than those of the two sides. Horizontally on the posts of the three rows, three big heavy beams are placed. Three cross beams are then fixed on both sides. These cross beams are tied with the middle and side posts. Then a number of rafters are tied over the cross beams. A frame work of split bamboo is then fixed over it. For roofing they generally use paddy straw or a special type of grass *(babai/ shaudi)*. This they get in abundance in the forest, and it is stronger and more durable than paddy straw. On the sides of each house is a platform measuring two to two and half feet in width and the length is equal to that of the room. The floor of the rooms and the platforms are made of beaten earth whereas placing branches between the post and plastering it with earth and cowdung, their thickness being about 18 inches make the walls. They apply a plaster of earth on the wall, and three feet from the floor, the smear a very dark greywash which is prepared by mixing cowdung with straw ash. There is only one door in each room and no other ventilation except a few holes high up on the wall. In front of the house the eaves of the roof are elongated which serves as a covered verandah. The floor of the house is a little raised above the ground, being filled up with earth. This is trampled down hard and made smooth with mud mixed with cowdung.

Every Santal house in which the head of the family lives has a *bhitar*. This is part of a room partitioned by a low wall, which runs out from the side wall some six to eight feet into the compartment. It is a kind of stall and is sacred to the ancestors. Here food is kept for the ancestors and no outsider is permitted inside. There is a separate shed for cattle, over it bamboo sticks and branches of trees and placed.

Inside the house on the verandah the '*dhenki*' (rice husker) is fixed and milt stone is kept. Except during cold season, when the inside is warm, the Santals generally stay outside on the courtyard both day and night. As they cook their food outside, but always have a fire place inside.

Debasthal (Gosnade)

The debasthal or raised platform is an important place for the Santals. They build this place generally, in the centre of the village, near the *majhi* (headman) house. There is no fixed size of debasthal. In a village Kabatghai, it is seven feet long, five feet broad and seven and half feet height. It consists of a raised mud mound over which a neatly thatched roof is constructed. Four upright wooden posts in the four corners of the debasthal support the roof. The floor of the debasthal is plastered with mud and cowdung.

The Jahirsthan (Sacred Grove)

The Jahirsthan is a place where the santal and the Ho perform religious ceremonies of the village. It is situated at the end of the village and it must be within the boundary of the village. A cluster of *sal* trees about twenty to twenty five in number is always required. Among these trees three are essential, and at the foot of each of these three trees a stone is put representing the deities, *Jaherera, Turukomonrako*, and *Marang-Buru*. The fourth one is an *Ashan* tree, which grows near these three sal trees and at the foot of this tree a stone representing the deity *Parganabonga* is kept. The fifth and the last one is a Mowa tree, which stands a little apart, and the foot of this tree, a stone is placed to represent Gosainera, a female deity. This mowa tree is known as the lady of the grove.

Household Materials

Earthen *chulla* (hearths) are used to cook food in every household. *Chulla* is made of mud in which fire is burnt and the food is cooked.

Chullas are found to be of two types, *Chulla* with one mouth and *ulchulla* with two mouths. It is made of thin plastered of mud. Firewood they collected from forest and cow-dung use for cooking. They cooked rice, dal and vegetables in a aluminum *dekchi* or iron made *kadai*. The utensils used for eating purpose are *thali* (dishes) and lota (*charu*). Dal or vegetables curry is taken in small plates. Water is stored in earthen pots of black colour. For bringing water from the *dhari*, aluminum or silver vessels are used.

Different types of baskets with or without lids made of wicker – work sometimes with rectangular strips form part of the domestic objects. The materials used are either bamboo strips or creepers. Hos and Santals commonly use ropes made of the creeper of siari tree or sabai grass. Broomsticks are made of a kind of grass plaited together at one end and wound into a pyramid shaped knob. Visitors are offered cots to sit upon. The cots are made up of wooden frame and stringed with the rope or coir.

Chakki (*Janti*)

It is a rotary quern found in every house and is usually installed in one corner of the house. It is constructed of two flat stone slabs, one fixed in the earth on which 8 inches iron bar stands which works as an axis to operate the next one. The next slab is fixed with a wooden handle *(kila)* six inches in height to rotate the slab on the fixed. There is a hole *(gala)* in the middle of the next slab, which the grains are put in and pounded by rotating it. The womenfolk do this work.

Shil and Nora

Shashank (shil) and *diri* (nora) is another device used to prepare *chatni* and pound chilly, coriander etc. *Shashank* is a rectangular stone slab of one and half feet in length and 10 inches breadth. The thickness of the *shashank* is usually 2 to 3 inches. The upper part of the slab is made rough by strokes of chisel and hammer. *Diri* is a pestle of stone. It is a circular slab of about six inches length and 2 inches radius.

Fishing

Fishing net they called *jalam/ gire*. Traps for catching fish in the river they called *dhunti, puan* etc., occasionally they catch fish in the river by *jalam*.

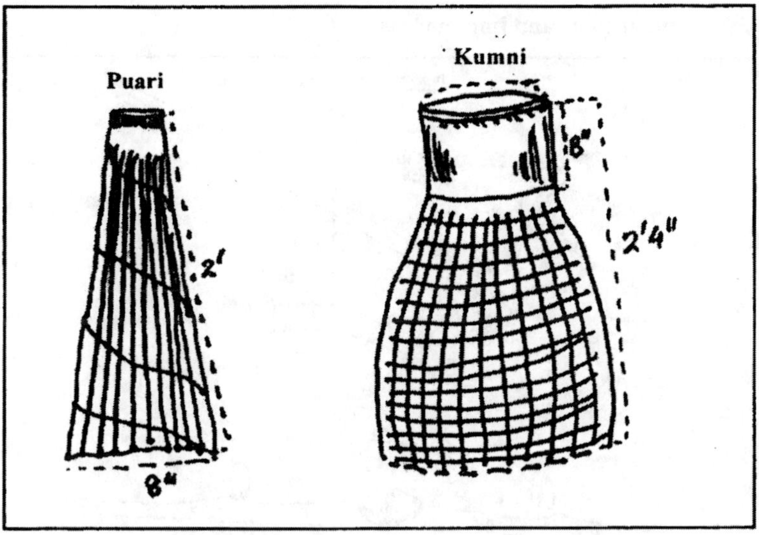

Fishing Trap made of Bamboo

Bells for Domestic Animals

Torki is made of wood and its size is bigger for cattle's neck prepared by Naik or Bathudi in Jenabil village. *Dhuna / ghanti* a bell smaller in size made of iron or bell-metal used also for goat.

Basketry

Different types of baskets with or without lids made of bamboo strips are used in their day-to-day work. *Kanchi, topli, challa* used for domestic work are made from bamboo strips or creepers. *Bandi* is big sized grain containers made by roped of sabai grass or paddy straw is commonly used.

There are two types of design of bamboo work found in the villages one is *check* and another is *twined. Check*, in which the wrap and weft pass over and under each other singly, as in woven cloth. *Twined*, when two or more wefts, pass alternately in front and behind each of the warps, crossing them obliquely. *Aankurabari*, a long bamboo sticks used in agricultural works at the time of threshing *(dauri)* paddy straw. All these articles from bamboo are useful in day to day work of the villagers.

Agricultural tools and implements

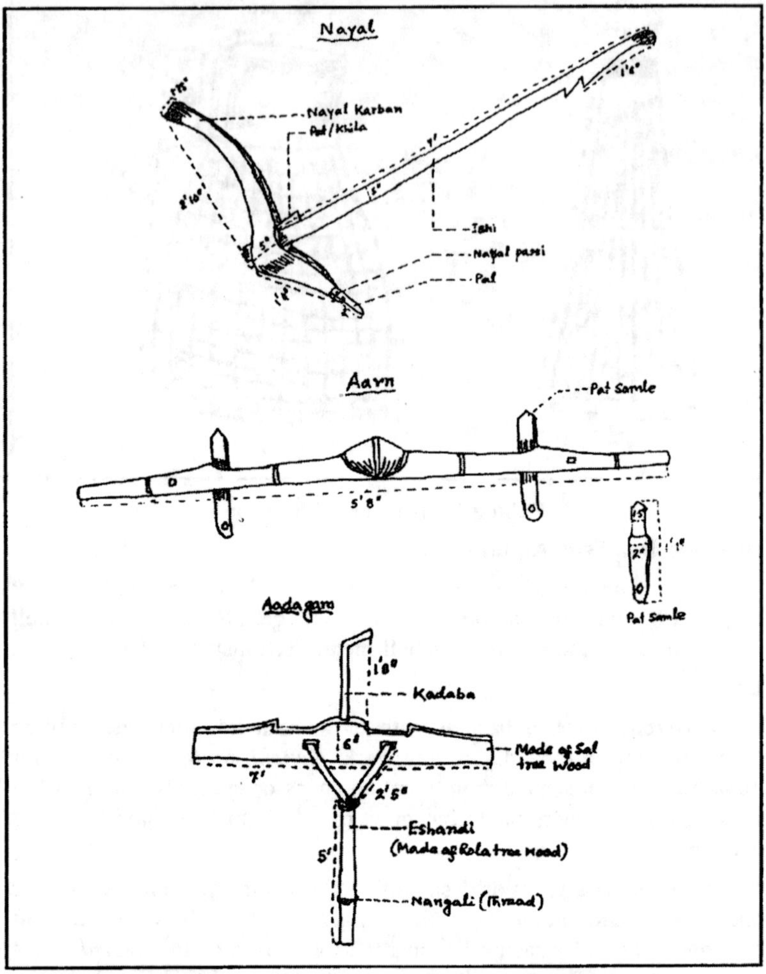

Land and cattle are the two main resources of the people of core area villages' economy. Bullocks have much importance in agriculture.

Tools used in agricultural operation are *nayal* (plough), *aarn, aadagam* (used for leveling the field), *kudlam, maden, gaita* (for digging), *datram* (pick-axe), *dauri* (used for separating straws from the grains at the time of threshing) etc

The common plough is *nayal* an ordinary wooden plough with an iron shave, which varies in weight. The heavier plough certainly produces better crop. The plough proper (*nayal*) is made of single piece of wood, which is slightly curved at the middle. It consists of mainly three parts, (1) the plough proper *(nayal)*, (2) the plough beam *(isi)*, (3) The yoke *(arar)*. The *nayal karban* is long and is made of wooden piece as the body. The beam *(isi)* is a long, flat and slender piece of wood. It has two notches, on its inner side, where it is to be fastened with the *aarnd*. *Pal*, the iron shave, projects about six inch beyond the wooden boot. The iron shave is fixed with wooden boot called *nayal passi*.

A bamboo tube *(por)* is fixed, with a broad wooden cup *(chadi)* at the top, which is fed by a boy or woman who follows the plough. Often the hole at the bottom gets clogged with earth, and for a few yards, till the boy finds the tube getting full of seed, the furrow is unfed.

The *aarnd /aarand* is curved out of a solid piece of wood. Near the two ends, there are two wooden pegs *(pal samle)* inserted through the holes. These pegs are meant to keep the bullocks in proper position.

Aadagam is used for leveling the field. It is a wooden piece made of *sal* tree wood about 7 feet length. One wooden handle is there called *kadaba* and *eshandi* connected with *aarnd*, which is fastened with the bullock. The villagers themselves make their own plough excepting the iron blade, which they purchased from the blacksmith of the local market.

These are the important implements of agriculture, but for other operations the cultivator is also equipped with tools of recognized shape and name. The *khoria* is the long bamboo stick, pointed at one end for the benefit of oxen, and fitten with flat piece of iron at the other to scrape the mud from the plough. The sowing basket is called *supra*. The reaping sickle known as *datram*. The implements of threshing are crooked fork, called *aankubari*, and the wooden rake *karha* use for collecting paddy after threshing. For winnowing the basket *(topli)* and the broom *(jharu)* are used. For carrying manure or mud they use *bharo*. They carry paddy straw by *chiguri* (wooden stick) from the field to their hut. Most of the households are having *dhenki* in their courtyard for husking purpose.

Preparation of Rice-field

They level the field as far as possible with the hoe and a heavy leveler *(karha)*. With the help of plough they plough the rice field several times before transplanting the seedlings.

Transplantation

In a rice field, during the months of May and June after making the earth loose they bring seeds in a basket and scatter them over the loose earth with the hand. Then the earth is again pressed down with the leveler and the seeds are thus buried in. When the seedlings grow to a certain height they are then pulled out and transplanted two or three together in the prepared field. They transplant the seedlings about one foot apart in rows. They have three kinds of transplanted paddy. The earliest varieties ripen in the month of September. The second one, which they transplant in high land, ripens in the month of October. The third variety becomes ripe between November and December and is transplanted in low land. The womenfolk generally do the transplanting.

Threshing

When the crops grow properly and become ripe they cut them with sickle and carry them to a place, which they clean with earth and cowdung. There in the centre they keep one big stone or wooden platform on which they strike the rice saplings. Sometimes they use bullock for threshing. A few bullocks, tied in a line, walk round and round on the spreading straw. With the above processes they separate the paddy from the stalk. Then with the winnowing fan they clean the paddy separating them from impurities.

Before husking, they treat the paddy in the following ways. First they soak the paddy in water by keeping in an earthen vessel for several hours, then they boil the soaked paddy. After boiling they spread the paddy on the courtyard for drying. They do not keep the paddy long in the sun because too much dried paddy will be brittle and during the time of husking the rice will break into pieces. Sometimes they do it not by boiling but only drying in the sun. After drying they store the paddy in baskets *(bandi)*, which prepare with the strands of straw. The paddy treated thus is husked either with a 'dhenki' (husking machine) or with the pestle and mortar. To get clean rice they husk the paddy more than once. Then with the winnowing fan *(kulo)* they remove the husk separated from the grain.

Husking implements

Cereals are husked in the villages either with mortar and pestle worked by hand or by a pestle, which is set at one end of a heavy wooden lever, which is worked at the other end by one or two persons treading on and off it. The mortar is made from a hard log of wood. One side of it is scooped out to form a deep cup like pit having a round bottom. The basal portion is a solid block of wood with a flat end, which looks like the stand of the mortar. There is a constriction outside between the hollow portion and solid stand.

The pestle is made of a heavy, long, solid, cylindrical, wooden rod with a constriction at the middle, which serves as the grip. The hand pestle and mortar are in extensive use. Mortars are made either of wood or stone or rarely of thick and heavy pottery.

Husking lever (Dhenki)

Dhenki

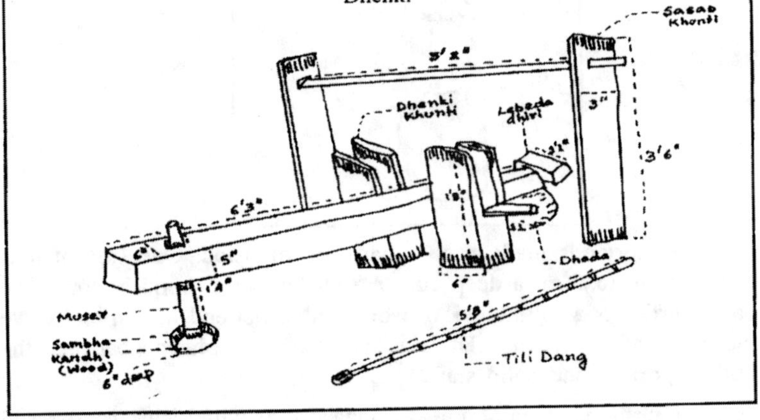

The implement consists of a horizontal lever and a short heavy pestle. The lever is made out of a long, heavy and solid wood which is about six feet long, roughly rectangular in cross-section. The anterior end of the lever is somewhat conical and near to this on its under surface a pestle made of a separate piece of wood is vertically fixed. The pestle is narrow, cylindrical in shape and about two and half feet long. A wooden mortar is placed in the earth in such a way that the pestle falls vertically in that when the implement is in action. The posterior part of the lever is comparatively thinner and lighter than the anterior one. A hole, which is perforated throughout the breadth of the lever, is present at about a third of its length from the anterior end and a wooden bar is inserted into it. Two vertical poles with deep notches, one in each, at its upper end is fixed into the ground on the either side of the lever. Both ends of the wooden bar rest on these notches. The whole arrangement is such that the pestle is raised along with the head of the lever by the pressure of the foot and falls on the grain by its own weight when the foot is removed.

Mortar and Pestle

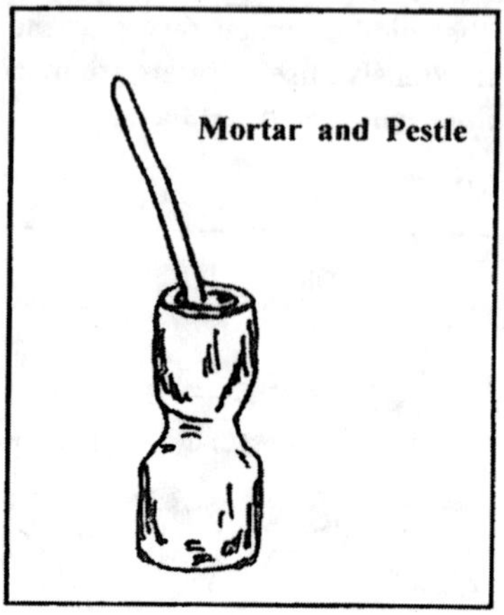

The mortar is made from a hard log of wood. One side of it is scooped out to form a deep cup like pit having a round bottom. The basal portion is a solid block of wood with a flat end, which looks like the stand of the mortar. There is a constriction outside between the hollow portion and solid stand.

The pestle is made of a heavy, long, solid, cylindrical, wooden rod with a constriction at the middle, which serves as the grip.

Sickle

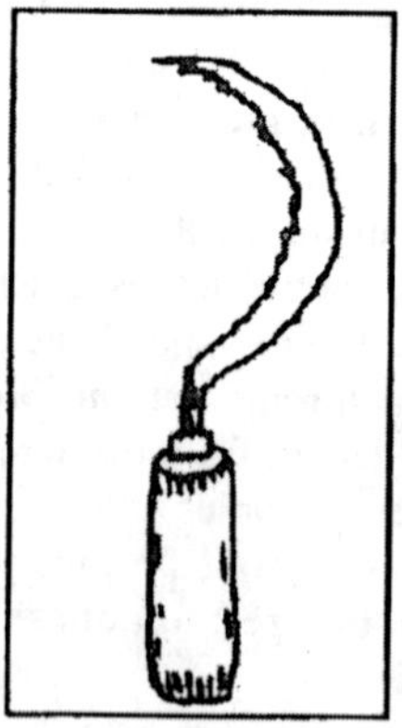

The implement is entirely made of iron. It consists of a blade and a handle made from the same piece of iron. The blade is curved and crescentic in shape. The anterior end of the blade is pointed. The concave edge of the blade is serrated obliquely and these teeth are directed inwards. The basal portion of the blade ends in a hollow cylindrical handle. The portion of the handle near the blade is narrow and it gradually becomes broad at the free end.

Scythe

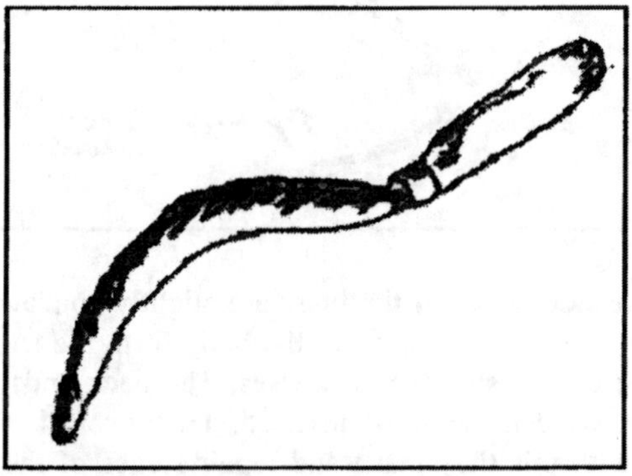

The implement consists of a wooden handle and an iron blade. The wooden handle is solid, cylindrical in shape and more or less uniform in diameter. At one end it becomes suddenly narrow to provide the insertion of the iron ferrule.

The iron blade is made from a single piece of iron. It is crescentic in shape with the outer side convex and a concave sharp inner edge meant for cutting. The blade ends in a long tang, which is introduced into narrow ends of the handle and strengthened by an iron ferrule, which protects the handle from the splitting at the time of insertion of the tang of the blade. The total length is 13 inches (5 inches handle and 8 inches blade) and the maximum breadth of the blade is about one inch.

Hoe (Kudi)

The implement consists of two parts- one is wooden handle and other is iron blade. The handle is about two feet and six inches long, roughly circular in cross-section about two feet in diameter and slightly thicker at one end where it is to be inserted into the socket of the blade. The blade is made of a flat sheet of iron. It consists of a body, which is more or less rectangular in shape, twelve inches long and eight inches

Kudi

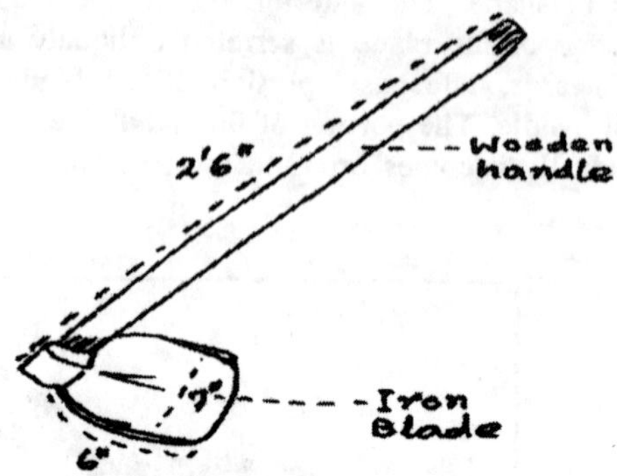

broad. The side edges of the body are slightly concave in outline. The broad projected rear portions of the body form the shoulders and from the middle of this side the neck rises. The neck ends with a socket in which the wooden handle is inserted. They use this for digging earth and also with this they make and repair ridges of the rice field.

Daily Life and Personal Care

The care of the teeth they use the twig of *nim, babul, jamun* etc. and ash of cowdung cake for cleaning of the teeth in the morning. Some people are not very particular about this; they wash their teeth by finger after taking meals.

Bathing Customs and Habits

Most of the people of the core villages have the habits of taking bath once in a day. Those who are agricultural labour or doing labour work prefer to take bath late in the evening after completing agricultural work. When they returned home hot, hungry and exhausted, after a short rest he goes to the neighbouring stream or river to have his daily bath. Most of the canals are outside their village settlements and they are used for washing clothes and bathing purposes.

Washing clothes

Rivers and canals are nearer to the core villages and it is a great advantage to the villagers to clean their clothes and bath time to time round the year.

Toilet practices

Most of the people go for their nature call outside their settlement a little far away. Women usually complete their daily routine either early in the morning or late in the evening.

Disposal of Garbage

The garbage's, cowdung and kitchen ash are stored in the outskirt of their residence by individual families as they are used as manures in the agricultural field. Those who do not have agricultural land, sell the same annually.

Drinking Water

They collect drinking water from river Khairy, canal and stream by *dekchi/ ghagra*. In core villages it is observed that just behind the river they dig three to four feet deep which they called *dhari* and fetch water *(da)* for drinking purpose.

Dress

The younger generation wears pant and shirt *(datta)*. The elder person of the villages wear dhoti *(Katcheha)* four to five yards in length which is worn up to the knees, banyan and shirt or *kurta.* The womenfolk wear *orapalak* /shadi and blouse which is shorter in length of standard shadi. Women wear no special dress during festivals except that new clothes are used. Small girl children wear fork.

Ornaments

Women mostly like ornaments. They wear both silver or brass or gold ornaments. They wear *pagra / lolo* in ears, bangles / *churi (lathi)* in hand, *phuli / chuti* in nose, *har /hiser* in neck, *pajab* in leg anklet. Nose rings are often worn on the left nasal and it is made of gold or brass. Necklace made of silver coins are the fashion among Hos. Anklets are made of the same materials and are oval in shape. They sometimes wear chappal *(chatki).*

Food habits

The food habits of the people of core villages are more or less common. Rice *(kanki)*, kudo *(khoda)* prepared as roti are staple food. Boiled rice kept overnight immersed in water with common salt they eat in the morning at 8 am. with some cooked vegetables like *mata-aara* leaf. In the daytime they eat rice *(mandi)* in the field at about 12.30 pm.

with vegetables. In the evening the main meal of the day composed of boil rice, potherbs, vegetables and a common salt.

They prepared food by firewood *(shan)* collected from forest by women folk. Sometimes they consume dal like *ramda* (black colour), chana dal *(morai jang)*. Cooking medium is mustard oil by very less quantity. Sometimes in the rainy season they prepared fish, first boil it then a few drop of oil with haldi, mirchi and masala they prepared fish item.

Position of women

The management of the house is entirely in the hands of women. Womenfolk of the house in a corner of the courtyard generally clean the utensils. The grinding stone *(chakki)* are kept on one side of the house. Womenfolk use *chakki* for grinding *kudo / kutki* etc. But in addition to her household work she has to take a large share in the work of the fields. Except in ploughig and sowing which are the monopolies of men, she helps in all the other operations of the field. During the off-season she works as a day-labour also.

Musical Instrument

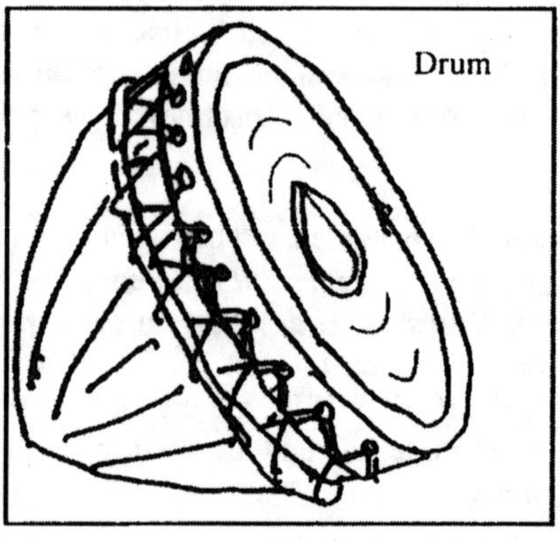

The Hos and Santals are very fond of music and dancing. The wind instrument commonly used is the bamboo flute. The drum is known as the *madal*. It is large and semispherical. Two sticks are used to play on it. They play these during marriage, birth and on the other festive occasions.

Change

Change is inevitable. Material aspects have much susceptibility for change due to various innovative processes, which go side by side with the technological advancement. These influence the mode of life of the people to a large extent who are on the way to urbanization. The earthen pots, brass and bell metal utensils have found a replacement in light and handy aluminium and stainless steel utensils. Thus, change is easily perceptible on the above-mentioned aspects of material culture.

DOCUMENTATION

No. 1

Local name: Datram

English equivalent: Sickle

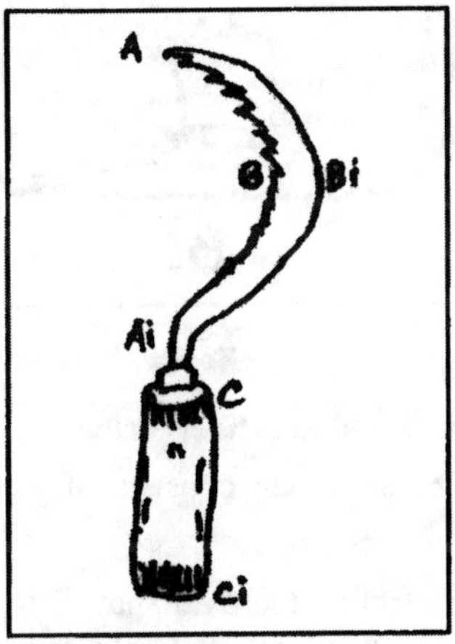

Datram

Material used: Iron for blade and bamboo for handle *(muthia)*

Description: It has two parts (i) a blade and (ii) a handle. The blade is crescent shaped with one concave toothed cutting edge. The opposite edge is thick, blunt and convex. The tip of the blade is pointed. Posteriorly the blade has a tang which is introduced into a handle. The handle is uniform in girth and straight.

Measurement: (a) Length of the blade (A-Ai): 12 inch.

(b) Maximum width of the blade (B-Bi): 1 ft 5 inch.

(c) Length of the handle (C-Ci): 5 inch.

No. 2

Local name: Torki

English equivalent: Wooden bell

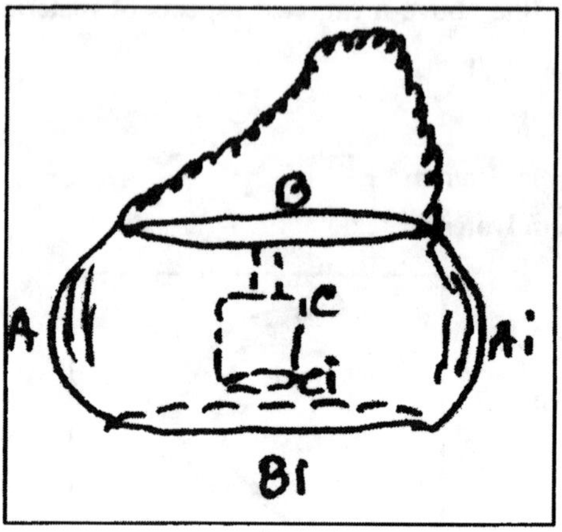

Torki

Material used: Wood and fiber string

Description: The specimen consists of the following parts

(i) The hollow wooden body

(ii) One wooden clapper, the oval shaped body was fashioned from a solid block of wood with the help of a chisel and hammer. The clapper (wooden) is suspended into the hollow cavity by means of strings.

Measurements:

(i) Maximum length of the wooden body (B-Bi): 7 inch

(ii) Maximum width of the same (A-Ai): 6 inch

(iii) Length of the clapper (C-Ci): 3 inch

Nature of use: It is tied to the neck of the cattle.

Method of use: It is suspended to the neck of the cow or bullock. As the object moves the clappers strike each other and produce a sound. To find out the cattle into the forest, the bell help a lot as the sound gives as idea of the cattle's where about.

No. 3

Local name: Dhuna / Ghanta

English equivalent: Bell

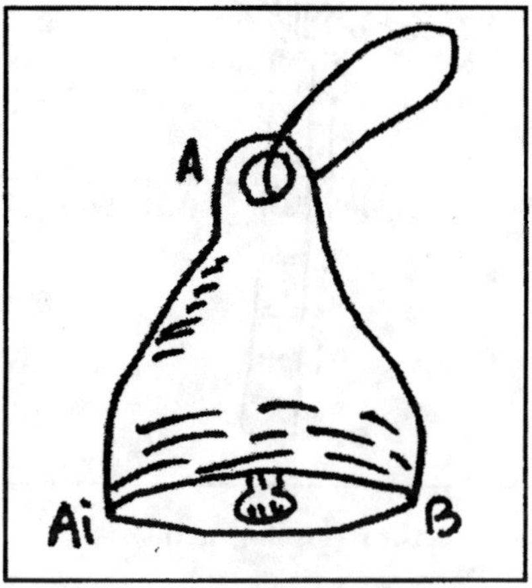

Dhuna

Material use: Iron / Bell-metal

Description: The specimen consists of two identical bells suspended from a cord. A hollow funnel shaped gong, which is widely open below, and a solid club shaped striker which is suspended in the gong from a hook at the top. At the top of the bell, a circular aperture is found through which the cord carrying the bell is passed.

Measurement:

(i) Height of the bell (A-Ai): 3 inch

(ii) Diameter of the open end (Ai-B): 1 feet 5 inch

Nature of use: Used for hanging round the neck of goat or some other animals.

No. 4

Local name: Gira

English equivalent: Fishing net

Material used: Bamboo

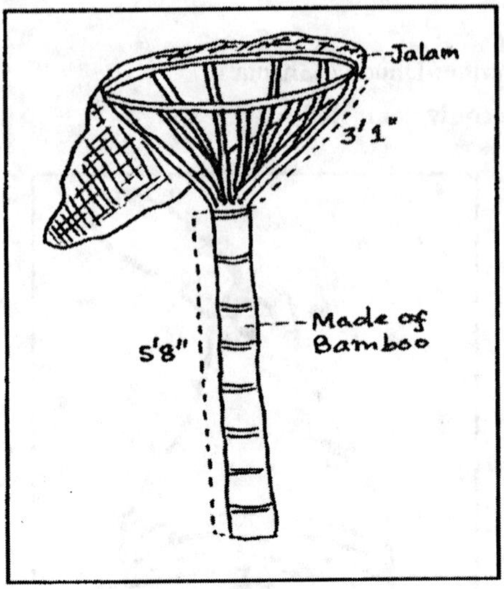

Gira (Fishing net)

Description: This is a funnel looks with a stand made of same piece of bamboo. Anterior part of the bamboo is split into several rays and they act as wraps funnel shaped conical. The open mouth is wide and circular. The stand is cylindrical and straight. A nylon net about four feet is fitted in the anterior portion of the circular mouth.

Measurement:

(a) Height of the stand: 5 feet 8 inch

(b) Conical shaped body length: 3 feet 1 inch

(c) Diameter of the basket mouth: 4 feet 5 inch

Nature of use: Used in the stream or river at the time of catching fish.

No. 5

Local name: Topli

English equivalent: Basket

Material used: Bamboo strips

Description: This is a bamboo strip small basket used for keeping materials like fish, vegetables etc. It has a small circular mouth and a round base. The basket is woven in check pattern of wickerwork. Four cords jointed one side and another side tied over the mouth of the basket.

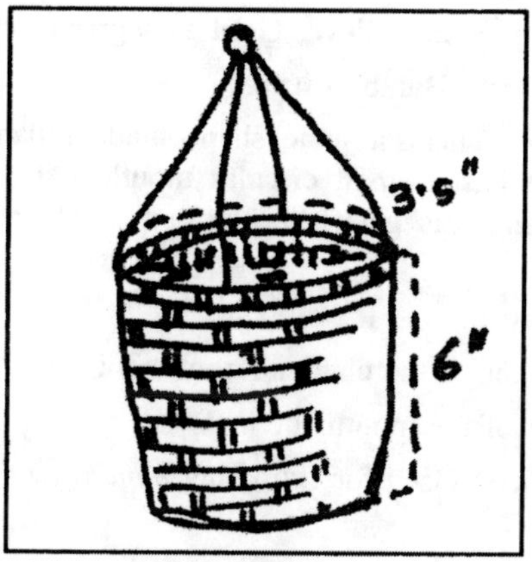

Topli

Measurement:

(a) Total height of the basket: 8 inch

(b) Diameter of the mouth: 3 feet 5inch

(c) Length of base: 5 inch

Nature of use: Used for container for fish or vegetables etc.

No. 6

Local name: Challa

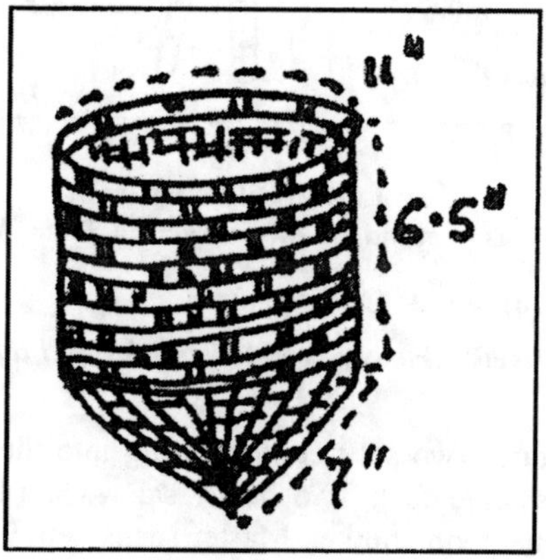

Challa
Use for extracting Handia

English equivalent: Sieve; Used as segregating the rice

Material used: Bamboo strips

Description: This is a funnel shape, made of bamboo strips looking small basket. It has a small circular mouth and posterior portion is conical shape and this part is perforated in character regarded as a sieve.

Measurement:

(i) Total height of the challa: 1 foot 1.5 inch

(ii) Diameter of the mouth: 11 inch

Nature of use: Used for extracting *handia* by segregating the rice.

No. 7

Local name: Putle

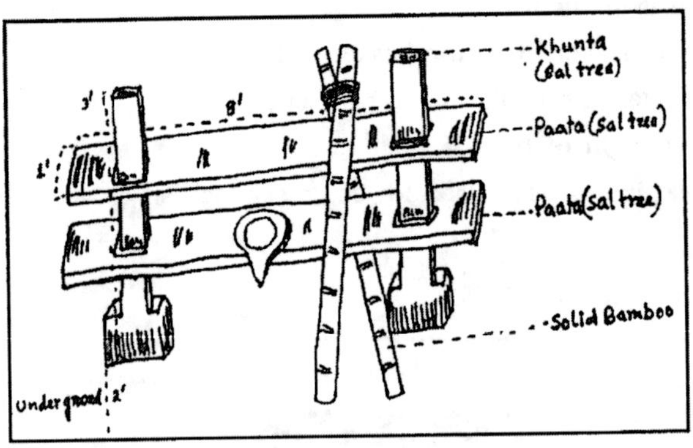

Putle (Oil pressure)
Indigenous method of oil extracting from Master seed

English equivalent: Oil pressure

Material used: Heavy wooden planks *(patta)* and *khunta* (post) of *sal* tree.

Description: Two sal tree khunta dig into the ground in two feet at a distance of five feet. Two heavy sal wood planks (patta) put on these khunta one upon another. Master seeds kept in a small bag (putli) and put in the round shaped area of the lower patta. A rope tied round one end of the pattas. By giving pressure another side of the pattas by two-sal balli, oil come out through lower patta

Measurement:

(a) Length and breadth of the two wooden patta: 8 feet / 1 feet

(b) Khunta's height from the gound level: 3 feet

(c) Oil extracting cutting area of the lower patta in a round shape: 10 inch

Nature of use: Indigenous method of oil extracting from master seed. The oil is collected below as it drips.

No. 8

Local name: Kulo

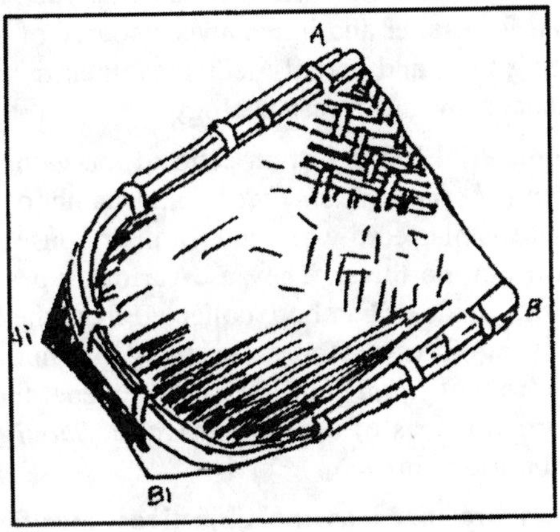

Kulo (Winnowing fan)

English equivalent: Winnowing fan

Material use: Bamboo strips

Description: The specimen is 'U' shaped with posterior part slightly broader than the anterior part. It has a raised rounded posterior end and a straight anterior end. The body of the winnowing fan is made of bamboo strips in twin pattern. Tying bamboo splints all around strengthens the margin.

Measurements:

(a) Length of the straight (A-Ai): 1 foot 6 inch

(b) Breadth of the winnowing fan (A-B): 1 foot 3 inch

(c) Breadth of the posterior end (Ai-Bi): 1 foot

Nature of use: Used during separating dusts and chaffs from grains.

Conclusion

The present endeavor aimed to document the material culture prevalent in the villages found in the core area of Similipal Biosphere Reserve. Hence this study is not claiming the material culture tradition of the entire Biosphere area, while it obviously depicts the traditional structure of material culture as it belongs to the core area. It is observed that forest plays an important role in their economic life both directly and indirectly. It supplies them various food items, fuel, raw material for cash earning and for shelter and it becomes a source of labour, provides them with grazing field and herbal medicines, their subsistence pattern is highly influenced by the forest ecology.

The entire material life is just an echo of the nature. The house is made of *sal* tree (*Shorea robusta*) wood and bamboo, collected from forests. The house is plastered with mud. In their houses, leaf plates are found collected from the forest. They use various types of baskets and winnowing fan etc. made of bamboo, collected from the bamboo groves of the neighbouring forests. Ropes for tying or binding of wooden pieces are made from of *sabai* grass. The *siari* creeper they make strings for bow or tie small pieces of bamboo or wood. *Sabai* grass profusely grows in neighbouring forest.

The study is organized as the objectives specified earlier. The material culture is analytically classified and documented as intended. It also establishes the fact that the entire core area shares common pattern in material culture in its production and use. Moreover, the obvious of ethnic bound craftsmanship in the area further underlines the truth that the similar ecological conditions mould similar material culture irrespective of ethnic boundaries. The observable similarity in the subsistence pattern (agricultural depend) is also proves that the ethic difference is not surfaced in the material culture rather ecology governs the character of material culture. It is significant to note that gender plays a substantial role in material culture practice as it is observed that some of the items are marked by gender in use irrespective of ethnic differences. Thus the study concludes that the traditional structure of material culture of core area is more eco-specific than ethno-specific.

13

Study on Society and Polity in the Core Area Villages

Tamal Maity

Introduction

Similipal Biosphere Reserve (SBR) is located on north eastern border of Orissa, adjoining the states of West Bengal and Jharkhand. Similipal presents a picture of rugged terrain dense forest and wildlife species. All these make Similipal one of the most important Biosphere Reserve, located on the southern extension of Chhotanagpur plateau or northern extent of the Deccan Plateau. Administratively Similipal Biosphere Reserve covers three talukas namely, Baripada, Karanjia and Bamanghati (Riarangpur) of the district of Mayurbhanj, Orissa.

Similipal was declared as the 8th.Biosphere Reserve of India with a total area of 5324 sq. kilometers of which 845 sq. kilometers is the core area. It is also the Tiger reserve and another 1904.25 sq. kilometers is the buffer area. There are four villages in the core area, 61villages in the buffer area and about 1200 villages in the transition area.

Area of study

The study is undertaken in core area (845 sq. kilometers) of which four villages are situated namely, (1) Kabatghai, (2) Jamunagarh, (3) Jenabil and (4) Bakua.

Types of Forest

There are mainly three types of forest in the area, viz. (1) Reserve Forest, (2) Protected Forest and (3) Tiger Reserve.

1. Reserve Forest: Where permission for collecting forest produces mainly nontimber produces is not supposed to be given as per rule, but as a concession the tribal living in the villages within it is allowed to collect fallen twigs and small dry branches from the ground in head load.

2. Protected Forest: Where collection of minor forest produces is permitted and selective residence is also allowed.

Society and Polity of the Core area villages

Society is a group of humans broadly distinguishes from other groups by mutual interests, participation in characteristics relationship, shared institutions and a common culture (Michael C. Howard, 1989).

Polity is considered here as a general term that refers to political organization of a social group. Against the above reference and theoretical back drop the entire gamut of the existing social and political milieu of the Santal, the Ho and the Bathudi tribes have been documented.

Methodology

Standard ethnographic methods have been adopted in collecting data through case studies, observation, schedules and informal interviews, both directed and undirected.

Ethnic composition of Core area:

The core area villages are dominated with scheduled tribes as Ho/ Kol, Santal/Majhi and Bathudi.

Table 1. Ethnic composition in core area of Similipal Biosphere Reserve

Sl. No.	*Village*	*Comm-unities*	*No. of family*	*No. of popul-ation*	*Total popul-ation*	*Male*	*Female*
1	Kabatghai	Santal Ho	13 4	20 60	80	42	38
2	Jamunagarh	Ho	18	86	86	39	47
3	Jenabil	Ho Bathudi	27 1	183 1	184	93	91
4	Bakua	Ho	21	97	97	44	53
	Total		**84**	**447**	**447**	**218**	**229**

From the above table, it is observed that the Santal are dominated only in Kabatghai village and only one member of Bathudi is residing in Jenabil village. Rest of the three villages, namely, Jamunagarh, Jenabil and Bakua are absolutely dominated by the Hos.

About village

Villages are generally located at the foot hills – at safe distances from the water courses. Low lands are generally avoided. Similipal forests being studded with sal trees, many of its villages are contiguous to *sal* forests. There is no well in the villages and water has to be fetched from the nearest canals and rivers. Usually a village consists of 17 to 28 households.

There is a central lane running through the village and on both sides of it huts are arranged. As a typical Santal feature the huts of Kabatghai village do not face the street but are built with their back to it. But in Jamunagarh, Jenabil and Bakua huts are scattered and build nearest to their agricultural field according to their convenience. The adjacent forests and paddy fields after the harvest are the common grazing grounds for the village cattle. In these villages, it is noticed that along with small and planted trees, a few large original forest trees such as the *nim* (Melia azadiraehta)and *tetul* (Tamarindus indica) could be seen around and are commonly found in the *sasans (cremation ground)* of the different *kilis (lineage)*. The *sasans* (*marghatti)* of Ho community are situated just at the boundary of the village and different *kilis have their specific burial spot.*

Santal Society

The Santal of Mayurbhanj district of Orissa is also known as Majhi. They are distributed in the districts of Keonjhar, Balasore and Mayurbhanj. According to 1981 census their population is 5, 30,776. Though their mother tongue is Santali, they are conversant with Oriya, the Indo-Aryan language.

The Santal are divided into twelve patrilineal totemic clans *(pari)*, namely, Hansda, Murmu, Hembram, Soren, Kisku, Marandi, Tudu, Baske, Pauria, Chonrey and Bedia, which are further divided into several *khunt* or sub clans.

Ten clan names have been found in Kabatghai village among the Santal. The names and their meaning are as follows.

S. No.	*Clan Names*	*Meaning*	*Sr. No.*	*Clan Names*	*Meaning*
1.	Hansda	Wildgoose	6.	Chonrey	Lizard
2.	Murmu	Nilgai	7.	Besra	Hawk
3.	Hembram	Betel palm	8.	Tudu	-
4.	Marandi	Sabai grass	9.	Baske	-
5.	Soren	Group of stars in the sky in a pattern	10.	Kisku	-

A good number of them have their own agricultural land and are engaged in agriculture. Santal belong to the Austro-Asiatic family of languages. They speak the Santali language at home. Rice is their staple food. They are non-vegetarian.

Family

Amongst the Santal generally two types of families can be seen. (1) The Nuclear family consisting of father, mother, sons and daughters. (2) The second consisting of grandfather, grandmother, sons, their wives, daughters (unmarried) and grandchildren, which is equivalent to the Hindus joint family.

There is third type, in which the son-in-law is taken as a son in the father-in-law's family. In a Santal family if there is no son but only a daughter, they marry their daughter off and give the son-in-law the position of a son, provided he lives with his father-in-law. If in a family there are grown up daughters and minor sons, the son-in-law works for his father-in-law without wages for five years. He then gets two bullocks, some rice and some agricultural implements, and set up a house for himself and his wife. This sort of son-in-law is called *Ghardi jawae.* In Kabatghai village, Santals are 13 families out of that seven nuclear, six are joint families.

Kinship

The Santal are a patrilineal people and the clans' descends from father to son. The son transmits it, in their turn, to their children again, and so on. The following table of kinship terminology is given below.

S. No.	*Kinship term*	*In Santali*
1.	Father	Baba
2.	Mother	Aayo
3.	Father's Father	Garom haram
4.	Father's mother	Garom budi
5.	Mother's father	Mamu baba
6.	Mother's mother	Mamu aaya
7.	Mother's brother	Mamo
8.	Mother's Br's wife	Mami
9.	Elder brother	Marang dada
10.	Younger brother	Bokayan
11.	Elder sister	Marang daye
12.	Younger sister	Hudin maye
13.	Elder brother's wife	Hili
14.	Junior sister's husband	Jamai
15.	Father's elder brother	Marangba
16.	Father's younger brother	Kaka/Hopanba
17.	Father's elder brother	Marango
18.	Father's young bro's wife	Kakki/Hopanaya

Position of Women

The Santal women have an important status in their society. Though they are shut off from certain religious and administrative functions, their economic status is very high. They generally go to market, do the bargaining – all the produce is in their hand and they make over the proceeds for the sustenance of the family to the male members. They are the repositories of the moral and religious tradition of the society.

Marriage

The Santali name for marriage is *'Bapla'*. It is strictly forbidden for any Santal to marry within his own clan/sept. He may marry into any other clan/sept or sub-clan to which his mother belonged. There are some septs/clans which never intermarry with one another in consequence of some ancient feud between them. A Hansda male or a female never marries a Murmu female or a Murmu male respectively. Similarly a Tudu male never marries a Besra female and vice versa.

Earlier no marriage was allowed between the two clans/septs. But now no such restriction can be found.

Birth and Naming Ceremony

When the Santal woman becomes pregnant, she and her husband observe certain taboos. The husband during his wife's pregnancy never kills any animal, nor participates in any funeral ceremony and does not come in contact with any dead body. The pregnant woman during the evening very rarely comes out of the house.

At birth, an experienced elderly woman of the village or of the adjoining village officiates as midwife. After birth, the umbilical cord is cut by the midwife with an arrow or sharp knife. The placenta is then buried in a pit in one corner of the courtyard of the house.

After the fifth day of child birth purification is performed. On that day the child is given its name. In case the child is boy, he should heir the name of his grandfather. The same procedure is followed for girls, the feminine relations being in the same order from the female side. The men, women and children of the village who have assembled at the house receive each a leaf cup full of rice water with which the leaves of *Nim* tree (Melia azadirachta, L) have been mixed. After this they all take home made *hariya* (liquor) and depart.

Divorce

Divorce is a common sequel to Santal marriage and is granted at the wish of either husband or wife. Following are the grounds for which the Santal men and women demand divorce. The husband can demand divorce if his wife is proved to be a witch, or is sexually immoral and if she does not obey him or lives all the time in her father's house. The wife can claim divorce if her husband cannot supply her sufficient food, clothing, ornaments, etc. Sterility is also an important ground for divorce.

Disposal of the dead

After death, the body is carried on a bedstead by the relatives and villagers who cover the body with a new shroud. They carry the dead body to the bank of a streamlet nearby. The pyre is prepared with the wood which is brought from the nearby jungle. Before setting fire to the body it is covered with a branch and four pieces of wood are put on it. After funeral they shave and bathe and before they enter the village smear themselves with *sal* resin. Coming back after cremating the body they breathed his last.

After five days there is a ceremony called *Tel nahan*. All the villagers assemble at the dead man's house and shave. Then they go to bathe, the men in one place, the women in another. The men take with them a little earth (used as soap), oil cake, oil, *sal* twigs (used as tooth brushes) and a couple of leaves.

Gods and their worship

The Santals have no temple or any shed for worship but they have a *sthan* (place) known as *Jahirsthan* where they worship the village deities. The *Jahirsthan* is a place where religious ceremonies of the village are performed by the Santals. It is situated at the end of the village and it must be within the boundary of the village. It should consist of cluster of *sal* trees about ten to fifteen or more in number, is always required. Among these trees three are essential and they must stand in a row. At the base of each of these three trees a small stone is placed representing the deities *Jaherera, Truku Muruko, Marang-Buru*. The fourth is an *Ashan* tree which grows anywhere near those three *sal* tree, and a stone is put on its base representing the deity *Pargana Bonga*. The fifth and the last is a *Mowah* tree; on its foot a stone representing the deity *Gosain Era* is kept. This is the most important one and is known as lady of the Grove.

In the Santal village there is a succession of festivals throughout the year, nearly all connected with agricultural operations.

Sl. No.	*Name of the Festival*	*Month*	*Details activity*
1.	*Sankrat*	*Pous* (Dec.-Jan.)	This is an imitation of the Hindu *Pous Sankranti* festival. It takes place on the last day of the month of Pous (Dec-Jan).
2.	*Baha Parab*	*Phalgun* (Feb.-March)	The festival takes place when the *sal* trees *(Shorea robusta)* begin to flower. Baha is a Santali word which means flower. The purpose of holding this festival is that the new year is well commenced. Drinking, dancing and singing go on in this festival. The festival takes place in the *Jahersthan* and the *sthan* are cleaned by plastering with cowdung.

Sl. No.	*Name of the Festival*	*Month*	*Details activity*
3.	*Erok-Puja*	*Asadh* (June-July)	This festival they perform before sowing seeds in the newly ploughed field. They celebrate it in the *Jahersthan.* Head of the village perform the function. Five fowls are sacrificed in the *Jahersthan.* After the function they cook these fowls and eat them.
4.	*Hariar Sim*	*Sraban* (July-Aug.)	The Santals hold the festivals at the time of the sprouting of rice, which takes place in the month of *Sraban.* Fowls are sacrificed to at the tribal and village deities and prayers are offered for a bountiful harvest.
5.	Janthar Puja (Nawai)	Agrahayan (Nov.-Dec.)	This festival is held when the first gathering of the winter rice crop is offered. Similarly the first gathering of the millets is offered in the sacred grove in the month of *Bhadra* (Aug.-Sept.) which the Santals call *(Irrigrudhi nawai).* Similarly offerings are made by each householder of the village to the spirits of their ancestors.
6.	*Magh-Sim*	*Magh* (jan.-Feb.)	When the jungle grass is mown; fowls are sacrificed to all the bongas by the village priest and his assistant. This festival marks the end of the Santal year.

Inter-ethnic Relationship

The Santals have cordial relationship with the Hos and other outsiders in the village. For their agricultural implements which are made of iron and baskets, they are dependent on Lohar and Mahli, which they purchase from the market. They have multifaceted relationship with the outsiders in their area. Their intercommunity linkage may be limited, but it cannot be denied that no tribe, other than the Santal, has contributed so much to enrich the culture of the neighbouring people, or for that matter, of the country as a whole.

Polity and Law

Santals will never settle alone in an uncultivated area; they go there in a body and settle with a leader and his assistants. The leader is known as *Manjhi* (headman), being chosen by the village people to administer the rights, rules and ceremonies of the Santal village community. No public sacrifice, no festival, no ceremony such as marriage can properly be done without the *manjhi* taking the lead or initiative. The *majhi* is the representative of the village both in external and internal affairs.

The traditional village council consists of headman *(manjhi)*, the messenger *(dakua)*, the priest *(dehiri)*. The headman *(manjhi),* who is assisted by the other members of the council, to look after the village affairs, the traditional village council has no specific name; it is headed by a headman *(majhi)*, a chief authority of the village. *Manjhi* is considered to be the overall in charge of the village. He looks after the law and order situation in the village like disputes regarding property, marriage and criminal assaults. If any one wishes to settle in the village, his request is discussed in the council. Inter-personal relationship between leaders of the village is pleasant. Kinship plays an important role in the political organisation of the village. The office bearers are either affinally or consanguineally related to one another.

Each village is represented in the Gram Panchayat consisting of tribes. Members are elected on the basis of adult franchise. The number of members from a particular village depends on the size of its population. This body is headed by the Sarpanch who is elected by the members of the Panchayat.

Ho Society

The Ho is from Orissa and are conversant with Oriya and little bit of Hindi. They also use the Oriya script. In Orissa, they are distributed in Kalahandi, Cuttak, Keonjhargarh, and Mayurbhanj and are numbered 44,496 persons (1981 census). The Ho of Mayurbhanj district of Orissa had migrated from Singhbhum district of Bihar.

Marriage Customs

The Hos are divided into a number of exogamous clans called *kilis.* The names of the clans are as Alda, Munda, Samad, Purti, Hembram, Hansda, Pingua etc. Marriage within the *kili* is strictly prohibited. Hos Marriage through negotiation is the common mode of acquiring mates. Marriage is usually adult in Ho society. The bride is not necessarily younger than the bridegroom. The consent of the bride is usually not taken and the match is generally arranged by the parents or guardians or through some common friends. Marriage by force or capture is recognized. The bridegroom may forcibly carry away the bride from a festival or from the bazaar. A free fight often takes place between the two parties and if the attempt is successful the suitor pays the bride-price and a formal marriage takes place.

The bride price has to be paid before the marriage day and is not attended with any special ceremony. The bridegroom has also to supply a fixed quantity of rice and rice-beer to be consumed in the marriage feast. For three days including the day of marriage, the bodies of the bride and bridegroom are anointed with oil and turmeric paste. This is then rubbed off and not washed off. Usually the bride is brought over to the bridegroom's place on the marriage day. But the bridegroom may go over to the bride's place- there being no special hard and fast rule. The bride's party stops at the outskirt of the village where it is met by the bridegroom's party. There is an exchange of abuses and hot words which ends in a mock fight from which the bridegroom's party emerges victorious and the bride is forcibly carried away. Both the parties then enter the village and proceed to the bridegroom's residence. All the villagers assemble at the house and *panch* presides over the ceremony.

The bride and the bridegroom, dressed in yellow clothes with red borders, their bodies anointed with mustard oil and turmeric, stand face to face (the bride facing east and the bridegroom west). The presiding elder joins the hands of the couple. The bridegroom then proceeds to mark with vermilion the parting of the hair of the bride for which any finger of the left hand may be used. Simultaneously the bride with her right hand marks the forehead of the bridegroom with vermillion. The ceremony is now complete and the couple together with the assembled guest breaks up to partake of the bridal feast. The ceremony may be performed during the day or the night. The feast sometimes continues for two or three days.

Poligyny is practiced by the Hos. A man may marry two sisters but in such cases he must marry the elder sister first.

Divorce is allowed in Ho society and divorced women are married according to the *sanga* form of marriage. Widows are allowed to remarry. Any of the younger brother's of the former husband has a priority of claim. But on no account can she be married to the elder brother who is tabooed. If the younger brothers forego their claim on the widow she may marry any other person who is not a member of her father's *kili.*

Birth Custom

After the birth of the child, the father enters the room and the husband and wife remain confined in the room for three days. During this period the husband cooks food for himself and his wife. After these-three days the father had a shave, his nails are pared (by a Ho belonging to a different kili) and he may now resume his usual work-though both the parents remain ceremonially unclean for one month. On this day the mother's nails are also pared.

No special rite is observed at birth. On the fifth or ninth or tenth day, a quantity of *hariya* is drunk, but not in the house of pollution. After a month, a Ho from a different kili again shaves the father and pares the nails of both the parents. After this they are regarded as ceremonially clean and they are allowed to resume their ordinary life. The child is usually named after their grandfather or may be named after the granduncle or uncle.

Death Custom

Among the Hos the usual practice is cremation. Burial is now coming into vogue through want of firewood-the state having strictly prohibited the felling of trees by the Hos on account of denudation of forests.

The *sasan* of the *kili* is generally situated within the village and near about the residents of the *kili.* Each *kili* has its own *sasan.* Due to restriction of firewood, the body is buried in the *sasan.* A grave is dug five or six feet in depth. The body is lowered into this and placed in the position with the head towards the south. All these are covered with a plank and the grave filled with earth. There are no separate rites for females- married, unmarried or widowed and children. But persons who have died of infections diseases such as cholera, smallpox, etc. are buried outside the village.

All the members of the *kili* living in the same village go into mourning for a period of nine days. During this period they are regarded a ceremonially unclean must not eat meat, eggs or fish. Nor are they allowed to anoint their body with oil. On the ninth day a Ho from a different *kili* shaves are pares the nails of the mourners and the period of mourning is ended.

The spirits of the departed are worshipped every year on the last day of the month of Paus (Dec-Jan.) when all the relatives meet in the *sasan* to pay their homage to the departed soul. They afterwards hold an annual feast where rice and *hariya* are consumed.

Kinship and Social Organisation

The Ho tribe is divided into a number of social groups known as *kilis*. The *kili* is primarily a unilateral kinship group which may be called 'clan'. All the members of a *kili* regarded themselves as the descendants of one common ancestor either actual or fictitious and are thus related unilaterally. The Hos are patrilineal people and their *kili* descents from father to children. It is noticed that the kilis are used as a surname by the Hos.

The clan name of four villages Kabatghai, Jamunagarh, Jenabil and Bakua have been found given below.

Sl. No.	*Clan*	*Sl. No.*	*Clan*	*Sl. No.*	*Clan*	*Sl. No.*	*Clan*
1.	Alda	12.	Lugan	23.	Hansda	34.	Chakia
2.	Mundu	13.	Hembram	24.	Barla	35.	Topno
3.	Goipai	14.	Sinku	25.	Suren	36.	Mundri
4.	Koda	15.	Birna	26.	Pingua	37.	Kichua
5.	Soye	16.	Deogam	27.	Bonaria	38.	Bhadora
6.	Nag	17.	Salukad	28.	Baunra	39.	Mal
7.	Samarth	18.	Semen	29.	Tiyu	40.	Dighi
8.	Hasa	19.	Tuti	30.	Danga	41.	Dohra
9.	Samad	20.	Bengra	31.	Tiyu	42.	Murum
10.	Purti	21.	Sanga	32.	Danga		
11.	Oliya	22.	Verla	33.	Gudia		

The family generally consists of the parents, the unmarried daughters, the sons and their wives and children. The Ho son does not, as among some of the primitive tribes, leave his parents as soon as married and set up a new house. He shares the same house with his parents and brothers and toils in the same field and adds his earnings to the common family fund administered by the father. The father is the recognized head of the family and as such has to officiate in some family rites and ceremonies. He generally represents the family in the village Panchayat. Here we see a difference with the Santal who after marriage sets his own house.

Family type in core villages of Ho

Family type	*Kabatghai*	*Jamunagarh*	*Jenabil*	*Bakua*	*Total*
Nucler	4	16	11	14	**45**
Joint	--	8	15	7	**30**
Broken	--	--	2	--	**2**
Total	4	24	28	21	**77**

The Hos are a patrilineal people and the clan's decends from father to son. The following kinship terminology is given below.

Sl. No.	*Kinship term*	*Ho*
1.	Father	Aaba
2.	Mother	Ma
3.	Father's father	Buda
4.	Father's mother	Nani
5.	Mother's father	Buda
6.	Mother's mother	Nani
7.	Mother's brother	Mamu
8.	Mother's bro's wife	Hatam
9.	Elder brother	dada
10.	Younger brother	Bokan
11.	Elder sister	Dai
12.	Younger sister	Bokan
13.	Elder bro's wife	Hili
14.	Junior sis's husband	Bokan
15.	Father's elder br.	Bada
16.	Father's younger br.	Kaka
17.	Father's elder br.'s wife	Badi
18.	Father's younger br.'s wife	Kaki

Gods and their Worship

The spirits of the ancestors are worshipped on every festive occasion by the owner of the house, who acts as priest, by sacrificing a cock and offering *hariya* but no vermillion is used on these occasion.

On the last day of the month of *Paus* (Dec.-Jan.), the departed ancestors are specially worshipped when all the relatives visit the *sasan* of the *kili* and pay homage to the dead.

The Hos worship *Sing Bonga* who appears to be their supreme deity and they identify him with the sun. He is worshipped on all important occasions and *Dussera* day in the month of *Aswin* (Sep.-Oct.) is specially associated with the worship of *Sing Bonga* when sweets and vermillion are offered and a cock is sacrificed. He is also worshipped on the day on which the Hos first partake of the new rice. On the day when the date of a marriage is fixed, *Sing Bonga* is worshipped with hariya and vermillion.

Another important deity is *Marang Bonga* or the 'great god' whom they associate with the hills and mountains. The usual offerings to *Marang Bonga* are sunned rice, vermillion and areca nut and often a goat is sacrificed.

An associated pair of deities is the *Bura Buri* who is ancestors-spirits. They are worshipped on many occasions, especially during the *Sarhul parab.*

The fourth deity is *Goram*. He is worshipped every month and the offerings are vermillion, rice and flowers, fowls and goats being also sacrificed.

The Hindu goddess Laxmi is worshipped on every Thursday in the month of *Agrahayana* (Nov.-Dec.). The owner of the house officiates as priest and the worship takes place in an open space in front of the granary. No special sacrifice is associated with the worship. After this, the paddy is taken home and *Sing Bonga* and other village deities are worshipped in the usual form. On the first day of the month of *Jyaistha* (May-June), before the early rice is sown, a cock and a goat are sacrificed in honour of *Dharam.*

In the month of *Asadh* (June- July), before the Hos are allowed to make rice cakes *Hero Bonga* is worshipped. His special offerings are boiled rice, rice-cakes and hariya.

Festivals

The Hos being mainly an agricultural people, their festivals are mostly associated with that pursuit. They have as many as nine such festivals these are as follows:

Sl. No.	*Name of the Festival*	*Month*	*Priest*	*Details*
1.	*Goma Puja*	*Srabana* (July-August) Full moon day	Headman of the village	The headman officiates as priest, and the Hos abstain from all work and spend the day in merriment.
2.	*Chitan Parab*	*Srabana* (July-August)	The owner of the house	Fowls and goats are sacrificed on this occasion. The festival has now fallen into disuse.
3.	*Gama Parab*	*Bhadra* (August- September)	No special puja is performed	But a cock and goat are sacrificed. The Hos feed their relatives on this day, drink *hariya* and make merry.
4.	*Shyama Puja*	*Kartik* (October-November) New moon day	*Deheri* act as a priest.	They celebrate the Bandana by sacrificing a cock and a hen to *Sing Bonga* and offering him sunned rice, vermillion, sweet, meat and flowers.
5.	*Gohal Puja*	*Kartik* (October-November)	*Majhi* act as a priest	All the deities are worshipped and a cock and a goat are sacrificed and prayers are made for all sorts of benefit.

Sl. No.	*Name of the Festival*	*Month*	*Priest*	*Details*
6.	*Paus Parab*	*Paus* (December-January)	The headman as usual, officiates as priest	The Hos worship the spirits of their departed ancestors and specially *Bura Buri.*
7.	*Magh Parab*	*Magh* (January-February)	The headman of the village officiates as priest.	When the granaries are full, the Hos celebrate the Magh Parab. A Cock and a goat are sacrificed to Sing Bonga. New clothes are put on.
8.	*Chait Parab*	*Chaitra* (March- April)		The god Shiva is worshipped. The people of the neighbouring villages take part and special dances are held.
9.	*Sarhul Parab*	*Chaitra* (March- April)		When the flowers of the *Sal* trees are blossom, basketfuls of these are brought home. The Hos decorate themselves with garland and wreaths of these flowers and after worshipping Bura Buri they spend the day and night in feasting and dancing.

Inter- Ethnic relationship

They have traditional intercommunity relations with peasants like Santal, Bathudi and serving people like the Kamar, Rajak and Napit. Their literacy rate is poor due to economic and social reason. According to 1981 census 12.75% of them are returned literate. They rely upon modern medicine when their indigenous medicines fail to work. They have benefited from some of the developmental programmes also.

Polity and Law

The traditional village council has no specific name; it is headed by a *Mukhia/Mura (panch),* a chief authority of the village. He was elected by the villagers and the post was hereditary in the family in the male line. The eldest son succeeded after the death of the father. He also decides minor criminal cases and boundaries disputes. Serious criminal offences are brought to the notice of the police at once.

There is a Panchayat in every village and the *Mukhia/ mura* on receiving a complaint-provided both the parties agrees to abide by his decision- sends his *dakua* to ask the householders to send their representatives and to produce the accused. Each household of the village must send at least one male member whether old or young to take part in the proceedings. Even a child may represent his household in the absence of an older member. The decision of the Panchayat must be unanimous and the *Mukhya/Mura* merely deliver judgement.

In addition to his judicial and executive duties the *Mura/panch* has to officiate in certain religious rites and ceremonies. He is the chief priest of the village. All religious rites and ceremonies are first performed in his home after which they may be carried out in other household.

Inter-personal relationship between leaders of the village is pleasant. Kinship plays an important role in the political organization of the village. The office bearers are either affinally or consanguineally related to one another.

Bathudi: Society and Polity

They are the original settlers of Mayurbhanj and Keonjhar districts in Orissa and the Singhbhum district in Bihar. As reported, the Zamindars of their community originally reigned in the Similipal range and established their *garh* in Joshipur. The probable homeland of the Bathudi was the Panchpir and Sadar subdivisions of Mayurbhanj district and later on migrated to Keonjhar in Orissa. According to 1981 census, their population in Orissa is 1,47,969 (males 73,241 and females 74, 728). They are mainly rural community and speak a dialect of Oriya and use the Oriya script.

The Bathudi are divided into a number of totemic clans like (1) Saluka (wood husking), (2) *Kushmardia* (wild grass), (3) *Nag* (serpent), (4) *Thariyar* (a kind of wild fruit), (5) *Panipuk* (water insects), (6) *Sankhiyar* (conch) and (7) *Dhanphul* (flower of the wild paddy). They use surnames such as Nayak, Giri, Bishal, Rout, Bansori, Patra, Mundiyar, Pradhan, Dakna, etc.

The Bathudi practice community endogamy and clan exogamy. The marriageable age for boys between twenty to twenty five years and for girls is between sixteen to twenty years. Marriage through negotiation is the common mode of acquiring mates. The minimum bride price was Rs. 250/- to Rs. 300/-, but it is now replaced by dowry. The rule of residence after marriage is patrilocal. Either spouse can seek divorce or the grown-up children of a divorced couple stay with their father while the minor children remain with their mother. Widow and divorced remarriage *(sangha)* is allowed. The marriage symbol of women is the vermilion mark and iron bangles. The Bathudi families are mostly of the nuclear type. The son inherits the parental property equally and the eldest son succeeds his father.

A predelivery ceremony (*sad bhaskhan*) ceremony is performed in the ninth month of pregnancy. A purification ceremony *(nartha)* is observed on the seventh or ninth day after child birth. The first rice feeding ceremony *(mu juthani)* is observed when the baby is seventh months old. Marriage is performed in front of a sacred fire and is conducted by a Brahman priest. The dead is either buried or cremated. Ancestor worship *(tita bhat)* is performed on the third day after a death. Another death rite *(ghat sradha)* is conducted by a Brahman priest.

The Bathudi are mostly marginal farmer who also work as labourers. The forest is their other economic resource. They now collect and sell forest produce. A few of them are also employed as teacher, and government officials.

The Bathudi have their own community council (*jati samaj)* controlled by the hereditary heads like pradhan, dakua and members. The *jati samaj* deals with intercommunity social affairs. In Mayurbhanj, a few persons have been elected as members of the gram *panchayat* and Legislative Assembly.

They profess both Hinduism and the traditional tribal religion and worship *sarna (salgroove)*. *Mangla Thakur* is their family deity. They also sing *Jhumur* and *Thusu* songs.

Discussion and Findings

In 1982 Hos and Santals of Similipal Biosphere Reserve forest was included under the Tiger Reserve project. As rules, "no human activity is permitted in core area." Thus the laws of this project curtailed food gathering activity. Even hunting small game is totally prohibited. The bow and arrow became symbols only, which the Hos and Santals even now carry Development work could not be taken up.

Tribal people in and around Similipal Biosphere Reserve indulge in mass hunting *(Akhand Shikar)* as a ritual. Earlier they practiced unrestricted hunting through the year except rainy season. But now it is restricted due to strong anti-hunting measures undertaken by the Park management. Previously they killed animals, burnt and ate inside the forest. Now, they practice as a mock hunting as name sake of traditional practice.

They are managing their natural resources, an integral part of their environment, e.g. land, water, forest by using their traditional knowledge. They nurture the nature in such a way that ecological balance and biodiversity are not disturbed. Their cultural ties with local environment are unique.

Material life is a fair index of the general level of knowledge and of the intellectual attainment of the people. Ecological factors play a great role in the various aspects of culture, especially in the case of material culture. Material life of a man is rendezvous of external environment and socio-cultural representation-manifestation of ecological relation.

Sal is mainly used as an ideal timber for house construction. Beside hard and strong, *sal* wood can withstand the attack of insects. White soil is used for plastering of the walls inside the house. They use soil for flooring.

The Santal and the Ho are now mostly bi-lingual (Oriya and their own language). A few of them picked up Hindi to a little extent due to their contact with Hindi speaking traders and some contractors in market places.

The rate of female literacy is miserably low among the Hos and Santals in the core area. They do not understand the importance of education, especially for their females. They consider participation in domestic chores is more important for their females than receiving formal education from any institution. In case of males they do not feel much inclination towards school education. Only with the advice of school teachers of Kabatghai village they are now sending the children to school. Out of four villages, Kabatghai village has only one primary school.

The Ho of Kabatghai village along with Santal works together in the P.W.D. which consists of digging the earth and spreading the same alsongwith *moorum* on the roads. They are also engaged by the forest contractors.

The Ho has no hesitation in accepting food and drink from the hands of Santal. But Santal would not accept cooked food, fried articles and puffed rice from the hands of Ho. Because, Santals are higher in status than the Ho, as reported. Even when a death is happen in Hos house; they visit the house of death person but do not enter into the house.

There are certain rites and duties of kinsmen during the birth, marriage and funeral ceremonies. The relatives or the kinsmen of the respective families must assemble on the ceremonial occasion.

Plants are considered as the principle source of all essentials of life, such as food, fodder, fiber, fuel and medicines. Many useful flowers, fruits and whole plants have been used as offerings in worship. This practice of the regular requirement of these plants for worshipping keeps them available in the neighborhood for recurrent use. Some plants such as *tulsi*, mangos are worshipped and protected in the form of sacred plant. Among them it is more of sacred plants than the sacred groove.

Settle agriculture is the main source of livelihood of the people of core area; hence land is the main economic resource to them. Land is individually owned and is inherited through the main line. They solely cultivate their own land.

Out of the four villages Kabatghai, Jamunagarh, Jenabil and Bakua, two villages namely, Jamunagarh and Jenabil are very near only a distance of 5 kms. The distance from Joshipur to Gurguria village is 27 kms., from Gurguria to Kabatghai core villages is 12 kms., from Kabatghai to Jamunagarh is 19 kms and Jamunagarh to Jenabil is only 5 kms but from Jenabil to Bakua is 18 kms.

It is observed from the genealogy collected from the core villages, that there are six marriage alliances between the Jenabil and Jamunagarh village. Because these two villages are situated within 5 kms of distance and their socio-cultural interaction between them is more than the other two villages. Village endogamy is found in Jenabil and Bakua village. Twelve marriages of Hos are held in Jenabil village between Hos and four marriages are held within Hos in Bakua village.

The people of core village are now very much politically conscious. They represent as a member in the Panchayat election to select their leaders of their own choice or sometimes advised by the different political leaders. It is evident that different higher political leaders motivate these local people.

The political organization of Ho and Santal of the core villages of Similipal Biosphere Reserve, which may be termed as traditional village political organization, inter-village political organization, formal statutory organization and inter-village organization of nominated members. The traditional village organization is represented by the council of respected elderly villagers. Inter-village political organization is represented by influential but informal local leaders. Formal statutory organization forms part of the Panchayat Raj system.

It is observed that in the core area of Similipal Biosphere Reserve, forest department has constituted a committee in the village Jenabil namely, Eco-Development committee. There are five members in this committee, four from the village Jenabil and one is Ranger of the village. They are named Ganu, Kanna, Dukhram and Ganga all from the Ho Society. A bank is situated at Joshipur where Rs. 25,000/- has been deposited by the forest department in the name of the committee. This committee's activities are to look after the forest's flora and fauna and to save animals from the poacher. And also to see whether anybody is lighting fire in dry leaves in summer. For keeping watch to the forest the villagers are given some facilities like if anybody of the village fallen sick he can get financial help as advance money from the committee. Even for marriage of a member's daughter they are provided finance (as advance) of Rs. 3000/- to Rs. 5000/- with 5 percent interest rate. This amount will be returned in monthly equal installments.

The people of the core villages are revenue tax payers from 1965 to the Government. As they reported, last 8 years Government has denied receiving tax from them. Because Government has occupied their entire land and advised them to vacate the village and shifted to Udara

/Ambodia village which is outside Biosphere Reserve. Government has offered them for rehabilitation of Rs. 1,50.000/- cash, those who possess 12 acres of land. Government also offered them agricultural land of 3 acres and Rs. 45,000/- for constructing a new house. There are about 12 families of Khadia tribe already shifted from Jenabil village to Udara and Ambodia village.

It is observed that forest plays an important role in their economic life both directly and indirectly. It supplies them various food items, fuel, raw material for cash earning and for shelter. It becomes a source of labour, provides them with grazing field and herbal medicines. Their subsistence pattern is highly influenced by the forest ecology.

Annexure – I

Table 1. Ethnic composition in core area of Similipal Biosphere Reserve

Sl. No.	*Village*	*Comm-unities*	*No. of family*	*No of Popul-ation*	*Total popul-ation*	*Male*	*Fe-male*
1	Kabatghai	Santal Ho	13 4	20 60	80	42	38
2	Jamunagarh	Ho	18	86	86	39	47
3	Jenabil	Ho Bathudi	27 1	183 1	184	93	91
4	Bakua	Ho	21	97	97	44	53
	Total		**84**	**447**	**447**	**218**	**229**

Table 2. Family type in Core villages

Family type	*Kabatghai*	*Jamunagarh*	*Jenabil*	*Bakua*	*Total*	*Per-centage*
Nuclear	11	16	11	14	52	43.68
Joint	6	2	15	7	30	25.2
Broken	-	-	2	-	2	1.68
Total	**17**	**18**	**28**	**21**	**84**	

Table 3. Village wise Marital Status in Core villages

S. No.	Village wise	Married			Un-married			Widow	Widower	Divorce		Total
		M	F	T	M	F	T			M	F	
1.	Kabatghai	22	21	43	20	16	36	1	-	-	-	80
2.	Jamunagarh	18	18	36	20	16	49	-	1	-	-	86
3.	Jenabil	41	40	81	50	46	96	4	1	1	1	184
4.	Bakua	26	26	52	17	26	43	1	-	1	-	97
	Total	**107**	**105**	**212**	**107**	**117**	**224**	**6**	**2**	**2**	**1**	**447**

Table 4. Village wise Literacy in Core villages

Village Name	*Illiterate*		*Literate*		*Total*
	Male	*Female*	*Male*	*Female*	
Kabatghai	25	26	17	2	**80**
Jamunagarh	27	46	12	1	**86**
Jenabil	90	88	3	3	**184**
Bakua	42	53	2	0	**97**
Total	**184**	**223**	**34**	**6**	**447**

Table 5. Age-group wise Population of four Core villages

Sl. No.	*Age - group*	*Kabatghai*			*Jamunagarh*			*Jenabil*			*Bakua*		
		M	*F*	*T*	*M*	*F*	*T*	*M*	*F*	*T*	*M*	*F*	*T*
1	0-4	3	5	8	7	8	15	19	17	36	5	9	14
2	5-9	4	9	13	6	9	15	12	15	27	3	5	8
3	10-14	8	0	8	5	5	10	8	5	13	4	5	9
4	15-19	3	6	9	1	7	8	3	6	9	2	10	12
5	20-24	3	3	6	3	3	6	9	8	17	5	0	5
6	25-29	3	4	7	2	1	3	8	10	18	2	4	6
7	30-34	3	4	7	0	6	6	5	6	11	3	2	5
8	35-39	6	1	7	5	1	6	9	4	13	1	4	5
9	40-44	2	4	6	2	2	4	2	5	7	4	5	9
10	45-49	2	1	3	0	3	3	6	4	10	4	2	6
11	50-54	4	0	4	3	2	5	3	4	7	3	3	6
12	55-59	0	1	1	0	0	0	3	3	6	4	1	5
13	60-64	1	0	1	5	0	5	2	3	5	2	1	3
14	65 +	0	0	0	0	0	0	4	1	5	2	2	4
Total		**42**	**38**	**80**	**39**	**47**	**86**	**93**	**91**	**184**	**44**	**53**	**97**

Table 6. Marriage distance from Kabatghai village

Sl. No.	*Name of the village*	*Distance in kms.*	*Tribe*
1.	Baliagora	40	Ho
2.	Phulgundi	42	-
3.	Basantpur	45	-
4.	Tagavilla	45	-
5	Cuttak	150	-
6.	Dudhiani	30	-
7	Badkachar	7	-
8	Jashipur	27	-
9	Kadiani	25	Santal
10.	Khorikudu	40	-
11.	Dudhiani Khalpar	35	-
12.	Dubha	30	-
13.	Joshipur	27	-
14.	Basantpur, Joshipur	35	-
15.	Podagarh	30	-
16.	Kenjra	80	-
17	Kumraban	15.	-
18.	Rumrabari	15	-
19.	Dardaro	19	-
20.	Kuchaikudu	42	-

There is no village endogamy found in Kabatghai village from the above table. It is observed that Hos and Santals of Kabatghai, marriage alliances are performed of 40 kms and above from Kabatghai village.

Table 7. Marriage distance from Jamunagarh village

Sl. No.	*Name of the village*	*Distance in kms*	*Tribe*
1.	Jenabil	5	Ho
2.	Edelkocha	40	-
3.	Podalia	55	-
4.	Deepasai	45	-
5.	Punduagora	60	-
6.	Udara	45	-
7.	Aanidi	60	-
8.	Asanboni	50	-
9	Rugudisai	45	-
10.	Bandribasa	40	-
11.	Kapand	45	-
12.	Kumari	15	-
13.	Noniagoda	50	-

It is found from the above table that marriage alliances are happened in and around 60 kms distance from Jamunagarh. In core area Jenabil and Jamunagarh the two villages are very nearer of 5 kms distances. Therefore, social interaction is more between the two village and marriage alliances is happened between them. From Jamunagarh, there are five girls married to Jenabil village and one in Kumari village which is 15 kms distance outside core area.

Table 8. Marriage distance from Jenabil village

Sl. No.	*Name of the village*	*Distance in kms.*	*Tribe*
1.	Sarat	75	Ho
2.	Jamunagarh	5	-
3.	Budabulam	17	-
4.	Jagogoda	30	-
5.	Jenabil	-	-
6.	Udara	45	-
7.	Phulbari	50	-
8.	Sortua	55	-
9.	Rania, Ranchi	154	-
10.	Jadugoda, Ranchi	154	-
11.	Poradia	52	-
12.	Bandibasa	35	-
13.	Mochia, Ranchi	154	-
14.	Gare Kunti	38	-
15.	Behrampur	48	-
16.	Pundurai	75	-
17.	Balina	52	-
18.	Sadba / Khonti	154	-
19.	Sayedba / Khunti	152	-
20.	Ranchi	152	-
21.	Baliyara Koradia	82	-
22.	Jatipani	45	-
23.	Chaibasa	154	-
24.	Delhisur, Chaibasa	154	-

Sl. No.	*Name of the village*	*Distance in kms.*	*Tribe*
25.	Balarampur	25	-
26.	Sarchua	75	-
27.	Raina, Ranchi	154	-
28.	Adolkucha	32	-
29.	Binipur	23	-
30.	Soraikori	80	-
31.	Sansasur	33	-
32.	Beldia	30	-
33.	Pahanda	90	-
34.	Bhatkaman	45	-

From the above table, it is observed that village endogamy is very much prevalent. There are twelve marriages materialize in the Jenabil village between the Ho tribe. It is also found that six girls are married to Jamunagarh village. Most of the marriage alliances are around 75 kms outside Biosphere reserve.

Table 9. Marriage distance from Bakua village

Sl. No.	*Name of the village*	*Distance in kms.*	*Tribe*
1.	Bakua	-	Ho
2.	Kudaghunta	15	-
3.	Khejuri	15	-
4.	Mokabari	20	-
5.	Kundagoda	50	-
6.	Barigaon	17	-
7.	Lambujhoran	35	-
8.	Kadachappal	15	-
9.	Lambujhoran	35	-
10.	Motiagarh	60	-
11.	Bilapora	28	-
12.	Bagdia	6	-
13.	Utrash	40	-
14.	Kumarivil	15	-

Sl. No.	*Name of the village*	*Distance in kms.*	*Tribe*
15.	Gorsai	9	-
16.	Naora	15	-
17.	Haldipur, Joshipur	50	-
18.	Nagarpattar	80	-
19.	Kumarigoda	20	-
20.	Valiagoda	60	-
21.	Barigram	12	-
22.	Jharkhand	90	-
23.	Kandivil	10	-
24.	Golnudu	18	-

In Bakua, village endogamy is also found from the above table. There are four cases of marriage alliances observed between the Hos of Bakua village. They prefer to marry their spouses around 15 kms distance villages like Khejuri, Kumarivil, Gosai, and Kandivil etc. which are outside the core area.

ANNEXURE II

Name of the Head of family of four villages in Core area of SBR

Kabatghai		*Jamunagarh*		*Jenabil*		*Bakua*	
House No.	*Head of family*	*House No.*	*Head of family*	*House No.*	*Head of family*	*House No.*	*Head of family*
1.	Shakra Ho	1.	Budu Ho	1.	Rasu Ho	1.	Raghunath Alda
2.	Lobia Ho	2.	Doman Ho	2.	Ganga Ho	2.	Menaki Alda
3.	Ghasiram Marandi	3.	Sagar Ho	3.	Kanna Ho	3.	Dubraj Purti
4.	Salge Marandi	4.	Ram Soi	4.	Budhuram Ho	4.	Jena Sinku
5.	Baiju Marandi	5.	Bhaiya Soi	5.	Suguna Ho	5.	Ramsing Suren
6.	Motilal Murmu	6.	Sanika (kanda) Hasa	6.	Sukhram Ho	6.	Mangal Suren
7.	Sadananda Murmu	7.	Jagdev Hasa	7.	Birse Ho	7.	Rupa Suren
8.	Ramdas Murmu	8.	Manga Hasa	8.	Ganju Ho	8.	Turam Purti
9.	Jhun Arjun Murmu	9.	Telenga Hasa	9.	Doman Ho	9.	Rengo Sinku
10.	Srihari Ho	10.	Bhaiya Hasa	10.	Soma Ho	10.	Lado Sinku
11.	Raitu Ho	11.	Chamru Hasa	11.	Sagar Ho	11.	Sidhu Sinku

Kabatghai		*Jamunagarh*		*Jenabil*		*Bakua*	
House No.	*Head of family*	*House No.*	*Head of family*	*House No.*	*Head of family*	*House No.*	*Head of family*
12.	Govinda Murmu	12.	Surendra Hasa	12.	Hadua Ho	12.	Dubraj Purti
13.	Kalidas Tudu	13.	Surendra Ho	13.	Ganu Ho	13.	Mongro Purti
14.	Bari Marandi	14.	Mangal Ho	14.	Bada Ho	14.	Soma Purti
15.	Ramrai Murmu	15.	Manika Ho	15.	Jarka Ho	15.	Dado Purti
16.	Govinda Murmu	16.	Guman Ho	16.	Harka Ho	16.	Mura Purti
17.	Arjun Murmu	17.	Guman Ho	17.	Londung Ho	17.	Mara Purti
	--	18.	Chamru Ho	18.	Budhu Ho	18.	Bajai Purti
				19.	Sanika Ho	19.	Doco Purti
				20.	Dani Ho	20.	Golaram Purti
				21.	Kande Ho	21.	Lago Suren
				22.	Santo Ho		
				23.	Samuel Ho		
				24.	Joro(nirmal) Ho		
				25.	Balku Ho		
				26.	Magdali Ho		
				27.	Lacco Bewa Ho		
				28.	Mohan Sundar Naik		

Photographs

Tourist Check Post: Similipal Biosphere Reserve

Water Falls: Similipal Biosphere Reserve

River Khairi : Life Line of Similipal

Tribal Family at a Core Area Village

Sacred Groove

Harvesting of Crops at Core Village

A Baul Singer of Similipal

Team Members with Director-in-Charge

References

Adhikari, Ashim K. 1984. Society and World View of the Birhor. Calcutta: Anthropological Survey of India; pp. 5-13, 60-90.

Agarwal, Anil and Sunita Narain (eds). 2003. Dying wisdom: rise, fall and potential of India's traditional water harvesting systems. New Delhi: Centre for Science and Environment.

Annual Report, 2004-05, Orissa Voluntary Health Association, Lokswasthya Bhawan, 165, Laxmisagar Square, Bhubaneswar 751006, Orissa, India.

Anonymous. 2004. Database on Similipal Biosphere Reserve, Orissa. Bhubaneswar: Regional Plant Resource Centre.

Anonymous. 2005. Agronica. Bhubaneswar: Directorate of Agriculture, Govt. of Orissa.

Anonymous. 2006. Information on Similipal Biosphere and Tiger Reserve. Baripada : CF and Director STR.

Bande, Usha. 2006. Ecotourism and Mountains. In S.B. Verma, S. K. Jiloka and Thryambakam (ed). Rural Tourism and Tribal Development. New Delhi: Deep and Deep Publications Pvt. Ltd.

Banerjee, Kumar. 1967. Dances of Orissa through the ages. *Calcutta Review.* 161 : 66-68.

Behura, B.K. and L.A.K. Singh 1979. Conservation of Crocodiles in India. Indian Forester. pp. 83-92.

Behura, N K. 2000. Status of Health Care Facilities and Ethno-Medicine in Orissa. (Mimeo). Bhubaneswar. N.K.C Centre for Development Studies.

Bhattacharya, S. 1968. *Ethnomusicology and India.* Calcutta: Indian Publication.

Bhowmick, P.K. Bhowmick. 1994. Primitive Tribal Groups in Eastern India: Welfare and Evaluation. New Delhi. Gyan Publishing House.

Binu, R. and K. Sashikumar. 2006. Ecotourism and Mountains In S.B. Verma, S. K. Jiloka and Thryambakam (ed.). Rural Tourism and Tribal Development. New Delhi. Deep and Deep Publications Pvt. Ltd.

Biranchi, Mishra, 2004. Ecotourism in Similipal: Prospects and Problems. Bhubaneswar. Tourism Department, Orissa Tourism.

Biswal, K.C. 2000. Folk classical elements in Mayurbhanj chhau in *Chaitraparva,* Baripada. Mayurbhanj chhau Nritya Pratishtan.

Biswas, P.C. 1956. Santals of the Santal Parganas. Delhi. Bharatiya Adimjati Sewak Sang.

Black, Marry, B. 1973. 'Belief system'. In J.J. Honigmann (ed.). Hand book of social and cultural anthropology. pp. 511-520. Chicago: Rand Mcnally and Company.

Bustard, H.R. 1975. Integrated scheme for the conservation and farming of gharial, crocodiles and sea turtle in Orissa. Source Implementation Report II, FAO Reports arising from the crocodile breeding and management project. FO: IND/71/033, October 1975. 1975. pp. 1-17.

Census of India 1931. Bihar and Orissa Part I Report. Volume VII. W. G. LACEY New Delhi. Govt. of India Press.

Census of India 1931. Bihar and Orissa Part II Tables, Volume VII. W. G. LACEY. New Delhi. Govt. of India.

Census of India. 1981. Catalouge of tribal specimens, Office of the Registrar General of India. New Delhi. Ministry of Home Affairs.

Chambers, R. 1993. Challenging the professions: frontiers for rural development. Review I:(I)34–41. London. Intermediate Technology Publications.

Chandra, R. (ed.) 2005. *Wild Life and Ecotourism, Issues and Challenges.* New Delhi. Akansha Publishing House.

Chawla Romila (ed). 2003. Wildlife Tourism and Development. New Delhi. Soni Publications.

Choudhury, S.R. 1999. *Khairi: The Beloved Tigress.* Natraj Publishers, Dehradun.

Das, R.N. 2000. Mayurbhanj Chhau and its Basic Techniques. *Ibid.* pp-21-22.

Das, T.C. 1931. The Bhumijas of Seraikella. Calcutta. Calcutta Univesity Press.

Danda, Ajit Kumar. 1977. Chhattisgarh: An Area Study. Calcutta. Anthropological Survey of India.

Danda, Dipali. G. 1978. *Among the Dimasa of Assam: An Ethnographic Study*. New Delhi. Sterling Publishers.

Dormall, J. Von. 1999. *A Conceptual Framework for Community Bared Health Insurance*, Lisbon, Portugal.

Firth, Raymond. 1990. *Tikopia Songs*. Cambridge. Cambridge University Press.

Gadgil, M. and F. Berkes. 1991. Traditional Resource Management Systems. *Resource Management and Optimization* 18:(3-4)127-141.

Gadgil, M., F. Berkes, and C. Folke. 1993. Indigenous knowledge for biodiversity conservation. *Ambio*. 22:151-156.

Guerci, A. (Ed.), 1997. Healing, Yesterday and Today. Tomorrow? Genova: Egra Multimedia Communication Links.

Hutton, J.H. 1944. *Caste in India its Nature Function and its Origin*. Delhi. Oxford University Press

Heywood, V.H. and R.T. Watson. 1995. *'Global Biodiversity Assessment, UNEP'*; New York. Cambridge University Press

Indigenous Peoples' Biodiversity Network (IPBN) 1 Nicholas Street, Suite 620 c/o Cultural Survival – Canada Ottawa, ON KIN 7B7 Canada.

Knight, D.H. 1975. A Phytological analysis of species rich tropical forest on Barro Colorado Islands, Panama. *Ecol. Monogr*. 45:259-289.

Kurath, G. P. 1964. *Iroquois Music and Dance*. Washington. Smithsonian Institution Bureau of American Ethnology.

Lomax, A. 1968. *Folk song Style and Culture.*Washington, American Association for the Advancement of Science.

Mahato, P. 2000. *Sanskritization vs Nirbakization, A Study of Cultural Silence and Ethnic Memocide in Jharkhand*. Calcutta : Sujan Publication.

Maity, T. 2007. 'Ecology and material culture of Pachmarhi Biosphere Reserve. pp. 261-292. In Sadhan Sengupta (ed.). Man in Biosphere: A Case Study of Pachmarhi Biosphere Reserve. New Delhi. Gyan Publishing House.

Maity, T. 2001.'Ethnic identity : A case study of Mawasi of Madhya Pradesh' pp. 73-77. In Samit Ghosal and K.R. Ravi. (ed.). *Ethnic Situation in Central India*. Special Issue - Jr. of the Anthropological Survey of India. Vol. 50. No. 3. Calcutta.

Martin Mclaugh 1995. Majority, Minority Differentials! Unemployment, Housing and Health, Chambers Press, London.

Mishra, R.K., *et.al.* 2003.Vegetation Diversity of Similipal Biosphere Reserve. *e-planet*. 1(1): 4-9.

Mohanty, R.K., *et.al.* 2005. Plant diversity assessment of shorea robusta dominated forest stands of Similipal Biosphere Reserve. *e-planet.* 84L21-29.

Nettl, B. 1956. *Music in Primitive Culture.* Cambridge, Harvard University Press.

Newell, Kenneth W. 1975. *Health by the People*, WHO Publication, Genewa.

Nirmal Kumar Bose. 1961. (ed.). Peasant life in India- A Study in Indian Unity and Diversity. Calcutta. Anthropological Survey of India.

Orissa District Gazetteers. 1967. Mayurbhanj. Nilmani Senapati and Nabin Kumar Sahu (ed.). Cuttack. Orissa Govt. Press.

Panda, S.C. 2004. Dryland Agriculture. Jodhpur : Agrobios.

Patro, S.N. and M.K. Misra. (*ed.*). 1985. Conservation of Similipal in its Wilderness. Bhubaneswar. Orissa Environmental Society.

Poffenberger, Mark. (ed). 1996. Linking Government with Community Resource Management. Research Network Report 9. Berkeley. CA: Asia Forest Network.

Posey, D.A. and G. Dutfield. 1996. Beyond intellectual property: towards traditional resource rights for indigenous people and local communities. Ottawa. International Development Research Centre.

Prasad, Narmadeswar. 1961. Land and People of Tribal Bihar. Ranchi. Tribal Research Institute, pp. 69-80, 90-103, 109-122, 143-148, 221-224, 237-243.

Reid, Held, 1993. A New lease on life : Chapter I. Biodiversity Prospecting: Using Genetic Resources for Sustainable Development, Washington, World Resource Unit.

Robins, Richard. H. 1973. Identity, culture and behaviour. In J.J. Honigmann (ed.). Hand book of social and cultural anthropology. p. 1201. Chicago : Rand Mcnally and Company.

Roy, S.C. 1925. The Birhor : A little known Jangal tribe of Chhotanagpur. pp.141-172. Ranchi : Man in India office.

Roy, S.C. and R.C. Roy. 1937. The Kherias. Ranchi. Man in India Office.

Sachchidananda. 2006; 'Development, Action and Applied Anthropology. in Swapan Kr. Paramanik and Samita Manna (ed.).

Sagar, S.R. and Singh, L.A.K., 1993. People and Ecodevelapment Scheme – A Pilot Study in Similipal Tiger Reserve. Abstract in : International Symposium on the Tiger, Delhi Ministry of Environment and Forest, February.

Senapati, Nilmani. 1967, *Orissa District Gazetteers, Mayurbhanj*, Cuttack. Govt. of India.

Sethi, Praveen. 1999. *Strategies for the Future of Travel and Tourism.* Delhi. Rajat Publication.

Sevetz, T.W. 1961. *American Economic Review*, Vol. 51, No. 1. Investment in Human Capital, pp. 1-17.

Sharma, Subhash Chandra. 2002. *Geography of Tourism.* Vol. 1. New Delhi. Rajat Publication Pvt. Ltd.

Singh, K.S. and P.P. Mahato. 1983. Mahato-Kurmi Mahasabha Movement in Chhotanagpur: A preliminary Appraisal. In K.S. Singh (ed). *Tribal Movement in India.* Vol. 2. p. 109, Delhi. Manohar Publication.

Singh, K.S. 1994. *The Scheduled Tribes. People of India*, National Series. Vol. III. Delhi. Oxford University Press.

Singh, K.S. 1998. India's Communities. People of India, National Series.Vol. IV, V, and VI, Delhi. Oxford University Press.

Singh, K.S. Singh. 1998. People of India-National Series on India's Community. Oxford University Press, Delhi.

Singh, L.A.K. 1997. Wildlife Research in Similipal. In: Tripathy, P.C. and Patro, S.N. (Ed.): *Similipal: A Natural Habitat of Unique Biodiversity.* Orissa Environmental Society, Bhubaneswar.

Stastistical Bulletin. 2005. Department of Tourism and Culture (Tourism), Government of Orissa.

The Orissa Gazette, No. 800, June, 1998. Extra Ordinary Published Authority, Department of Tourism and Culture.

Tiger Reserve Status Report. 2001. Ministry of Environment and Forest, Bhubaneswar.

Van Boten, H and Van Dormael. 1999. "Health Service Professional and Uses," International Social Science Journal. No. 161, September.

Vatsyayana, K. 1976, *Traditions of Indian Folk Dance.* New Delhi: Indian Book Co.

Vitousek, P.M. Ehrlich, P., Ehrlich, A. and Matson, P. 1986. Human Appropriation of the products of Photosynthesis. *Bio Science* 36: 368-373.

Voget, Fred.W., 1973. History of Cultural Anthropology. p. 35. In J.J. Honigmann (ed.). 'Hand Book of Social and Cultural Anthropology'. p. 35. Chicago: Rand Mcnally and Company.

Wahab, Salah and John J. Pigram. (ed.). 1997. Tourism Development and Growth: the Challange of Sustainability. London. Routledge.

Woods, C.M. 1977. Alternative Curing Strategies in a Changing Medical Situation. *Medical Anthropology.* 3:25-53.

World Bank Report on Madhya Pradesh Forestry: On Indigenous People, 1994. World Bank Report on Madhya Pradesh Forestry: on Indigenous People. Bhopal. Government of Madhya Pradesh, Satpura Bhavan.

Yadav, Santosh. 2006. Ecotourism: Problems and Prospects. In S.B. Verma, S.K. Jiloka and P. Thryambakam (ed.). *Rural Tourism and Tribal Development*. New Delhi. Deep and Deep Publications Pvt. Ltd.

Index

A

Adhanti Songs of Marriage, 256
Adivasis (Tribals), 18, 39, 121
Administrative Zones, 207
    Buffer, 207
    Core Zone, 207
    Transition Zone, 207
Adult Franchise, 71
Agricultural Land, 48, 358
    And Soil Types, 333
Agriculture and
    Cropping Pattern, 98
    Rice Culture, 96
Agro-Climate
    Classification, 98
    Zones, 112
Alda—Pingua, 465
Alma Ata, 212
    Declaration, 211, 212
Tribal and Christians, 73
Anthropology and Geography, 17
Ashram Hostels, 203
Astakunar Gram Panchayat, 119, 120
Athrareuar, 137
Austro-Asiatic
    Group, 132
    Origin, 44
Auxiliary Nurse Midwife (ANM), 215, 236

B

Bakua Village, 351, 354, 363, 405, 427, 434
    Land Use, 355, 357
Balasore, 115
Banan String Instruments, 256
Bangripasi Block, 163
Barahipani Waterfalls, 218
Baripada, 184
Basic Heath Services, 211, 212
Basuki Thakurani (Earth Goddess), 133
Bathing Customs and Habits, 446
Bathudi, 166, 259
    And Kharia Population, 368, 392
    Festivals, 126
        Asharipuja, 136
        Dhuliapuja, 136
        Rashpurnima, 136
        Sarolpuja, 136
        Sitalapuja—Shivachaturdasi, 136
    Original Settlers of Bhanjubhum, 436
    Society and Polity, 473
    Totemic Clans, 431, 474
Bay of Bengal, 349

Below Poverty Line, 208
Benevolent and Malevolent Spirits (HO), 134
Bhadra, 115
  Shukla Ekadsi, 259
Bhore Committee, 211
Bhumij, 137, 167, 257
Bhumji,
  Belong to Austro-Asiatic Family of Languages, 137
  Divisions in Gotras, 137
  Jadur Ritual Dance, 257
  Marriage by Negotiations, 137
Bihar, 18, 79, 115
Bihor, 140
Biodiversity, 18, 78, 81, 88, 125, 126
Biological Reserves, 161
Biosphere
  Conservation, 122
  In India, 268
  Management, 98
  Programmes by UNESCO, 162
  Reserve (BR), 17, 72, 73, 78, 125, 18, 126, 194, 207, 267, 323, 325, 327, 347, 478
    Core Area, 324
    Defined, 162
    Similipal, 218
    Villages, 236
  Zone, 268
    Villages, 39
Birhor, 140, 169, 182
  People, 128
  Primitive Tribal Group, 140
BJP, 71, 73
Breast Feeding, 212
Bride Price, 135, 137
British
  Administrator, 163
  Rule, 131
Buddhupal Goddess of Rainfall, 133
Buffer
  And Transitional Zone Villages of Joshipur, 39
  Area, 268
  Area Villages, (BAV), 17
  Area Villages Ethno-Medical and Health Culture of-, 209-252
  Region of Similipal Biosphere Reserve, 174
  Zone, 79, 162, 323, 347, 429
  Zone Villages (BZV), 21, 72, 73, 79, 117, 125-207
    Basilakocha, 79
    Chekundkocha, 79
    Gad Sinutipal, 79
    Gudgudia, 79
    Gudgudia—Basilakocha, 21
    Kolhjari, 79
    Kumari, 79
    Lanjighosra, 79
    Saruda, 79
      Impact of Development Plans/Programme in-, 161-207
Building Materials, 48
Buru Bonga, Forest Deity, 133
Buru Song, God of Bhumij, 257
Bathudi, Origin Settlers of Bhanjabhum, 136

C

Canal at Kumari Village, 184
Case Study of Childcare, 250-252
Cash Crops, 131

Caste
Community, 176
Groups, 31, 131, 141
Teli, Goud, Lohar, Bindhani, Kudmi, Mahato, 131
People, 305
Goud—Lohar, 305
Changu Song, 258
Charabil Forest, 351
Check Dam, 184
Cheroananbi Forest, 351
Chhau Dancers of Mayurbhanj, 262
Chhotanagpur, 457
Biotic Province, 79
Child Death Rate, 215
Chitta Gyani Charitable Trust, 237
Christian Burial Grounds, 136
Christianity, 159
Chua, A kind of Well, 106
Clan Exogamy, 70, 136
Coastal Ecosystem, 323
Common Trees, 269
Kadamba—Salai, 269
Communication Diseases, 212
Community
Development Blocks, 214
Health Care, (CHC), 211, 215, 236, 242
Congress (I), 70, 73
Conservation
And Management of
Natural Resources, 113
Water Resources, 109
And Management, 17, 120
Of Biodiversity, 323
Of Jahiras, 71, 73
Of Rainwater, 107
Of Water, 114
Of Watershed/Water Resources, 78
Of Wildlife, 98, 304
Core Area
Ethnic Composition, 429, 458
Santal—Bathudi, 431, 432
Of Similipal Biosphere Reserve, (CASBR), 315, 327, 328
Agricultural Colony, 321
Bukua, 327
Cropping Pattern, 329
Forest Labour Colony at Barakamra, 315
Jamunagarh, 315
Jenabil, 327
Kabatghai, 315
Villages, 268, 315-326, 347, 457
Literacy, 320
Livestock, 320
Population Structure, 315
Society and Polity, 457-490
Core Zone, 79, 162, 323, 347, 351, 429
Cosmology, 128
CREFTDA, NGO, 286
Crop Varieties, 104
Cropping Pattern, 98, 329
Crude Birth Rate, 215
Cultivation
Crops, 333
Of Sugarcane, 112
Cultural
Anthropology, 127
Identity, 161
System, 114
Traditions, 256

Traits, 158

Culture, 21

D

Damu Changu, 259

Dance,

Chhau Dance of Mayurbhanj, 260, 261

Dhang and Langi, 257

Jhika, 257

Decentralized Development Programme, 210

Denudation of Forests, 116

Destruction of Forest/Wildlife, 126

Development of Education in Studied SBR Villages, 198

Dharm Devta (Shiv), 134

Sidhi Dam (Abode), 134

District

Balasore, 459

Keonjhar, 459

Mayurbhanj, 71, 79, 89, 135, 132, 162, 218, 269, 347, 429, 457, 459, 473

Mayurbhanj and Keonjhar (Orissa), 136

Midnapur, West Bengal, 218

Purulia, West Bengal, 131

Singhbhum, 131, 132, 218, 473

DPEP (District Primary Education Programme), 203

Drainage

Basins, 78

Outlet, 78

Drought Prone Parts, 115

Dry Deciduous Hill Forests, 349

Duarsuni Forest, 363

Durang Marriage Song, 258

Durdura Makdiya Tanda, Transitional Village, 79, 188, 203, 207

E

Earth's Ecosystem, 77

Eastern Ghats, 79

Ecological

Balance of Area, 31

Balance, 304

Change, 210

Conservation, 268

Diversity, 268

Ecology

And Material Culture, 433

And Tourism, 268

Economic

Changes, 210

Development, 46, 174

Committee (EDCs), 164, 195, 196, 339

Programme, 109, 113, 164, 195, 208

Liberalization, 217

Sector, 267

Transactions, 210

Economy, 21, 48, 175, 184, 352, 392

Ecosystem, 121, 126

In India, 268

Ecotourism, 268, 270, 304

Education

And Literacy, 304

Guarantee Scheme, 203

Educational Status, 64

Employment

Generation, 115

Opportunity, 267

Endogamous Groups, 141

Jaharua Goud, 141

Krishna Goud, 141
Mogda Goud, 141
Nand Goud, 141
Endogamy, 70
Environment and Economy, 81
Environmental Degradation, 267
Erosion for Deforestation, 116
Ethnic
Composition, 44
Groups, 18, 39, 45, 48, 52, 71
Identity, 46
Structure, 433
Ethno
Biologists, 210
Medical, 209
Ethnoscape, 21, 39, 71
Evil Spirits, 257
Exogamous Clans, 431

F

Family Welfare Programmes, 213
Faunal and Floral Species, 268
Festival,
Ashare Parav, 257
Atharo Deol, 259
Baha Utsav, 257
Chaitra Parav, 262, 263
Chhau, 261
Dassai, 257
Fagu Parav, 257
Karam, 257-260
Sohrae, 257
Festivals
In Similipal Biosphere Reserve, 151
Of Ho, 133
Feudal Rule, 80
Forest Resources, 80
Flathead Indians, 253
Folk Music of-, 253
Flora and Fauna, (FF), 17
Folk
Awareness, 126
Ecological Knowledge, 209
Knowledge, 209
Music and Dance, 253-265
Or Domestic Medicare, 214
Tales—Hunting, 138
Taxonomy, 126, 128, 152, 153, 154, 158
Forest
Dependency, 177
Dwelling Commuhities, 209
Ecology, 48
Ecology and Environment, 18
Economic System, 135
Management, 362
Policy, 268
Preservation, 268
Resources, 47, 48, 88

G

Gender Empowerment, 46
Geography of Tourism Development, 267
Global Warming, 125
God Similipal, 119
Irrigation Canal, 184
Gond, Caste Group, 80
Gondwana Land, 82
Goud, 131, 141, 172
Endogamous Group, 141
Settlers of Singhbhum District of Jharkhand, 141

Gram
  Panchayat, 21, 22, 119, 121
  Panchayat of Joshipur, 22
  Saraj of Baripada, NGO, 286
Grassland and Savannah, 349
Green House Gases, (GHG), 125
Ground Water
  Harvesting, 106
  Recharge, 115
Growth Monitoring, 212
Gudgudia
  And Makdiya, 189
  Panchayat, 203
  Village, 182, 203, 236

H

Handia, Rice Beer, 44
Harsha Chullah, 198
Harvesting Rainwater and Groundwater, 106
Health
  Culture, 209
  For All, 211
  Policies, 212
  Reform Policies, 211
Higher Peaks, 269
  Khaireburu—Dhudruchampa, 269
Hills in Similipal Biosphere Reserve, (HSBR), 114
  Aamjhera Pahad, 114
  Athera Demlli Pahad, 114
  Badargoda Pahad, 114
  Bakua Pahad, 114
  Bhadirakad Pahad, 114
  Chandan Pahad, 114
  Daukocha Pahad, 114
  Sarua Pahad, 114
Hills, 347
  Baampahar, 347
  Balideha, 347
  Chahala, 347
  Dhudruchampa, 347
  Gorumahasani, 347
Hindu
  Area, 31
  Pantheons, 71
    Shiva—Kali, 71
  Society, 31
Hinduism, 31, 475
HO, 132
  Of Mayurbhanj, 431
  Polity, 133
  Population, 44
  Religious Head (Deuri), 134
Ho's
  And Santals, 448
  Birth Customs, 467
  Bride Price, 466
  Death Customs, 467
  Distribution, 465
    Cuttack, 465
    Kalahandi, 465
    Kunjhargarh, 465
  Divorce, 466
  Festivals, 470
  Gods and Worship, 469
  Inter-Ethnic Relationships, 473
  Marriage Customs, 465
  Marriage through Negotiations, 431
  Mayurbhanj, 465
  Polity and Law, 473
  Polygyny, 466
  Society, 465

Tribe and Kinship and Social Organisation, 468

Horn Works, 218

Human

Civilization, 77, 78, 109

Settlement, 89, 120

Hunting, 338

Husking Implements, 442

I

Ill Health and Illiteracy, 211

Illiteracy, 64

Immunization, 239

In Migration, 46

Indian Parliament, 125

Indigenous Knowledge, 324, 325, 433

Parameters, 325

Indira Abash Yojna, 182, 184

Infection or Immunization, 212

Infections with Plasmodium, 236

Information

Education and Communication, (IEC), 217

For Tourists, 272, 273

Infrastructure Development, 184

Integrated Tribal Development Agency (ITDA), 140, 189

Inter-Ethnic Relationship, 464

International Conference on

Population and Development at Cairo, 212

Primary Health Care, 212

Irrigation, 108, 113

Facility, 184

Structures, 117

J

Jamunagarh, 427, 482

Land Use, 392, 393, 394, 395

Jashipur, 187, 351, 352, 392, 405, 426, 473

Block, 163

Community Health Centre, 242

Jenabil

Hill, 363

Land Use, 364, 365, 367

Village, 427

Jenaburu Hill, 392

Jharkhand, 18, 45, 79, 115, 457

And West Bengal, 157

Jhumar,

And Thusu Songs, 475

In Mayurbhanj—Jharkhand State, 263

Song, 258

Song in Chhau Dances, 265

Song in Marraiges, 264

Song Style, 263

Joshipur Blocks, 44

Jute Mills, 218

K

Kaibarta, 142, 171

Karam

Songs, 256

Tree, 136, 260

Kendu Leaves, 340

Kharia, 135, 170, 177

And Bhumij, 259

As Food Gatherers and Hunters, 135

Dependent on Minor Forest Produce (MNP), 135

Exogamous Clan, 135

Of Mayurbhanj, 258

Population, 135

Killis, 133
  Identified Details, 141
  Names, 137
    Prefer Hinduism, 137
Kobalghai Village, 405, 411
Kol/Ho, 167, 259
Kolha/Ho Group, 44
Kolha—Kaibarta, 46
Kudmi Mahato, 44, 45, 141, 172
Kukurbhuka, 184

L

Land
  And Forest Resources, 323-345
  Records, 47
  Resources, 328
  Revenue Officer, Baripada, 351
  Use Changes, 347-428
Landless Communities, 115
Large Scale Deforestation, 268
Laterite Soil, 349
Life Expectancy Rate, 216
Literacy Rate, 202
Local Handicrafts, 304
Lohar Bindhan, 171, 142
Lord Krishna, 143
Low Cultivated Land Categories, 407

M

Mahali, 139, 169
  Concentration, 139
    Jharkhand—West Bengal, 139
  Sub Castes, 139
Mahanadi Bio-Geographical Region, 79
Mahatma Gandhi Nation Rural Employment Guarantee Programme, (MGNREGP), 164, 194
Mahato, 176
Mahurbhanj State (Now in Orissa), 18
Makdiya, 182, 189
  And Kharia, 177
  Or Birhor, 52
  Settlements, 196
Man and Biosphere Programme (MAB) by UNESCO, 323
Management of
  Forest Resources, 337
  Natural Resources, 328
Manbharias Traditional Money Catchers, 269
Mankhadiya (Makdiya), 44
Maoist, 73
Marginal Farmers-cum-Owner Cultivators, 81
Marital Status, 59
Market Economy, 48
Marriage
  By Exchange, 135
  Ceremonies, 187
  In Negotiations, 70
  With Cross Cousin, 135
Material Culture in Core Area, 429-456
Maternal Mortality Rate, 215
Mayurbhanj, 45, 115
  Chhau, 262
  State, 79, 80, 129
Medicinal Plants, 196
  And Animals, 209
Mega
  Health Mela, 237
  Water Projects, 121
Meghasini
  Hills, 347

Mountains, 218
MFP, (Honey, Resin, Paro, Sal seed, Sal Leaf, Firewood), 48, 135, 176
Migration, 44, 131
Minerals of Area, 219
  Chromites, 219
  Iron Ore, 219
Manganese, 219
Ministry of
  Environment and Forests, 162
  Rural Development, 194
Modern
  Education System, 203
  Health Infrastructure, 213
Medicines, 209
Monocrop, 184
  Area, 112
Monogamy, 70, 135, 136, 143
Morality Rate, 214
Mudaliar Commission, 217
Multi-purpose Health Workers, 215
Munda, 138, 168
  Agricultural Tribe, 138
  Are Patrilocal, 138
  Details, 138
  Exogamous Clan, 138
  Religious Festivals, 138
  Settled in Chhotanagpur, 138
Munda-Saruda Hill, 195
Music and Dance, 448
Musical Instruments, 448
  Of Chhau, 263

N

National
  Forest, 268
  Park, 18, 78, 324
  Resources Forest, 324
  Rural Employment Guarantee Act (NREGA), 164, 194
  Rural Employment Guarantee Programme, (NREGP), 208
Natural Calamities, 96
Nature to Supernatural Power, 150
New Draft Health Plan, 211
Non-Governmental Organisations (NGOs), 48, 196, 212
Non-Tribal Population, 270
Northern Tropical
  Most Deciduous Forest, 349
  Semi-Evergreen Forest, 349

O

Odissi Dance, 262
Old Age Population, 31
Oral Traditional Medicines, 242
Orissa Tourism, 272
Out Migration, 45

P

Panchayat System, 71
Panchayati Raj, 164
Panchayats, 44
  Astakunar, 44
  Barheipani, 44
  Dhalabani, 44
  Gudgudia, 44
Pandhat, Traditional Dress, 257
Pane Sangkranti, 46
Paramedics, 215
Parba Traditional Panchayat, 138
Participating Rural Appraisal (PRA), 210
Patrilineal Rule of Inheritance, 70
People's Biodiversity Register (PBR),

Programme, 209, 210

Perennial

Streams, 349

Water Sources, 89

PESA (Extension of Panchayat Act in Scheduled Areas), 164

Phagu Parab, (Holi), 138

Pharmaceutical Industry, 210

Planning, 213, 270

Poaching, 339

Podu Cultivation, 48

Pollution, 125

Polygamy, 70

Population

Growth, 54

Structures, 31

Position of Women, 448, 461

Pradhan, 133

Prenatal Mortality Rate, 217

Primary Health Care, (PHC), 211, 212, 214, 215, 238

Primitive Tribal Groups, 214

Public Health Engineering Department, 96

R

Railways and Mining, 88

Rainfall, 81

Red Soil, 349

Religious

Structure, 31

Traditions, 461

Reproductive

And Child Health (RCH), 213

And Child Health Programme, 212

Health System, 213

Tract Infection (RTI), 213

Reserve

Forest, 207

Forests in India, 269

Revolving Fund, 196

Rice Beer, 132

River

Basins, 141

Budhabalonga, 218

Daldali, 392

Jamuna, 392

Jenabil, 363

Khairi, 363

Roads and Bridges, 184

Rohali Buru Forest, 363

Role of Ecology, 433

Rope Making Project, 182

Rural Development, 78, 115

S

Sabuj Bahini, Protector of Wildlife, 339

Sadgop, 143, 173

Biswas, 143

Chowdhuri, 143

Distribution Districts, 143

Burdwan—Singhbhum of Jharkhand, 143

Exogamous Gotras, 143

Gotras, Sandilya—Alambayan, 143

Khan, 143

Majumdar, 143

Mallick, 143

Mondel, 143

Roy, 143

Sarkar, 143

Tellis, 143

Sal
- Forests, 349
- Most Prominent Timber, 84
- Trees, 358

Saluka Clan, 136

Sambandh, NGO, 196

Sanctuary, 129

Santal (Santhals), 139
- And Ho, 326, 475
- And Kol Communities, 326
- Birth and Name Ceremony, 462
- Disposal of the Dead, 462
- Divorce, 462
- Gods and Worship, 463
- Mutiny, 131
- Parganas of Jharkhand—Tea Garden of Assam, 141
- Polity and Law, 465
- Population, 44
- Women, 196, 462
- 12 Patrilineal Totemic Clans, 139

Santal/Majhi, 169
- Group, 44

Santal's Society, 459

Santali Marriages, 461

Santals (Santhals), 256
- And Munda, 134
- Of Mayurbhanj, 459
- Totemic Clans, 430, 459
- Tunes, 257
- Women, 257

Sarhul Parab (Flowers), 138

Scheduled
- Areas, 214
- Tribe, 219
  - Communities, 214
  - Population Composition in Orissa, 214
  - Traditional Forest Dwellers, (Recognition of Forest Rights) Bill, 2006, 125

Sedimentation, 115

Self-Destruction, 267

Self-Employment Programme, 164

Self-Help Groups (SHGs), 48, 164, 195, 208, 340,

Settled Agriculture, 48

Settlements, 44

Sexually Transmitted Diseases, 213, 222

SGSY (Swaranjyanti Gram Swarojgar Yojna), 164, 208

Shifting Agriculture, 48

Siati Leaves, 340

Siati Uses, 341

Similipal
- Biosphere, 109, 112, 116, 341
- Buffer Area Village, 219
- Folk Music and Dance, 256
- Forest Area, 305
- Forests, 104
- Hill Range, (SHR), 18, 79, 81, 89, 129, 218
- Region, 44
  - Multilingual, Multicultural, 45
- Reserve Forests, 194, 324
- Richest Watershed in Orissa, 269
- Soils-Acidic, 82
- Tiger Reserve (STR), 79, 196, 198, 272
- Tourism, 286

Similipal Biosphere Reserve (SBR), 18, 21, 31, 71, 79, 81, 83, 96, 120,

122, 126, 129, 157, 162, 163, 177, 187, 188, 194, 196, 198, 205, 208, 218, 222, 237, 238, 256, 255, 268, 275, 285, 286, 324, 326, 347, 351, 411, 429, 456, 457, 475
Agriculture Pattern, 98
Buffer Zone Communities, 158
Buffer Zone Villages Religion, 143
Clans, 132
Adwaspu—Tutti, Munda, 132
Dense Forest Area, 79
Forest Resources, 84
Forests, 48, 164
House of Richest Watersheds, 79
Land and Soil Resources, 82
Multicultural Zone, 70
Of Orissa, 77
Prominent Biosphere in India, 18
Rainfall, 81
Region, 18, 48, 130
Ethnic Composition, 130
Sabuj Bahini, 339
Socio-Cultural Background, 70
Socio-Economic Background, 48
Studied Villages, 135
Benedia—Suar, 135
Taluks, 457
Villages, 71, 104, 131, 135
Water Resources, 89
Zones, 18
Sing Bonga (Sun God), 71
Prime God of Santals, 132, 133, 137, 140
Singhbhum District Jharkhand, 131
Sitala Goddess, 136
Social Structure, 127, 253
Socialist Unity Centre of India (SUCI), 71, 73, 285
Role, 118, 119
Socio-
Cultural and Economic System, 48
Economic, 21
Economic Development, 98, 212
Economic Empowerment, 46
Soil
Classification, 83
Conservation Measures, 116
Erosion, 122
Types, 98
Spinning Mills, 218
Spirits and Deities, 135
Benevolent and Malevolent, 135
Srivastav Committee, 211
Structural Adjustment Programme, 217
Study Area, 429
Bakua, 429
Jamunagarh, 429
Jenabil, 429
Kabatghai, 429
Study Villages, 22, 457, 459
Kabatghai—Bakua, 459
Sub-metamorphic Layers, 82
Subsistence Agriculture, 48
Economic System, 329
Sustainable Development, 116, 267

T

Taluks, 347
Baghmali, 347
Baripada, 347
Panchim, 347
Tamak and Tamdak, Drums of Santals, 256
Techniques and Devices, 108

Teli, 142
Teli Goud, 173
Terrace Cultivation, 107
Tiger Reserve, 18, 129, 324
  Baripada, 275
  Status Report, 347
Timber Species, 84
Totemic Clans, 126, 141
Tour Operators, 274
Tourism, 267-305
  Development, 268
  In Similipal, 267-305
  Industry, 267, 268, 304
  Promotion, 48
Tourist
  Map, 275
  Movement Routes, 276, 277
  Spots, 280
    Barheipani Waterfalls, 282
    Brundaban Gali, 283
    Chahala, 283
    Chandipur Beaches—Khiching, 283
    Gudgudia, 281
    Jorandu Waterfalls, 282
    Kabani, 280
    Kalika Prasad, 281
    Khairinivas, 280
    Mahabruthshasal, 281
    Oski Water Falls, 282
    Ramtirth, 280
Traditional
  Healers, 210
  Medicines Details with Diseases, 243-249
  Village Council, 430
  Water Harvesting System, 106
  Watershed Management, 113
Transition Zone, 323, 347
Transitional
  Area, 268
  Villages Bordering, (TVB), 44
    Jharkhand District, 44
    Kunjhar District, 44
  Zones, 17
Transplantation, 442
Tribal
  And Caste Communities, 135
  And Caste Population, 21
  And Jumuras, 262
  Community (ies), (TC), 80, 135, 158, 174, 175, 213, 326
    Bathudi, 80, 175
    Bhumij, 175
    Goud, 175
    Kaibarta, 175
    Khadia, 80
    Kol, 175
    Kolha, 80
    Mahato, 175
    Makdiya (Birhor), 80
    Munda, 80, 175
    Santal, 80, 175
    Teli, 175
Tribal Groups (TG), 44, 45, 131, 132, 158
  Bathudi, 44, 131
  Bhumji, 44, 45
  Bhuniya, 44
  Khadia, 44, 45, 131
  Kol/Ho, 131
  Kolha, 45

Mahali, 45, 131
Makdiya/Birhor, 131
Mankhadia, 45
Munda, 44, 45, 131
Santal, 45
Santal/Majhi, 131
Tribal
Panchsheel (Nehru), 164
Population, 44, 125, 132, 203
Problems, 163
Societies, 253
Songs, 256
Sub Plan, 164
Villages, 21
Bakua, 21
Jamunagarh, 21
Jenebil, 21
Kabatghai, 21
Tribes of
Mayurbhanj, 256
Of the Districts, 219
Bakunth—Saunh, 219
Bhumija, 219
Kolha, 219
Munda, 219
Santal, 219
Turi and Rashi (Til), 336
Tussar Silk, 218
Types of Forests, 457
Protected Forests, 458
Reserve Forest, 458
Tiger Reserve, 458

U

UNICEF, 212
Child Survival Programme, 212
Uni-Ethnic, 21

V

Vector Born Diseases, 212
Vegetation, 349
Village
Communities, 131
Community Council (VCC), 70
Deities, 138
Endogamy, 483, 484
Gudgudia, 184, 187
Judiciary, 133
Vishvakarma, 171
Volcano Sedimentary Rocks, 82

W

Water
And Biodiversity, 162
Harvesting Structure, (WHS), 96, 115
Harvesting, 106
Issues, 78
Global Problem, 78
Watershed
Classification, 78
Man Made, 78
Natural, 78
Development, 115, 117
Development Basic Principles, (WDBP), 115
Generation, 115
Resource Conservation, 115
Utilization, 115
Explained, 78
Management, 77-124
In Similipal, 77-124
Management Programme, 116
Watersheds, 113
West Bengal, 18, 79, 115, 445, 457

And Jharkhand, 135
Widow and Divorce Remarriage, 136
Wild Animals, 269
Wildlife, 18, 80, 269
And Vegetation, 71
Habitat, 89
Management, 96
Population, 304
Protection Act, 268
Sanctuaries, 78
Species, 457
Witch Doctors or Sorcerers, 213
Women Self-Help Group, 46
World Health Organization, (WHO), 212
World Tourism Organization (WTO), 267

Y

Yadur (Song) of Spring, 256
Youth Hostel, Jashipur, 290